Handbook of Agriculture

Handbook of Agriculture

Edited by Michelle Jacobs

MURPHY & MOORE
www.murphy-moorepublishing.com

Published by Murphy & Moore Publishing,
1 Rockefeller Plaza,
New York City, NY 10020, USA
www.murphy-moorepublishing.com

Handbook of Agriculture
Edited by Michelle Jacobs

International Standard Book Number: 978-1-63987-273-2 (Hardback)

Cataloging-in-Publication Data

Handbook of agriculture / edited by Michelle Jacobs.
p. cm.
Includes bibliographical references and index.
ISBN 978-1-63987-273-2
1. Agriculture. 2. Agricultural systems. 3. Land use, Rural. I. Jacobs, Michelle.
S493 .H36 2022
630--dc23

TABLE OF CONTENTS

PREFACE

The science and practice of food production is known as agriculture. It encompasses livestock farming as well as crop cultivation. Modern agriculture is influenced by advances in scientific understanding in the domains of agronomy, agrochemicals, plant breeding, animal husbandry, and selective breeding of animals. It can be broadly categorized into industrial agriculture and subsistence agriculture. Industrial agriculture is primarily commercial in nature with higher level of input and output per unit area of agricultural land. This book brings forth some of the most innovative concepts and elucidates the unexplored aspects of agriculture. The various advancements in crop cultivation and animal husbandry are glanced at and their applications as well as ramifications are looked at in detail. This book will prove to be immensely beneficial to students and researchers in this field.

After months of intensive research and writing, this book is the end result of all who devoted their time and efforts in the initiation and progress of this book. It will surely be a source of reference in enhancing the required knowledge of the new developments in the area. During the course of developing this book, certain measures such as accuracy, authenticity and research focused analytical studies were given preference in order to produce a comprehensive book in the area of study.

This book would not have been possible without the efforts of the authors and the publisher. I extend my sincere thanks to them. Secondly, I express my gratitude to my family and well-wishers. And most importantly, I thank my students for constantly expressing their willingness and curiosity in enhancing their knowledge in the field, which encourages me to take up further research projects for the advancement of the area.

Editor

1

Assessment of Land Management Practices in Food Crops Production among Small Scale Farmers in Kwara State, Nigeria

Abdulazeez, Muhammad-Lawal[1*], Omotesho, Kemi Funmi [2], Adekola, Olabisi Fatimat [3] and Adekunle, Dayo[1]

Abstract

The study carried out an assessment of agricultural land management practices in food crops production among small scale farmers in Kwara Sate, Nigeria. Specifically, the study determined the socio economic characteristics of crop farmers in Kwara State; ascertained the cropping patterns common among them; highlighted the soil conservation methods adopted by the farmers; examined the relationship between selected socio-economic characteristics of the farmers and their adoption of major agricultural land management practices; and investigated the constraints to adoption of sustainable agricultural practices among crop farmers in Kwara State, Nigeria A three stage random sampling technique was used in selecting a total of one hundred and forty four small scale food crops farmers. Descriptive statistics, binary logistic regression model and four point Likert-type scale were used to analyse the data for the study. The study revealed that food crops production in Kwara State is dominated by middle aged men who are poorly educated and have poor access to agricultural extension services. Half of the respondents (48.60%) adopted cereal-based cropping systems. 48.6% of the farmers adopted a minimum of three management practices. Crop rotation was mostly adopted by all the farmers while irrigation was the least adopted by only 29.3% of the respondents. Furthermore, the study revealed that farm size, age, education status, number of contacts with extension agents, household size and number of farm plots of the respondents were the significant factors affecting their adoption of land management practices. The study also revealed that the major constraints to the use of sustainable crop management practices among the farmers included inadequate supply of fertilizer, inadequacy of labour and credit, poor knowledge of improved agricultural practices, poor transportation, low produce prices and high cost of production. The study recommended the need for training programme on the use of appropriate sustainable practices that will take into account the cropping systems adopted by the farmers.

Keywords:
Sustainable practices, Land management, Small Scale farmers, Food crops

[1] *Department of Agricultural Economics and Farm Management, University of Ilorin, P.M.B. 1515, Ilorin, Nigeria.*
[2] *Department of Agricultural Extension and Rural Development, University of Ilorin, P.M.B. 1515, Ilorin, Nigeria.*
[3] *Department of Agronomy, University of Ilorin, P.M.B. 1515, Ilorin, Nigeria.*
* *Corresponding author's email: lawaz71@yahoo.com; lawalazeez@unilorin.edu.ng*

INTRODUCTION

A unique characteristic of rural societies in many developing countries is the almost total dependence on agriculture for livelihood. In Nigeria for instance, rural communities are completely agrarian in nature with crop farming, fishing, or livestock keeping being the major occupation of the majority. Since almost all rural households depend directly or indirectly on agriculture, and given the large contribution of the sector to the overall economy, agriculture is a key component of growth and development (IFPRI, 2007). As a major source of income, agriculture plays a significant role in rural and economic development process (Thirtle *et al.*, 2003). As such, agricultural development is an integral part of all rural development approaches. This fact is well captured in the description of agriculture as the engine room of rural development (Ashley and Maxwell, 2001). It is therefore imperative that any attempt at rural development must focus on agriculture. It has however been observed that agriculture has changed dramatically, especially since the end of World War II. Food and fiber productivity soared due to new technologies, mechanization, increased chemical use, specialisation and government policies. Although these changes have had many positive effects and reduced many risks in farming, there have also been significant costs. Prominent among these are topsoil depletion, groundwater contamination, the decline of family farms, continued neglect of the living and working conditions for farm laborers, increasing costs of production, and the disintegration of economic and social conditions in rural communities (Agricultural Sustainability Institute, ASI, 2012).

The Nigerian agricultural sector is made up of four major sub-sectors: crop production, livestock production, forestry and fishery. According to Daramola *et al.* (2007), their contributions to the Nigerian agricultural GDP in 2007 were 85%, 19%, 4% and 1% respectively. The dominance of crop production in the agricultural sector in Nigeria therefore places it in a pivotal position in issues of sustainable agriculture in Nigeria. Food production in Nigeria, as in many developing countries is linked with small scale agriculture. The small scale farmers in this system are responsible for a large share of the total agricultural output, cultivated land and the farming population (Brown and Wolf, 2005). The importance of small scale farmers to sustainable agricultural and rural development is derived therefore from their dominance of the agricultural sector of the country.

Sustainable land management (SLM) has been defined as the adoption of appropriate land management practices that enables land users to maximize the economic and social benefits from the land while maintaining or enhancing the ecological support functions of the land resources (FAO, 2009). It is the key point for improving land resource resilience and productivity within the context of the potentially devastating effects of climate change in Sub-Saharan Africa, bridging the needs of agriculture and environment, with the twin objectives of maintaining long term productivity and ecosystem functions (land, water, biodiversity); and Increasing productivity (quality, quantity and diversity) of goods and services (including safe and healthy food). The major goal of SLM therefore is to develop economically viable agro-ecological system and to enhance the quality of the environment, so that farm lands will remain productive indefinitely. Since SLM includes the maintenance over time of soil productivity, it therefore requires the combination of soil fertility treatment such as application of mineral and organic fertilizers with soil and water conservation measures including implementation of agronomic, soil management and physical measures, such as contour ridging, terracing, tied ridges or providing ground cover through mulching, use of plants and leaving crop residues (Durno, 1992; Woodfine, 2009).

The production systems which follow a predominantly ecosystem approach offer a range of productivity, socio-economic and environmental benefits to producers and to society at large based on a sustainable basis. According to Kassam and Friedric (2011), they are based on five overall objectives: (i) simultaneous achievement of increased agricultural productivity and enhanced ecosystem services; (ii) enhanced input-use efficiency, including water, nutrients, pesticides, energy, land and labour; (iii) judicious use of external inputs derived from fossil fuels (such as mineral fertilizers and pesticides) and preference for alternatives (such as recycled organic matter, biological nitrogen fixation and integrated pest management); (iv) protection of soil, water and biodiversity through use of 'minimum soil disturbance' and maintaining organic matter cover on the soil surface to protect the soil and enhance soil organic matter and soil biodiversity; and (v) use of managed and natural biodiversity of species to build systems' resilience to abiotic, biotic and economic stresses, with an underlying emphasis on improving soils' content of organic matter as a substrate essential for the activity of the soil biota. Furthermore, they noted that the farming practices required to implement these objectives will differ according to local conditions and needs.

Sustainable production practices involve a variety of specific and general approaches. Specific strategies must take into account topography, soil characteristics, climate, pests, local availability of inputs and the individual grower's goals. Despite the site-specific and individual nature of sustainable agriculture, several general principles can be applied to help growers select appropriate management practices: selection of species and varieties that are well suited to the site and to conditions on the farm; diversification of crops and cultural practices to enhance the biological and economic stability of the farm; management of the soil to enhance and protect soil quality; efficient and humane use of inputs; and consideration of farmers' goals and lifestyle choices (Agricultural Sustainability Institute, ASI, 2012).

A common philosophy in sustainable crop management is that a "healthy" soil is a key component of sustainability. This implies that a healthy soil will produce healthy crop plants that have optimum vigour and are less susceptible to pests. On the other hand, crop management systems that impair soil quality often result in greater inputs of water, nutrients, pesticides, and/or energy for tillage to maintain yields. In sustainable crop management systems, the soil is viewed as a fragile and living medium that must be protected and nurtured to ensure its long-term productivity and stability. Methods to protect and enhance the productivity of the soil include using cover crops, compost and/or manures, reducing tillage, avoiding traffic on wet soils, and maintaining soil cover with plants and/or mulches (Agricultural Sustainability Institute, ASI, 2012).

One of the most severe threats to the sustainability of agricultural crops production in Nigeria is declining productivity as a result of the loss of soil fertility. Soils in Nigeria suffer deficiency common to the tropical soil such as low percentage organic matter and nitrogen, shallow depth and high acidity which predispose about 63% of agricultural soils in Nigeria to low productivity (Lekwa and Whiteside 1996). The Federal Ministry of Agriculture and Rural Development (FMARD, 1987) blamed this situation on the inappropriate land and agricultural management practices in many communities across the country. Such agricultural practices have been fingered in natural vegetation transformation causing loss of soil water content, soil structure and porosity. As such, soil depletion in Nigeria can generally be attributed to the following: increase population density and pressure on agricultural lands; over grazing due to increased stocking rate; wide spread deforestation due to reduced fallow period; incompatible and unsustainable agricultural practices; and excessive use of fertilizers and other agro-chemicals (Emma-Okafor *et al.*, 2010).

In spite of the wide spread knowledge (some even indigenous) about cropping patterns such as fallowing and crop rotation known to significantly contribute to soil sustainability and other soil water and nutrient conservative measures which could help to remedy soil condition, the situation has continued to be grim. It is uncertain if farmers in Kwara State, like in many other parts of Nigeria are taking full advantage of the farming practices required for improved and sustainable agricultural production. Studies by Okoye (1998), Marshall (2004) and Rezvanfar *et al.* (2009) have shown that socio-economic characteristics including age, educational background, family size and farm size influence farmers' adoption of land management practices Therefore, the ability of crop farmers in Kwara State to adopt sustainable agricultural practices is probably affected by their socio-economic conditions. As such, this study sets to answer the following research questions:

what are the socio-economic characteristics of crop farmers in Kwara State, Nigeria?

what cropping patterns are common to them?

what soil conservation methods do they practice?

What relationship exist between selected socio-economic characteristics of crop farmers in the State and their adoption of sustainable agricultural practices?

what militates against successful adoption of sustainable agricultural practices among farmers in Kwara State?

Objectives

The main objective of the study was to assess sustainable agricultural practices in food crop production among small scale farmers in Kwara state, Nigeria. The specific objectives of the study were to:

1- Determine the socio economic characteristics of crop farmers in Kwara State;

2- Ascertain the cropping patterns common among them;

3- Highlight the soil conservation methods adopted by the farmers;

4- Examine the relationship between selected socio-economic characteristics of the farmers and their adoption of major agricultural land management practices; and

5- Investigate the constraints to adoption of sustainable agricultural practices among crop farmers in Kwara State, Nigeria.

Literature review

The adoption of land management practices is multidimensional with numerous factors affecting the willingness of farmers to use various conservation practices (Rezvanfar *et al.*, 2009). The effects of most socio-economic variables on the adoption of land management practices have not been conclusive and have been noted to vary with location given the divergent reports available from existing literature.

Higher educational attainment is characteristically linked with increased awareness and hence better adoption. It has been reported that educational attainment has a positive and significant relationship with the adoption of sustainable land management practices among farmers (Okoye, 1998; Deininger *et al.*, 2003 and Pender *et al.,* 2003). While Clay *et al.* (1998) reported educational attainment to be insignificant to the adoption of land management practices, Warriner and Moul (1992) presented education as having a negative relationship with adoption. According to Okoye (1998), age is positively correlated with the adoption of sustainable land management practices; however, Abd-Ella *et al.* (1981) reported a negative relationship between the two. Pender and Gebremadhin (2004) concluded that male headed households were more likely to use contour plowing and manure on a regular basis, Mulat *et al.* (1997) held a contrary opinion from a study carried out in the Ethiopian highlands in which male –headed households were found to most likely use less soil management practices. A high household size could be expected to imply more labour to carry out land management practices, however, Shiferaw and Holden (1998) confirmed a negative correlation between family size and land conservation practices. Holden and Yohannes (2002) however found no relationship between family size and land management practices.

Farm size has been positively linked to the adoption of land management practices (Cramb, 1999; Hagos, 2003; Demeke, 2003 and Teklewold, 2004). Deininger *et al.* (2003) reported a negative relationship between farm size and some land management practices such as tree planting. Land tenure affects adoption of land management practices primarily by influencing transferability of property rights thereby affecting the reversibility of land investments and ability to use land as collateral. Ayalew *et al.* (2005) opined that where land tenure is insecure, chances of sustainable land management practices are slim. Deininger *et al.* (2003) observed that access to agricultural extension services has a positive relationship with the adoption of land management practices. This view was also held by Marshall (2004) and Okunade (2006).

MATERIALS AND METHODS

The study area is Kwara State, Nigeria. With a total of sixteen Local Government Areas, the state has a land area of about 36,825 km^2 (Federal Office of Statistics, 1999) and a population of 2,371,089 (Kwara State Planning Commission, 2007) Major tribes in the State are Yoruba, Nupe, Fulani and Baruba. The state is located between latitudes 7^{0}45'N and 9^{0}30'N and longitude 2^{0}30'E & 6^{0}25'E. The topography is mainly plain to slightly gentle rolling lands. The mean annual rainfall ranges between 1000mm and 1500mm. Average temperature ranges between 30^{0}C and 35^{0}C. The rainy season in the state falls between March and October with a short break in August. The dry season is between November and February. The climatic condition favors the cultivation of arable crops including millet, cassava, yam, cowpea, maize, and rice. As obtainable in other States of Nigeria, Agricultural extension service delivery has been largely public and administered by the Agricultural Development Projects (ADPs) under the supervision of the State Ministries of Agriculture. A number of special agricultural development schemes/projects embarked upon by the Government and some non-governmental organizations (NGOs) over time have also addressed agricultural extension in the State.

Small scale crop farmers in the study area constituted the population for this study. A three stage random sampling technique was used to select the sample for the study. The first stage involved the random selection of four (4) Local Government Areas: Edu, Asa, Kaiama, and Ifelodun out of the sixteen Local Government Areas in the State. The second stage involved the random selection of three (3) villages from each of the selected Local Government Areas using the Kwara State Agricultural Development Programme (KWADP) village listing. The third stage was the random selection of twelve (12) small scale farmers from each of the villages selected. Information on the number of the total farmers in each village was obtained from the village council. This was used in preparing the list of the farmers that form the sampling frame from which the random selection of the farmers was carried out. Every other farmer was thereafter randomly selected from the list. A total of one hundred and forty four (144) respondents were surveyed but responses from only one hundred and thirty five (135) farmers were found to be useful for the study.

Data for the study were obtained by the administration of a structured questionnaire by well trained enumerators. Descriptive statistical tools employed to analyze data collected were frequency, percentages and means. A four point Likert-type scale was employed to investigate constraints to the adoption of sustainable agricultural practices. The scale was graded as follows: Not severe = 1, Less severe = 2, Severe = 3 and Very severe = 4. Respondents rated various possible constraints based on their perception of the severity of the constraints to them. To examine the causal relationship between the use of soil management practices and some selected socio-economic variables, binary logistic regression model was adopted. The regression model in its linear form is presented as follows:

$Y = f(X_1, X_2, X_3, X_4, X_5, X_6, u)$

Where Y_i = use of ith soil management practice (used =1, 0 otherwise)

X_1 = Total farm size (ha)

X_2 = Age of farmer (years)

X_3 = Education level

X_4 = Number of visits by extension agents in the year.

X_5 = Number of farm plots

X_6 = Household size

X_7 = Error term

RESULTS AND DISCUSSION

Socio-economic characteristics of the respondents

The socio-economic characteristics of the respondents including age, household size, sex, marital and educational status are as presented in table 1.

Table 1 shows that with a mean age of 50.72 years and close to 60% of the respondent being 50 years and above, the youth who are known to possess the physical strength required for crop farming are largely not as involved in agriculture as expected. This is particularly significant giving the fact that youths have been reported to constitute about 60% of the population of the Nigeria (Vision 2010 Report, 2005) and that they account for 70% of the State population (Akinyemi, 2011). This may not be unconnected with the fact that agriculture is increasingly being rejected among the youths particularly because of preference for white collar jobs. The Majority of the respondents have household sizes of between 5 and 10. This is of particular importance because of the implication of household size on the availability of farm labour. The dominance of male in crop farming in the study area is expressed in the fact that 92.6% of the respondents were male. Only 28.9% of the farmers had a minimum of primary school education. Education has been known to influence adoption positively. However, the farmers are poorly educated on the average and this may affect their level of adoption of sustainable agricultural practices.

Occupational characteristics

In view of their importance on the use of sustainable crop and soil management practices,

Table 1: Socio-Economic characteristics of the respondents.

Characteristics	Frequency	Percentage
Age of Respondents		
>40	12	8.9
40-49	44	32.6
50-59	69	51.1
60 and above	10	7.4
Total	135	100
Household Size		
<5	4	3
5-10	125	92.6
>10	6	4.4
Total	135	100
Sex of the Respondents		
Female	10	7.4
Male	125	92.6
Total	135	100
Marital Status		
Married	127	94.1
Divorced	2	1.5
Widower	6	4.4
Single	0	0
Total	135	100
Educational Status		
No formal education	51	37.8
Adult education (out of school)	10	7.4
Quranic education	35	25.9
Primary education	24	17.8
Secondary education	4	3
Tertiary education	11	8.1
Total	135	100

Source: Field survey, 2012

this section is based on the discussion of the occupational characteristics of the farmers. Details of such characteristics including the place of farming as a source of income, mode of land acquisition, source of capital for farming, farm size, number of farm plots and contact with extension agents are presented in table 2.

About two thirds of the respondents have farming as their primary occupation. Only 35% of the farmers have full ownership titles by having purchased their farmland. The ownership structure is important as farmers may not be willing to expend effort towards conservation practices on land temporarily held by them. The fact that 83% of the respondents rely on their meager personal savings may have implications for the adoption of agricultural practices that are financially demanding. Mean size of farmland under cultivation was 2 ha. Some agricultural practices such as fallowing are known to require more land area. Inadequate land area may therefore pose a problem in the adoption of such practices. An abysmally poor level of extension contact was recorded among the respondents with only 6.7% of them having extension contact. Thus the use of the sustainable practices might have been affected negatively due to low level of enlightenment on the need for farmers to adopt them by the extension agents.

Cropping patterns among respondents

Certain management practices are dependent on the type of cropping systems adopted by the farmers. This is based on the diversity of cropping potentials in the country at large and the study area in particular. Table 3 therefore presents the distribution of the respondents according to their cropping systems.

The most common cropping pattern among the respondents is the maize/cassava mixture. Table 3 reveals that 19.1% of the respondents followed maize/ cassava intercropping on

Table 2: Distribution of respondents according to occupational characteristics.

Characteristics	Frequency	Percentage
Farming as primary occupation		
Yes	89	65.9
No	46	34.1
Total	135	100
Mode of land Acquisition		
Borrowed	7	5.2
Inherited	63	46.7
Communal	17	12.6
Purchased	48	35.6
Total	135	100
Sources of capital		
Credit	11	8.1
Internal source (friends & family)	12	8.9
Personal saving	112	83
Total	135	100
Farm size (Ha.)		
<2	76	56.3
2-2.99	19	14.1
3-3.99	20	14.8
>3.99	20	14.8
Total	135	100
Contact with Extension Agent		
Yes	9	6.7
No	126	93.3
Total	135	100
vi. No of Farm Plots		
1	47	34.8
>1	88	65.2
Total	135	100

Source: Field survey, 2012

21.87% of the entire farm land. Further analyses revealed that half of the respondents (48.60%) adopted cereal-based cropping systems comprising maize/cassava and maize/melon intercropping as well as maize and, sorghum mono-cropping which cover about half of the entire farm land (51.48%). Furthermore, heavy nutrient drawers such as cereals (maize, sorghum) and tubers (cassava and yam) planted either individually or in various combinations were planted on 58% of the entire farm land by about 3/5 of the farmers (59%). This is in sharp contrast with the cultivation of legumes with only 27.7% of the respondents growing leguminous soil conservation crops (soya beans and melon) on about 24% of the total land area. While the food needs of the farmers is very important in dictating the cropping system adopted, it is also possible that the farmers were not properly informed about the importance of legumes in nutrients fixation as well as their cover cropping characteristics which are beneficial in soil conservation.

Agricultural land management practices adopted by respondents

Various agricultural practices adopted by the farmers which have effects on agricultural sustainability including mulching, use of cover crops, crop rotation with or without legumes, irrigation, composting, use of organic manure, and fertilizer application are presented in table 4.

As shown in table 4, about half of the respondents (48.6%) engaged a minimum of three management practices. Crop rotation was the major single crop management practice adopted by all (100%) of the farmers. On the other hand, irrigation was the least adopted agricultural practice with only 29.3% of the respondents involved in it. This may be connected to the high cost of irrigation. As a sole management practice, crop rotation alone was adopted by 22.2%

Table 3: Cropping patterns among respondents.

Cropping System	Frequency	Percent	Farm Size (ha)	% Farm Size
Maize/ Cassava	53	19.1	67.2	21.87
Maize/ Melon	32	11.5	37	12.04
Cassava	39	14	39	12.69
Maize	34	12.2	36.5	11.88
Soya Beans	31	11.2	26	8.46
Yam	22	7.9	18	5.86
Sorghum	16	5.8	17.5	5.69
Cassava/ Soya beans	14	5	10.5	3.42
Others	37	13.3	55.6	18.09
Total	278	100	307.3	100

R^2=0/82 D.W=2.02

of the farmers. Further analysis however revealed that only 31.20% of the respondents practice crop rotation with legumes. This implies that the majority of the respondents wrongly applies crop rotation with non leguminous plants. The majority of the respondents (52.40%) apply inorganic fertilizer in production while only 16.80% use organic manure and/or composting. Further analysis of the data revealed that only 5.40% (1.6+1.9+1.9) of the respondents apply both inorganic fertilizer and compost or organic manure(The implication is that while it is desirable to apply various management practices, about 94.60% of the respondents who did not include both inorganic and organic fertilizers in their management practices can be said not to apply sustainable land management practices in production. This is because use of both inorganic and organic manure constitute what is generally termed integrated land management system.

Effects of selected socio-economic factors on usage of soil management practices among respondents

This section presents the effect of selected socio-economic characteristics on soil manage-

Table 4: Distribution of respondents according to agricultural land practices adopted.

Management Practices	Frequency	Percentage
Mulching, fertilizer, cover crops, crop rotation	6	1.9
Mulching, cover crops, crop rotation	5	1.6
Mulching, cover crops, crop rotation with legumes	1	0.3
Mulching, crop rotation	13	4.1
Mulching, crop rotation with legumes	3	0.9
Cover crops, crop rotation, crop rotation with legumes	12	3.8
Cover crops, crop rotation, irrigation	2	0.6
Crop rotation	70	22.2
Cover crops, crop rotation, composting	1	0.3
crop rotation, crop rotation with legumes, organic manure	15	4.7
crop rotation, organic manure	5	1.6
crop rotation, organic manure, irrigation	10	3.2
crop rotation, composting, irrigation	5	1.6
crop rotation, irrigation	14	4.4
Fertilizer, cover crops, crop rotation	14	4.4
Fertilizer, crop rotation	57	18
Fertilizer, crop rotation, crop rotation with legumes, composting	5	1.6
Fertilizer, crop rotation, crop rotation with legumes, irrigation	42	13.3
Fertilizer, crop rotation, composting, irrigation	6	1.9
Fertilizer, crop rotation, irrigation	15	4.7
Fertilizer, crop rotation with legumes, composting, irrigation	6	1.9
Fertilizer, crop rotation with legumes, irrigation	9	2.8
Total	316	100

Source: Field survey, 2012

Table 6: Distribution of respondents according to constraints faced in the adoption of sustainable agricultural practices.

Constraints to Use of Sustainable Practices	Frequency of Respondents Based on Perception					Likert Mean Score	Rank
	Very Severe	Severe	Less Severe	Not Severe	Total		
Inadequate Supply of Fertilizer	96	24	9	5	134	3.57	1
Non-availability of Labour	62	63	9	0	134	3.4	2
Non-availability of Credit	57	70	9	0	134	3.4	2
Inadequate Knowledge of Modern Technique	70	45	19	0	134	3.38	3
High Labour Cost	54	73	4	3	134	3.32	4
Transportation Problems	29	76	24	5	134	2.96	5
High Cost of Irrigation Water	28	79	19	8	134	2.95	6
Low Produce Price	40	44	50	0	134	2.93	7
High Cost of Soil Management	17	67	49	1	134	2.75	8
Inadequate Improved Variety of Seed/Planting Material	20	53	52	9	134	2.63	9
Insufficient Land Availability	20	12	47	55	134	1.98	10
Insufficient Extension Services	18	14	22	80	134	1.78	11

Source: Field survey, 2012

ment practices among the respondents. Table 5 shows the result of the binary logistic regression to determine the socio-economic factors that affects the adoption of soil management practices among the respondents.

The results presented in table 5 show that 51.2%, 32.70%, 15.60%, 55.40% and 39.8% of variation in the use of crop rotation, cover cropping, bush fallow, mulching and use of fertilizer respectively were explained by the independent variables as implied by their pseudo R-square (R^2) values. The variables that had significant effects on usage of the soil management practices are farm size, age, education, number of contacts with extension agents, number of farm plots, and household size of respondents.

Farm size positively influenced the use of mulching at 1% level of significance implying that respondents with large farm size had higher probability of making use of mulching. While farm size negatively influenced the use of crop rotation at 10% level of significance which implies that as farm size decreases the farmers may likely increase the use of crop rotation.

The age of the household head was positive and significant at 10% probability level for the use of crop rotation, while the situation was different for the use of mulching which was negatively influenced by age of the household head at 5% probability level. This implies that an increase in the age of the household head will likely increase the use of crop rotation. Hence while the older

Table 5: Effects of socioeconomic factors on adoption of soil management practices.

Variables	Soil Management Practices				
	Crop rotation	Cover Cropping	Bush Fallow	Mulching	Use of Fertilizer
Farm size (X_1)	-0.398***	0.251	-0.012	0.727*	0.069
Age (X_2)	0.05***	-0.024	-0.001	0.20**	-0.0065
Education (X_3)	0.103	0.034	0.001	0.089	0.150**
Contacts with Extension Agents (X_4)	-0.649*	-0.288*	-0.33*	0.569*	0.321**
Number of Farm Plots (X_5)	-1.675*	-1.358**	-0.567	2.612*	-0.58
Household size (X_6)	-0.219***	-0.078	-0.021	0.134***	0.086
Constant	4.569	2.787	-0.356	-4.869	-1.546
R^2	0.512	0.327	0.156	0.554	0.398

Source: Data Analysis, 2012.

* P ≤ 0.01; ** P ≤ 0.05; *** P ≤ 0.1

Note: Logistic regression could not be computed for irrigation, composting and organic manure because there were very few respondents who adopted them in the syudy and were therefore eliminated in the regression.

and more experienced farmers will likely adopt crop rotation, the younger farmers will likely prefer to use mulching. This may be as a result of the fact that younger farmers are more innovative and better adopters than older farmers. Besides, preference for such practice may be due to the fact that it will reduce drudgery associated with highly labour intensive practices.

The educational level of the farmers also positively influenced the use of fertilizer by the farmers. This was found to be significant at 5% level of significance. The more educated the farmers are, the higher their likelihood for the use fertilizer.

While the coefficient of number of contacts with extension agents was negative for crop rotation, cover cropping and bush fallow, it was positive for mulching and the use of fertilizer. This implies that the use of Mulching practice and use of fertilizer will likely increase as the number of contacts with extension agents increases. On the other hand, as the farmers contact with extension workers reduce the farmers' probability of using crop rotation, cover cropping, bush fallow will likely increase. This implies that the farmers will likely rely more on their indigenuos knowledge and practices with reduced level of extension contact.

The number of farm plots cultivated positively influenced the use of mulching. As such, farmers will likely use more of mulching as their farm plots increase in number. This may be due to increased labour requirements in higher number of farm plots which the farmers may not be able to afford. Crop rotation and cover cropping practices were however negatively influenced by the number of farm plots. This indicates that the use of crop rotation and cover cropping is likely to reduce with increase in number of farm plots. Similar findings have been reported by (Aromolaran, 1998).

Constraints faced by farmers in the adoption of sustainable agricultural practices

Farmers are often faced with various challenges in the use of sustainable practices in crop production. An assessment of the constraints faced by the farmers is presented in table 6.

Inadequate supply of fertilizer, with a mean score of 3.57 was reported as the most severe constraint. 89.5% of the respondents identified with this constraint. This often results in low level of application due to inadequate access to fertilizer among the farmers. Other problems considered to be severe are inadequacy of credit and labour (both with mean scores of 3.4); inadequate knowledge of improved methods and high cost of labour were also rated high with average scores of 3.38 and 3.32 respectively. The inadequate knowledge of improved methods may be a result of the poor level of extension contact as earlirer reported. Other constraints included cost of transportation, high cost of adoption, low produce price and poor transportation. Poor transportation may be fingered in a number of other constraints particularly those involving high costs. The low produce price reported may make it difficult for the farmers to afford the various costs associated with sustainable agricultural practices. Contrary to expectation, the least severe problem reported by the respondents was insufficient extension service. Even though, the negative impact of inadequate extension contact is already implied in inadequate knowledge of improved method of land management, response by the farmers may suggest the fact that farmers have not had the opportunity to appreciate the need for extension service in improved and sustainable agricultural production.

CONCLUSION AND RECOMMENDATIONS

The study assessed sustainable agricultural practices in food crop production among small scale farmers in Kwara State, Nigeria. Findings from the study revealed that while farmers engaged in a number of sustainable agricultural practices, the majority who constitute about 95% of the farmers do not apply sustainable practices in food crops production (see above). In order to improve on this situation, this study makes the following recommendations based on its findings:

1- There is the need for better information on the need for complementary use of organic manure and inorganic fertilizer in food crop production among farmers in Kwara State, Nigeria;

2- Increased extension contact is of paramount importance for small scale crop farmers with specific emphasis on sustainable agricultural practices. This may call for a stakeholder conference, including farmers, on the appropriate land management practices that would suit their cropping systems. Besides, there may be the need for regular joint traiing programmes for the farmers on relevant agricultural land management practices;

3- Farmers should make concerted efforts to avail themselves of training opportunities on appropriate use of farm input especially farm land for optimum production that would take account of costs of adoption of the various land management practices adopted by the farmers. Such training programmes should be used as a springboard for updating the knowledge of the young farmers on appropriate land management practices for their cropping systems;

4- Agriculture should be accorded more priority in youth empowerment scheme of the Kwara State government. This is with a view to encouraging the youth to taking better interest in agricultural food crops production using sustainable practices; and

5- Land fragmentation which often results from high number of farm plots operated by the farmers should be curtailed. This therefore calls for consolidation of land holdings among the farmers which may be achieved through rejuvenation of farmers' cooperatives and other farmers' associations.

REFERENCES

1- Abd-Ella, M., Eric, O.H., & Warren, R.D. (1981). Adoption behavior in family farm systems: An Iowa study. Journal of Rural Sociology, 46(1), 42-61.

2- Agricultural Sustainability Institute ASI, (2012). Sustainable Agricultural Research and Education Program Retrieved on 30th September, 2012 from http:// sarep. ucdavis. edu/ sarep /about/def.

3- Akinyemi, D. (2011). Unemployment: A life-line for youths in Kwara. Vanguard, December 6th Edition Retrieved on 26th April, 2013 from http://www.vanguardngr.com/2011/12/unemployment-a-life-line-for-youths-in-kwara/.

4- Aromolaran, A.B. (1998). Economic analysis of soil conservation practices in South-western Nigeria. Issues in African Rural Development, Monograph Series; Monograph # 11 African Rural Sciences Research Networks

5- Ayalew, D., Dercon, S., & Gautam, M. (2005). Land tenure insecurity and investment in ethiopian agriculture: Evidence from panel data. Mimeo.

6- Ashley, C., & Maxwell. S. (2001). Rethinking rural development. Development Policy Review, 19(4), 395–425.

7- Brown, L.R., & Wolf, E.C. (2005). Reversing africa decline. World Watch Courier (June) Environmental and Development No 87 in Odoemenem, I. U. & V. Adebisi (2011) Sustainable Agriculture For Small Scale Farmers In Niger State, Nigeria Journal of Sustainable Development in Africa, 13(2), 225-235. Retrieved on 30th September, 2012 fromhttp://www.jsd-africa.com/Jsda/V13No2_Spring 2011 _B/PDF/Sustainable%20Agriculture%20for%20Small%20Scale%20Farmer.pdf.

8- Clay, D., Reardon, T., & Kangasniemi, J. (1998). Sustainable intensification in the highland tropics: Rwandan Farmers' Investments in Land Conservation and Soil Fertility, Economic Development and Cultural Change, 45 (2), 351-378.

9- Cramb, R., (1999). Smallholder adoption of soil conservation technologies evidence from upland Projects in the Philippines, University of Queensland.

10- Daramola, A.G., Ehiu, E., Ukeje, E., & Mclntire, J. (2007). Agricultural Export Potential in Collier, P. and Pattilo, C. (eds) Economic Policy Options for Prosperous Nigerian. London: Palgrave Macmillan, 2007.

11- Deininger K., Jin, S. and Adenew, B. (2003). Market and non-Market transfers of land in ethiopia: Implications for Efficiency, Equity, and Non-Farm Development. World Bank Policy Research Working Paper 2992. World Bank, Washington, DC.

12- Demeke, A.B. (2003). Factors influencing the adoption of soil conservation practices in northwestern ethiopia. Discussion Paper 37. Institute of Rural Development, University of Goettingen.

13- Durno, J. (1992). Integrated farming in the lowlands resources. London: Zed books Ltd.

14- Emma-Okafor, L.C., Ibeawuchi, I.I., & Obiefuna, J.C. (2010). Biodiversity conservation for sustainable agriculture in tropical rainforest of nigeria. New York Science Journal 2010, 3(1), 81-88. Retrieved on 30th September, 2012 from http://www.sciencepub.net/newyork/ny0301/15_1086_Biodiversity _conservation_ny0301.pdf.

15- FAO. (2009). Country support tool – for scaling-up sustainable land management in sub-saharan africa. Food and Agriculture Organization of the United Nations, Rome, Italy.

16- FMARD. (1987). Federal ministry of agriculture water resources and rural development, Fourth National Development Plan 1981-1985, National Planning Office, Lagos, Nigeria.

17- Federal Office of Statistics (FOS, 1995). Abstracts of Statistics, FOS Abuja, Nigeria .

18- Hagos, F. (2003). Poverty, institutions, peasant behavior, and conservation investment in Northern

Ethiopia. PhD thesis No. 2003:2. Department of Economics and Social Sciences, Agricultural University of Norway, As.

19- Holden, S.T., & Yohannes, H. (2002). Land redistribution, tenure insecurity and input intensity: A Study of Farm Households in Southern Ethiopia. Land Economics, 78: 573-590.

20- IFPRI 2007. The role of agriculture in development: Implications for sub- Saharan Africa Research Report 153.

21- Kassam, A., & Friedric, T. (2011). Conservation agriculture: Principles, sustainable land management and ecosystem services. Società Italiana de Agronomia XL Convegno Nazionale, Università degli Studi Teramo. Retrieved on 26th April, 2013 from http://www.fao.org/ag/ca/CA-Publications/CA_Teramo_Kassam_Friedrich.pdf

22- Kwara State Planning Commission, KWSPC, 2007. Kwara State Year Book. Kwara State, Nigeria.

23- Lekwa, G., & Whiteside E.P. (1996). Coastal plain sand of southeastern Nigeria. Soil science society of American Journals 50: 154-160. In Emma-Okafor, L. C., Ibeawuchi, I. I., & Obiefuna, J. C. (2010). Biodiversity Conservation for Sustainable Agriculture in Tropical Rainforest of Nigeria. New York Science Journal 2010; 3(1), 81-88. Retrieved on 30th September, 2012 from http://www.sciencepub.net/newyork/ny0301/15_ 1086 _Biodiversity_conservation_ny0301.pdf.

24- Marshall, G.R. (2004). From words to deeds: Enforcing Farmers' Conservation Cost-sharing Commitments. Journal of Rural Studies, 20: 157-167.

25- Mulat, D., Ali, S., & Jayne, T. (1997). Promoting fertilizer use in ethiopia: The implications of improving grain market performance, input market efficiency, and farm management. Working Paper 5, Grain Market Research Project, Ministry of Economic Development and Cooperation, Addis Ababa.

26- Okoye, C. (1998). Comparative analysis of factors in the adoption of traditional and recommended soil erosion control practices in Nigeria. Soil and Tillage Research. 45: 251-263.

27- Okunade, E.O. (2006). Factors influencing adoption of improved farm practices among women farmers in Osun State. Journal of Human Ecology, 19 (1), 45-49.

28- Pender, J., & Gebremedhin, B. (2004). Impacts of policies and technologies in dryland agriculture: Evidence from Northern Ethiopia In S.C. Rao (ed.), Challenges and Strategies for Dryland Agriculture. American Society of Agronomy (ASA) and Crop Science Society of America (CSSA) Special Publication 32. Madison: ASA and CSSA.

29- Pender, J., Gebremedhin, B., & Haile, M. (2003). Livelihood strategies and land management practices in the highlands of tigray, presented at the conference on policies for sustainable land management in the East African Highlands, April 24-26, 2002, United Nations Economic Commission for Africa, Addis Ababa (revised). International Food Policy Research Institute (IFPRI), Washington, DC.

30- Rezvanfar, A., Samlee, A., & Faham, E. (2009). Analysis of factors affecting adoption of sustainable soil conservation practices among wheat growers. World Applied Sciences Journal, 6 (5), 644-651.

31- Shiferaw, B., & Holden, S.T. (1998). Resource degradation and adoption of land conservation Technologies in the Ethiopian Highlands: A Case Study in Andit Tid, North Shewa. Agricultural Economics, 18: 233-247.

32- Teklewold, H. (2004). Risk and time preferences on soil conservation decision in the central Ethiopian Highlands of Ethiopia. M.Sc thesis. Addis Ababa University.

33- Thirtle, C., Irz, X. Lin, L. McKenzie-Hill, V., & Wiggins. S. (2001). Relationship between changes in agricultural productivity and the incidence of poverty in developing countries. DFID Report 7946.

34- Vision 2010 Report (2005). Youth, Development and Empowerment in In Adekunle, O. A., Adefalu, L.L., Oladipo, F.O., Adisa, R.S., & Fatoye, A.D. (2009). Constraints to youths' involvement in agricultural production in Kwara State, Nigeria. Journal of Agricultural Extension, 13 (1), 102 - 108

35- Warriner, G.K., & Moul, T.M.(1992). Kinship and personal communication network influences on the adoption of agriculture conservation technology. Journal of Rural Studies, 8 (3), 279-291.

36- Woodfine, A. (2009). Using sustainable land management practices to adapt to and mitigate climate change in Sub-Saharan Africa- Resource Guide Version 1.0, TerrAfrica, Regional Sustainable Land Management.

2

Risk Analysis and Strategy of Rice Farmers in Swampland in the Face of Climate Change Impact (Case in South Kalimantan Province-Indonesia)

Yudi Ferrianta [1*]*, Muhammad Fauzi Makki* [1] *and Suprijanto1 and Rifiana* [1]

Abstract

The impact of climate change does not only affect the production aspects but also on household income and other socioeconomic aspects. The research aims to determine the household economic behaviors of rice farmers in decision-making with relation to the production risks due to climate change and determine the farmer's adaptation strategies in the face of climate change. The approach taken is the production risk analysis using the coefficient variance. The result shows that the farmers' behavior to manage rice farming in the tidal area has neutral risk behavior. This means that if there is a risk or possibility of risk recurrence towards rice farming, thus the decision maker (farmer) still does not reduce or enlarge the scale of their business. Adaptation strategies adopted by farmers in swampland consists of short, medium and long term strategies. The improved adaptive capacity on farm level should be through planned adaptation whose development is conducted by the government to be synergistic and rely on autonomous adaptation that has been a tradition in the farmers community.

Keywords:
Swampland, Risk, Farmers' adaptation, Climate change

[1] *Lecturers in Faculty of Agriculture, Lambung Mangkurat University, Indonesia.*
* *Corresponding author's email: ferrianta@gmail.com*

INTRODUCTION

Climate change is almost certainly one of the most important development challenges faced by every country in the 21st century. Global warming has cause more unstable climate, such as changes in rainfall pattern and increased frequency and intensity of the extreme weather events; and it has led to increase the average of global sea level (ADB, 2009; Finger, and Schmid, 2007). The threat to the food agriculture sector due to the impact of climate change, of course, occur in the centers of swampland rice in various areas of Indonesia, including South Kalimantan (Makki and Ferrianta, 2012). The impact of climate change on rice farming was not only affected the production aspects, but also on household income and other socio-economic aspects. This matter evident when there is flood that hit North Hulu Sungai Regency as the center of food production in the swampland in 2008; the rice production on that regency is decrease by 22% due to crop damage (Daily of Banjarmasin Post; on June 2008). Economic growth in South Hulu Sungai is decrease due to the growth of agricultural sector contracted -1.52 in 2010 due to flood in the swampland.

Since the impact of climate change on swampland environment in South Kalimantan is increasingly felt the influence during the last six years; the effort such as policy both anticipation and mitigation implemented are still reactive; temporary; instantaneous, partial and tend to be oriented project so that it does not touch the problem root and not related (Makki *et al.*, 2009). If it allowed, not only affect on food security, but also a process of impoverishment that will increase the number of poor people. One effort that can be done to reduce and anticipation the impacts are adaptation of climate change through adaptation strategies in agriculture business management in swampland by farmers and economic strengthening their household by first performing on risk analysis.

The research article aims to determine the household economic behavior of rice farmers in swampland in terms of decision-making in relation to the production risks due to climate change and farmers adaptation strategies in the face of climate change. The approach taken is the production risk analysis using the coefficient variance.

MATERIALS AND METHODS

Data and sampling techniques

This research is conducted in two swampland agro-ecosystem in South Kalimantan both non-tidal swampland and tidal swampland. To represent the agro-ecosystem of tidal swampland is selected two regency namely Kuala Barito Regency and Banjar Regency. Represents the type of non-tidal swampland is selected North Hulu Sungai. This regency is purposively selected with the consideration that the area of swampland is the largest in South Kalimantan.

Representing the agro-ecosystem of tidal swampland is selected two regency namely Kuala Barito Regency (Batola) and Banjar Regency. Each regency selected two sub-districts purposively, that is Barambai District and Cerbon District in Batola Regency, and Aluh-Aluh District and Beruntung Baru District in Banjar Regency. The total number of villages selected in this research is 12 villages. From each of these villages, the farmers sample will be determined by proportionate random sampling. Overall the total samples of farmers as the primary data source is 240 respondents consisting of 180 farmer respondents as the sample for tidal swampland type, and 60 farmer respondents as the sample for non tidal swampland. The main requirement of sample is these farmers conduct rice farming of minimum four plant season in swampland.

Data analysis

The design of this analysis model attempted to determine the risk posibility that will occur over the use of production input in the certain farm land size. Particularly relevant to this analysis is shown with reference to the empirical model of production risk analysis in accordance with the rules of Singh (1980). The benefit of this model is to identify the subjective probability of each farmers sample on rice farming activities in tidal land in different risk situation. General design model used in the analysis of these data presented in Table 1.

The next analysis stage, after sought opportunities to each farmer example, then proceed with the cash flow analysis as recommended by Horne (1993).

To calculate and measure the magnitude of the

Table 1: Calculation model of subjective probability and possible of cost Incurred to purchase input production in accordance with the circumstances experienced by farmers.

Situations	Subjective Probability	Possibility of Input Cost (IDR/Hectare)			
		X1	X2	X3	Xn
A	P (A)	ΣX1A	ΣX2A	ΣX3A	ΣXiA
B	P (B)	ΣX1B	ΣX2B	ΣX3B	ΣXjB
Expectation Value	ΣP (n)	ΣX1n	ΣX2n	ΣX3n	ΣXijn

Description: A = normal situation B = Flooded (one of the impacts of climate change)

production risk possibility that will occur in instances on sample farmer group, followed by calculating the probability distribution of net cash flow, by first calculating the standard deviation as:

$$\vartheta_x = \sqrt{\Sigma(R_{xt} - E(R_{xt}))^2 P_{xt}} \quad (1)$$

Where, Rx_t = *net cash flow* of sample farmer of the x at t time

Px_t = probability that occur in the net cash flow of sample

Farmerof the x at t time

$E(R_{tx})$ = value expectation of *net cash flow* at t time

Expectation value of the net cash flow in t period sought by the general formula:

$$E(R_{xt})+\Sigma R_{xt}.P_{xt} \quad (2)$$

The next step is to compare sample group of farmers who have larger standard deviation, which can be used as indicator which can states the spread magnitude of rice production probability that earned by sample farmers. To decide the degree of risk size between the samples farmers compared, it can be determined by calculating the coefficient variance (CV,) by:

$$CV_2= \vartheta_x / E(R_{tx}) \quad (3)$$

Where: CV_2 = coefficient variance in the sample farmers of the x

θ = standard deviation of the sample farmers of the x

E (R_{tx}) = expectation value of the net cash flow at t time and the sample farmers of the x

In this case, if the value of coefficient variance (CV) of sample farmer-x in instances greater than the others, then the sample farmers decided to have higher degree of risk (Harwood *et al.*, 1999; Horne, 1983). The coefficient variance is the standard deviation distributed to each sample farmers and it can be expressed in percent. The larger the CV value, the greater the risk variability that will occur.

RESULTS AND DISCUSSIONS

Production risk analysis of rice farming in swampland

Rice farming in swampland in particular tidal is faced with the problem of production risk. Indication of production risk of rice farming in the tidal area is indicated by fluctuation or the amount output variation generated due to the possible use of input production under normal condition and under condition of high rainfall, thus also causes the income fluctuation and profits shown on farming the cash flow. Input use and the expenditure of production cost in the tidal area can be seen from the comparison in normal condition and high rainfall condition.

In addition to the output variation due to input variations and variations in income; discussion of risks also relates the occurrence of an event of the probability that can be measured. In this research, the probability that lead to failure or risk based on the subjective probability of each season the assumed value. Basic assumption is the possibility of production failure or loss possibility due to lack of farmers' ability to predict the changes of weather and climate. Because there are two comparison of subjective probability between the normal situation and situation of rice farming faced heavy rainfall, then when the one probability value is X then the the other probability is (1-X). In the case of rice farming in swampland, especially tidal area; the probability of normal situation rated 0.6 so assuming the other value is only 0.4.

The existence of the probability value, then it can be done the production risk assessment which can be calculated using the Coefficient Variance (CV). CV is the ratio of standard deviation with return expectation value of rice farming in the expected net cash flow derived by the average farmer. With the size of the CV, business activity analysis has been carried out with the same size, which is the risk for every net cash flow.

The result of the risk analysis of rice production in the tidal area gets the value of coefficient variance (CV) of 0.45 to 0.48 or can be expressed in a percentage of 45% - 48%. In other words, the risks covered by rice farmers in the tidal area especially in Banjar Regency is 45% - 48% of the value of return (net cash flow) obtained. This means that every IDR 1 return (net cash flow) received by rice farmers in the tidal area will get risk of IDR 0.45 to 0.48. This value is the ratio of the standard deviation of the net cash flow (in normal climate and climate with very high rainfall).

When linked with the farmėr's behavior in two different seasons, then the CV value of 0.45 to 0.48 indicates that the rice farmers in tidal area in Banjar Regency have neutral risk behavior. It is appropriate that proposed by Debertin (1986). CV value is relatively closer to the CV value when farmers face high rainfall and also similar to the normal situation. This means that if there is a risk or possibility of risk recurrence towards rice farming, thus the decision maker (farmer) still does not reduce or enlarge the scale of their business. The data shows that the area of rice farming between the seasons is not different. This matter due to rice farming is a business that seems to "shall" be implemented and as much as possible the wetland should not suffer harvest failures, no matter how great the risks faced.

Strategies of farmers adaptation and anticipation of climate change

The agricultural sector is highly vulnerable towards climate change because it affects the cropping pattern, time of planting, production, and quality of results. Thus the necessary effort is relatively fast and able to reduce the negative effects of climate change. One effort that can be done through adaptation of rice plants by local farmers or farmer groups.

The general target of adaptation is to minimize vulnerabilities, develop resilience, and it developing them if the circumstances allow. Vulnerability of farmers to climate change is determined by the interaction potential impacts of climate change and adaptation capacity of farmers (Boer, 2007). The potential impact is the resultant of farmers from sensitivity farmers and exposure due to variations in sharp climate. On the other hand, farmers' adaptive capacity is determined by the internal conditions of farmers and external supporting factors. Internal conditions include the farmers' knowledge and mastery of farming technology, the ability of capital, and managerial skills, while the most important supporting factors are the availability of infrastructure, innovative technology packages, and institutional.

According to Boer (2007) and Irianto (2010); adaptive capacity is defined as the degree of adjustment that occurs in practice, process, or structure that can alleviate or overcome potential damage / loss or benefit from the opportunities that may exist. Factors that influence the adaptive capacity are socioeconomic, technology, infrastructure, and government policy factors.

The situation and condition of the factors that influence the adaptive capacity are vary so adaptive capacities between regions, between communities, and even between individual farmers also vary. The implication, critical nodes of increase strategy of adaptive capacity between regions, or between communities also vary. Information about diversity and the factors that influence it is very useful to support policy formulation, programming and implementation strategies of adaptation towards climate change (Brooks *et al.*, 2005).

Policy effectiveness and adaptation program are not only determined by the accuracy design and its instrument but also determined by the accuracy of the strategy adopted to implement them (FAO, 2007; Finger and Schmid, 2007; IPCC, 2007). This is an implication of the following conditions. Effectiveness of adaptation is determined by a combination of: (i) the selected instrument, (ii) the methods applied, (iii) the level and quality of public participation (target groups), and (iv) the provision efficiency of supporting factor (especially infrastructure and institutional), and (v) the consistency of policy and programs.

Although the shape or type is varied but the

target of adaptation to climate change in principle is to minimize the vulnerability, build resilience, and develop the ability to utilize the favorable opportunity of the situation and the conditions posed by climate change (Brooks and Adger, 2005). Vulnerability is the degree of easy or not subjected, damaged, lost, or weakens its existence. Unlike the vulnerability, resilience refers to the design ability to survive, recover, or even evolved from conditions created from the result which appear related to climate change (ECA, 2009; Harmony, 2006). Keyword to address this challenge is to strive for the vulnerability of farmers to unfavorable climate conditions can be reduced. In other words, farmers should be conditioned to be more resistant, robust, and resilience to confront the climate change.

Historically, every individual or community of farmers has always faced with the challenges to adapt with their environment, either physical or socio-economic environment. Therefore, the essence of policy and adaptation programs should be reoriented to accelerate the process of improving their adaptive capacity. Therefore, the main actor adaptation to climate change in the agricultural sector is the farmer, thus the forms of adaptation which has been independently developed by farmers or farmer communities (autonomous adaptation) is an important basic capital (ADB, 2009). The implication, increasing adaptive capacity through planned adaptation that the development is conducted by the government or at least should be resting synergistically with autonomous adaptation that has been a tradition in the farmer community (Boer, 2007; Irianto, 2010).

Adaptation can be done in the form of land and water resource management in an optimal and sustainable, crop management and crop adapted to local climate conditions, the use of agricultural production infrastructure that effective and efficient, and the application of appropriate agricultural technologies that adaptive. Agricultural environmental management strategies can be done through a variety of planning efforts, adjustments, between agricultural practices, resource management and application of agricultural technologies to address climate change impacts and anomalies. The strategy adopted consists of short, medium and long term strategies, include:

1. Short-term strategies

a. Determine the crop calendar based on the valid data elements of climate and longer data series.

b. The selection of superior plant varieties tolerant to drought exposure, and short-lived as the anticipation of El-Nino phenomenon.

c. Monitoring of pests and diseases that commonly occur when a long rainfall season and the change of seasons.

d. Development and empowerment of institutional farmers, such as the integration of climate field school to the integrated crop management field schools and integrated pest control field school.

2. Medium-term strategy

a. Continuous monitoring of the phenomenon of climate elements change, especially rainfall, air temperature and humidity.

b. Repair and maintenance of facility and folder infrastructure and irrigation that already exist.

c. Application of soil and water conservation techniques, such as check dams, and ponds in drought-prone areas.

d. Establishment of institutional managers and water users

e. Empowerment of farmers through coaching and mentoring to confront climate change and climate anomalies of the agricultural business.

3. Long-term strategy

a. Development planning of the agricultural sector is more integrated, systematic and comprehensive by considering the various aspects related with the performance of the agricultural sector, especially the agroclimatology aspect.

b. Making rice roadmap in swampland in the era of climate change

c. Involvement the participatively community in every planning of agricultural development.

d. The pattern of good coordination among government agencies, especially those directly related to the agricultural sector through the synchronization and harmonization work program.

e. Monitoring area that often affected due to the climate change and anomaly periodic and continuous.

f. Dissemination of weather and climate information quickly, accurately and actually.

In addition to above, the strategy of economic development in the agricultural sector can be done by agribusiness approach. Agricultural development through agribusiness approach is a step in the right and proper because this approach vertically integrates upstream and downstream activities and horizontally various sectors so as to create a reasonable profit for the farmer. Agribusiness institutions that need to be developed is a group of farmers, farmers' water user associations, cooperatives and rural financial institutions, production facilities and infrastructure providers, marketing, and tool and machine of agriculture services. In addition to these two institutions, the empowerment of field extension also needs to be done because they are dealing directly with farmers. Empowerment of farmer organization aims to increase farmer participation in institutional farm. Community institutions such as the traditional institution is participate to moves the communities in together activities, grow and enhance the community's role in the activities initiated by the local government, as well as increasing self-reliance of farmers on agricultural development. Meanwhile, village unit cooperatives play to help the farmer members in obtaining credit, production fasility, and agricultural tools and accommodating and market outcomes.

CONCLUSIONS

Extreme climate change such as heavy rainfall in the rice farming in swampland either tidal swampland and non tidal swampland has negative impact on production. based on the value of the coefficient of variance (CV) behavior of farmers in the face of risk in two different seasons worth 0.45 to 0.48. This suggests that farmers of rice farming in tidal area have neutral risk behavior. CV value is relatively closer to the CV value when farmers face high rainfall and also relatively similar to the normal situation. If there is a risk or possibility of recurrence risk towards rice farming the decision maker (farmer) still does not reduce or enlarge their Adaptation strategies adopted by farmers in swampland consist of short, medium and long term strategies. Short-term strategies include: (a) Timing of planting (crop calendar) based on information of valid climate data that and based on the experience of the farmers themselves, (b) Selection of superior plant varieties tolerant to drought exposure, and short-lived as the anticipation of El Nino phenomenon; (c) Monitoring of pests and diseases that commonly occur when the long rains season and the change of seasons. (d) Institutional development and empowerment of farmers, such as climate field school integration to the integrated crop management field schools and integrated pest management field school.

RECOMMENDATIONS

Anticipation of climate change should referrals from various aspects, such as climate change adaptation, diversification of food production, social and cultural development of society, economy and institutional strengthening of farmers, and policies in favor of agriculture. In non tidal swampland, adaptation of climate change on farm level should be followed by the availability of input factors of production earlier. Seed subsidy policy strongly supports the adaptive capacity of the economy. In the tidal area, in addition to the availability of production factors as time of planting; also beyond the ability to substitute the non-family hired labor with family labor will be able to reduce the decline of income. Supporting factor such as role of breeding and seed certification center is to produce the seed that is capable of adapting to climate anomalies. Similarly, the provision of business credit to be able to support farmers businesses and it also must calculate the change adaptation of planting time. There must be the weather information that can be accessed by farmers' groups and an increase in the ability of extension worker in the field to be able to help farmers in the group level to adapt the climate change. In this era of climate change; diversification activities in swampland on-farm level (non-rice farming) need more attention. Fisheries and poultry activities especially ducks have great opportunities in an effort to increase household income and household economic strengthening of farmers from non-rice farming.

Income factor of non-farm level has a role in strengthening of household economic of rice farming affected by climate change. Therefore, the maximization of income from non-farm activities can be done through increasing skills and employment opportunities. Adaptive capacity of farmers in swampland should be based on strengthening the synergies between adaptation that has historically been developed independently by farmers with the planned adaptation introduced by the government.

REFERENCE

1- Asian Development Bank (ADB). (2009). *Economics of climate change in South East Asia: A Regional Review.* Manila. Philippines.

2- Boer, R. (2007). *State of arts agroclimate for strategy and agricultural adaptation to climate change. Centre for climate risk and opportunity management in south east asia and pacific.* Bogor Agricultural University. Bogor.

3- Brooks, N., & Adger, W.N. (2005). *Assessing and enhancing adaptive capacity, in adaptation policy frameworks for climate change: Developing strategies, policies and measures*, Cambridge University Press, Cambridge.

4- Brooks, N., & Adger, W.N. (2005). *The Determinants of vulnerability and adaptive capacity at the national level and the implications for adaptation.* Global Environmental Change 15 (2005) 151-163. Elsevier Ltd. All rights reserved.

5- Debertin. (1986). *Agricultural production economics.* Macmillan Publishing Company, New York.

6- ECA, (2009). *Shaping climate-resilient development: A framework for decision- making, a report of the Economics of Climate Adaptation (ECA) working group, climate works foundation, global environment facility, european commission*, McKinsey & Company, The Rockefeller Foundation, Standard Chartered Bank and Swiss Re.

7- Finger, R, & Schmid. S. (2007). *Modelling agriculture production risk and the adaptation to climate change.* Paper Presented for Presentation at the 101st EAAE" Management of Climate Risk in Agriculture. Berlin. MPRA Paper No. 3943

8- Harmony, A. (2006). *Socioeconomic impacts of climate change.* Gunadarma University. Jakarta.

9- Harwood, J., Heifner, R., Coble, K., Perry, J., Somwaru, A. (1999). *Managing risk in farming: concepts; research and analysis.* Agricultural Economics Report No. 774 US Department of Agriculture.

10- Horne, J.C.V. (1983). *Financial management and policy.* Stanfort University. Prentice Hall of India.

11- IPCC (2007). *Climate change 2007: Impacts, adaptation and vulnerability.* Contribution of working group ii to the third assessment report of the intergovernmental panel on climate change. JJ. Mc Carthy, O.F. Canziani, N.A. Leary, D.J. Dokken and K.S. White (Eds). Cambridge University Press. Cambridge.

12- Irianto, G. (2010). *Adaptation to Climate Change in Agriculture.* Department of Agriculture. Jakarta.

13- Makki, M.F; Suslinawati & A.J. Kirnadi. (2009). *Optimization of Farming Business System Model Based on Rice in Swampland South Kalimantan.* Competitive Grant Reports. Lambung Mangkurat University Research Institute. Banjarmasin.

14- Makki, M. F., & Ferrianta, Y. (2012). Impacts of Climate Change on Productivity and Efficiency Paddy Farms: Empirical Evidence on Tidal Swamp Land South Kalimantan Province–Indonesia. *Journal of Economics and Sustainable Development*, 3(14), 66-72.

15- Singh, I. J. (1980). Farm decision under uncertainty, improving farm management teaching in Asia. The Agricultural Development Council, Inc.

3

Major Barriers to Application of Good Agricultural Practices (GAPs) Technologies in Sustainability of Livestock Units

Marzieh Panahzadeh Parikhani [1], Fatemeh Razzaghi Borkhani [1], Hossein Shabanali Fami [1], Naser Motiee [1] and Abotaleb Hosseinpoor [2]*

Keywords:
Good Agricultural Practices (GAPs), Barriers, Livestock systems, Sustainability, Safety food

Abstract

The main purpose of this study was to investigate the major barriers to application of Good Agricultural Practices (GAPs) technologies in sustainability of livestock by ranchers of Meshkinshahr. A sample of 120 farmers was selected by using proportional random sampling method. Data were collected by means of a questionnaire. Validity of questionnaire was determined through Agricultural Jihad exports of Meshkinshahr County and some faculty members at University of Tehran, Department of Agricultural Management and Development. Cronbach's alpha was used to estimate the reliability. The reliability was found to be acceptable. The results of Factor Analysis showed that infrastructure barriers, informational- educational barriers, institutional-support barriers, personal barriers, economical barriers were the five barriers to application of Good Agricultural Practices (GAPs) technologies in livestock unites. These factors explained 67.23 percent of the total variance.

[1] *Department of Agricultural Management and Development, Faculty of Agricultural Economics and Development, University of Tehran, Karaj, Iran.*
[2] *Assistant Professor in Agricultural Development Faculty of Agricultural, University of Mazandaran.*
* *Corresponding author's email: hfami@ut.ac.ir*

INTRODUCTION

There is growing demand for sustainable agricultural development in response to the environmental impacts of conventional agriculture (Rasul and Thapa, 2004). In recent years, maintaining the product, regarding environmental considerations coupled with reduce production costs, producing a healthy crop and empowerment of farmers, is caused extension of widespread sustainable agriculture. One of the major goals of sustainable agricultural systems is decreasing use of inputs in agriculture; Good Agricultural Practices (GAP) is one approach which deals with these issues. Good Agriculture Practices standard included human health and the environment, safety food and access to sustainable agricultural development, (achieve environmental, economic and social sustainability) which attention the sustainability of on-farm activities to certain safety and quality of food and non-food agricultural crops.

Good Agricultural Practices according to study done by Banzon *et al.* (2013) refer to approaches sustainability agriculture, safety and quality food and enables farmers to absorbent new market vantages by improving supply chain control, improving natural resource utilization, workers health, and working conditions, consumers and farmers families' health and creating new market opportunities for farmers in developing countries.

Livestock is source of income at least 20 million households who live in pastoral. As well as livestock provides the main income source at least 200 million small-holder farmers in Asia, Africa and Latin America (FAO, 1994). Livestock plays an important part in rural development in third world countries. This section is as producer of food, promoting of agricultural products level and providing of services and surplus economic goods and earn cash. Integrating livestock and agriculture make annual sustainable employment. As well as sale of livestock production provided necessary capital for the purchase of agricultural inputs and generally any kind of investment in the field. The animal is often considered as the main source of financing for agricultural households (Stainfeld and Mack, 1995) pay attention to there is the demand for healthy livestock and agriculture products in the world by consumers and buyers, Thus application of safety product regulations and international and national requirements is expanding to provide satisfactory of agricultural and livestock production consumers with production safety and healthy agricultural crops. one of this standards is Good Agricultural Practices (GAP) approach.

According to this definition, we can say that a healthy product or can be the result of application the guidelines and methods for organic production (agricultural biotechnology), which in this case is applied organic product. Or crops following the actions and practices of other methods of production can be obtained that contaminants and toxic ingredients in the product is Maximum Residue Limit (MRL). One of the guidelines of good agricultural practices that lead to GAP product (Koohsar, 2012).

GAPs for beef cattle farming are used to help farmers increasing their production of Livestock, with good quality, investment, safety production, the most using of existing resources, sustainable beef cattle production, and non polluted environment (NBACFS, 2005).

Now, the question this is: why despite the emphasis on healthy and organic produce, in practice, these methods are less used? It seems that due to the barriers and problems in the application of safe production technologies such as good agricultural practices. Undoubtedly, identifying of these challenges could facilitate adoption of good agricultural practices and provided development of successful GAP programs.

According the literatures of adoption technology lack of knowledge and awareness of stakeholders and producers is considered major constraint on application of good agricultural practices. (Banzon *et al.*, 2013) with increasing of knowledge of GAP probably application GAP intensified in the food safety programs. It seems that agricultural extension facilitates enhancement of awareness and participation of farmers.

Finally, among the exits variables as documented in the literatures major constraints according to social, cultural and economic condition of application of technologies are presented in this research. In order to have a more effective impact on promoting wide spread adoption of GAP technologies, that concentrating on barriers to adoption can be more effective.

Quality of below the standard of domestic beef due to the problem on the efficiency of production since beef cattle raising farmers lack knowledge and understanding on correct beef cattle raising. Extension of inadequate technology and poor efficiency in beef cattle – raising (Suppadit, 2003) are presented concentrating on barriers to application of good agricultural practices.

NuruIslam *et al.* (2012) stated that reasons for farmers being unable to adopt GAP standard included the lack or scarce access to credit for investment, lack of technical support. Therefore Malaysian GAP certification scheme require upgraded, extended and monitored to ensure the quality of the produce.

Razzaghi Borkhani *et al.*, (2010) in their studies concluded that barriers to adoption of IPM technologies included the infrastructure barriers, management barriers, economics-social barriers, institutional-support barriers and training-skills barriers.

Erbaugh *et al.*, (2010) found that knowledge was the major factor in the adoption of IPM technologies.

Moradi and Omidi Najafabadi (2010) in their studies concluded that barriers to application of GlobalGAP standard included the institutional structural barriers, attitude and awareness barriers, researches barrier, economic barriers, marketing and trade barriers and private part and mass media barriers.

Swinnen and Maertens (2007) indicated that challenges of developing countries in adaption with food safety standards including financial, technical, structural constraints, lack of organization potential and ability to control and support of standards.

Kleinwechter and Grethe (2006) based on a theoretical framework of a adoption process (information stage, decision stage, implementation stage), of the Europe GAP standard by mango exporters in Piura, Peru, stated that the major constraints to adoption is lack of access to information on the standard GAP. And indicated that this problems relevance of the socioeconomic constraints. That the important public problem caused by the standard is the increasing costs. Other barriers are of little importance and practical conformation of the farm to the standard and constraints in understanding the regulations.

According to study done by Tawadchai *et al.*, (2006) In Thailand different levels of education may affect perceptions and learning ability on correct beef cattle – rising. Some practices need high level of knowledge such as disease healing and drug using. Result revealed that beef cattle raisers who had higher income had higher tendency to accept GAPs for beef cattle farming. This might be because they can afford to spend money on some expensive equipment or tools needed for a high standard of beef cattle farming. Thus, difference in family income may cause different capacities to buy expensive equipment and tools.

Rodriguez Baide (2005) in his studies concluded that major barriers in adoption of technologies were included the economics barriers , education and information barriers, constancy to change, application of technology, social challenges, infrastructural, landlessness and personal characteristics.

Hobbs (2003) indicated that the major disincentive for the adoption of Good Agriculture Practices (GAPs) were Economics, Institutional and Regulatory and Legal, Human Capital. Economic disincentive including: increase variable production costs (e.g. labor), reduce output/increase average costs, increase fixed production costs (e.g. equipment), asset specific investments. Institutional and Regulatory disincentive e.g. reliance on institutional infrastructure, lack of public institution for monitoring GAP. Human capital limitations included limits on the farmer's ability to apply the prescribed production and management protocols and maintain the appropriate level of documentation (literacy), limitation of labor, limitation of time management, inappropriate and poor public extension institution. Table 1 displays the summary of major barriers to application of GAP technologies.

Generally, according to the aforementioned introduction and according to the fact that it is important to find factors influence non-adoption technologies by farmers. Main purpose of this

Table 1: Summary of major barriers to application of GAP technologies.

Factors	Source
Infrastructure Barriers	Rodriguez Baid (2005), Hobbs (2003), Moradi and Omidi Najafabadi (2011), Razzaghi Borkhani *et al.*, (2010), Swinnen and Maertens (2007).
Awareness – Informational Bbarriers	Rodriguez Baide (2005), Drost (1996), Erbaugh *et al.* (2010), Karimi, (2009), Kleinwechter and Grethe (2006), Moradi and Omidi Najafabadi (2011), Nowak (1991), Razzaghi Borkhani *et al.*, (2010), Vanclay (1992).
Institutional-Support Barriers	Rodriguez Baide (2005), Hobbs (2003), Moradi and Omidi Najafabadi.(2011), Nowak (1991), Razzaghi Borkhani *et al.*(2010), Swinnen and Maertens (2007).
Personal Barriers	Rodriguez Baide (2005), Hobbs (2003), Karimi, (2009), Tawadchai *et al.*, (2006), Vanclay (1992).
Economical Barriers	Rodriguez Baide (2005), Drost (1996), Hobbs (2003), Karimi, (2009), Moradi and Omidi Najafabadi. (2011), Nowak (1991), Razzaghi Borkhani *et al.*, (2010), Tawadchai *et al.*, (2006), Swinnen and Maertens (2007), Vanclay (1992).

study was to investigate the major barriers to application of Good Agricultural Practices (GAPs) technologies in sustainability of livestock systems by ranchers Meshkinshahr. The objectives of the study were:

- Identifying the demographic characteristics of respondents;
- Priority setting of respondents' view about barriers to Application of Good Agricultural Practices (GAPs) technologies of livestock unites;
- Factor analysis to indicate barriers to Application of Good Agricultural Practices (GAPs) technologies of livestock unites.

MATERIALS AND METHODS

Population and sample

This study was a descriptive-correlation research, carried out in Meshkinshahr County. The population of the study consisted of ranchers of traditional and industrial livestock systems. (N=366) in 2 district of (traditional livestock systems, N=360 and industrial livestock farms, N=6). By calculation Cochran's formula, a sample of 120 ranchers was selected by using proportional random sampling method. This formula is:

$$d = t\frac{s}{\sqrt{n}} \qquad n = \frac{N(t.s)^2}{Nd^2 + (t.s)^2}$$

In this formula, (n) is the number of sample, (N) is the number of population, (s) is standard deviation, and (t) is equal to 2. Table 2 displays the statistical population and sample size of this study.

Instrument

A questionnaire divided into three parts was used to collect data from the target group. Part one, asked farmers to specify their demographic and technical information such as age, educational level, and livestock experience, membership of in local associations (two groups: membership and non- membership) and total income.

Part two was assessed major barriers to application of GAP technologies of livestock unites using Likert-type scale (1="very low", 2= "low", 3= "intermediate", 4= "high" and 5= "very high").

Finally part three, assessed level application of Good Agricultural Practices (GAPs) technologies in livestock systems was measured in

Table 2: Statistical population and sample size of the study

County	District	No. of Ranchers per district	Sample size
Meshkinshahr	Traditional livestock farms	360	116
	Industrial livestock farms	6	4
	Total	366	120

Table 3: Demographic characteristics of respondents

Demographic Characteristics	Frequency	Percent	Mean	SD
Age (year)			53.35	12.67
< 31	6	5		
31-40	14	11.7		
41-50	28	23/3		
51-60	42	35		
> 60	30	25		
Educational level				
Illiterate	37	30.8		
Literate (not primary school)	51	42.5		
Primary school	15	12.5		
Secondary school	3	2.5		
High school	7	5.8		
Post high school	7	5.9		
Livestock experience			20.74	10.84
livestock Units ownership				
Personal	119	99.2		
rental	1	0.8		
Income from livestock (percent)			55.20	23.41

six parts including application of GAP practices for very high") installations and structures (seven statements), application of GAP practices for equipment and facilities (11 statements) application of GAP practices for health (17 statements), application of GAP practices for transportation (eight statements), application of GAP practices for nutrition) seven statements, application of GAP practices for recording events (8 statements), (accordance to studies done by (GlobLGAP, 2010 and Tawadchai *et al.*, 2006), All these parts were measured on a Likert-type scale ranged from 0 to 5 (0=No, 1=low, 2=intermediate and 3=high).

Validity and reliability

Validity of the instrument was obtained by Agricultural Jihad experts of Meshkinshahr County and the members of committee of thesis, Supervisor and advisers of thesis in Department of Agricultural Management and Development, University of Tehran. Reliability of the instrument was measured by calculating Cronbach's Alpha coefficient, a measure of internal consistency. The reliability for various parts was more than 0.7, which showed the acceptable level. Data were collected through face to face interviews with ranchers at their farms.

Data analysis

Data were analyzed using Statistical Package for the Social Sciences (SPSS). Descriptive and inferential statistics were used to analyze the collected data. Descriptive statistics included frequency, percentage, mean, and standard deviation and so forth and inferential statistics included factor analysis.

RESULTS AND DISCUSSION

Characteristics of the sample

According to the findings about 88.53% of respondents were men, and respondents were on average 53 years old. About 35 % of respondents were between the age of 51 and 60years. While, 5%, 11.70%, 23.30%, and 25% of respondents were <31, 31-40, 41-50, and >60 respectively. Most of the respondents were literate (69.20%) and 30.80% were illiterate. Respondents' experience in livestock activities was 20 years on average. Findings showed 99.02% respondents had personal ownership livestock unites. The average production of milk was 21.50 Liter daily. The average income of ranchers from livestock job was 58.20 percent of total income. According to the findings, anybody of the respondents in traditional livestock systems had access to internet and computer (Table3).

Priority Barriers to Application of Good Agricultural Practices (GAPs) for livestock

Table 4 shows that No guarantees prices for safety products has first priority of barriers to, because of having the highest extent of mean (M=2.275 Lack of veterinary clinics and early detection of diseases (M=2. 958), Delay in payment of compensation by insurance (M=2.866), Lack of support for organic milk producers (M=2.841), respectively have allocated priorities from second to fourth. In addition Low levels of literacy (M=1.808), The high age of most ranchers (M=2.808), The absence of avoiding laws excessive use of antibiotics and hormones in dairy farm (M=2.100) with the lowest extent of mean have allocated last priorities to themselves.

Factor Analysis

In order to indicate barriers to Application of Good Agricultural Practices (GAPs) by Ranchers factor analysis was conducted. To determine the appropriateness of data and measure the homogeneity of variables entered to the analysis, the Kaiser-Meyer-Olkin (KMO) and Bartlett' Test of Sphericity (BTS) were applied. KMO was 0.857 and BTS was 1125.557 ($p < 0.01$), indicating that the data were appropriate for factor analysis.

While performing the factor analysis, there are some decisions to be made: the method of factor extraction, the number of factors and the type of factor rotation. There are several factor extraction methods. The methods used for the final solution were chosen primarily on the interpretability of the resulting factors. In this study unweighted least squares factoring was used as the extraction method. Another decision to be made when conducting factor analysis is to determine the number of factors. One rule of thumb is to use

Table 4: Barriers to application of good agricultural practices (GAPs) for livestock.

Statement	Mean	SD	Priority
No guarantees prices for safety products	2.975	1.184	1
Lack of veterinary clinics and early detection of diseases	2.958	1.226	2
Delay in payment of compensation by insurance	2.866	1.076	3
Lack of support for organic milk producers	2.841	1.092	4
Lack of government support on livestock	2.808	1.937	5
Ranchers accustomed to the use of chemical drugs and method because of Immediate effects and lesser cost	2.766	0.993	6
Texture of old livestock buildings	2.750	1.317	7
Lack of access to extension-education services	2.741	0.957	8
Lack of access to safety and health food	2.741	1.111	9
Lack of familiarity with hygienic and health problem	2.708	1.032	10
The lack of Stalls health	2.666	1.079	11
Lack of sufficient knowledge about organic and safety production	2.600	1.110	12
Low risk of ranchers for application GAP	2.508	0.840	13
Lack of using efficient decontamination in livestock sites	2.466	1.003	14
Lack of equipments and facilities for using GAP technologies	2.383	0.842	15
Lack of access to affordable and quality vaccines	2.357	1.059	16
Lack of adequate information on GAP for livestock	2.225	0.911	17
The absence of avoiding laws excessive use of antibiotics and hormones in dairy farm	2.100	0.901	18
The high age of most ranchers	2.041	0.863	19
Low levels of literacy	1.808	1.055	20

Table 5: Eigen Values, variance percentage and The cumulative variance percentage of extracted determinants

Factors	Eigenvalue	% of variance	Cumulative % of variance
1	4.170	21.950	21.950
2	2.407	12.668	34.618
3	2.185	11.498	46.115
4	2.033	10.425	56.814
5	1.981	10.425	67.238

Table 6: Items Loaded In The Factors Using Varimax Rotated Factor Analysis

Factors	Variable	Factor loadings
Infrastructure	Lack of equipments and facilities for using GAP technologies	0.718
	The lack of Stalls health	0.535
	Lack of using efficient decontamination in livestock sites	0.737
	Lack of veterinary clinics and early detection of diseases	0.832
	Lack of access to safety and health food	0.833
	Texture of old livestock buildings	0.568
	Lack of access to affordable and quality vaccines	0.672
Informational – educational	Lack of adequate information on GAP for livestock	0.744
	Lack of sufficient knowledge and technique about organic and safety productior	0.651
	Lack of familiarity with hygienic and health problem	0.579
	Low access or lack of access to extension-education services	0.6710
Institutional-support	Delay in payment of compensation by insurance	0.550
	Lack of support for organic milk and dairy production	0.593
	Lack of avoiding laws on excessive use of antibiotics and hormones ir livestock farming systems	0.744
Personal	Low levels of literacy	0.889
	The high age of most ranchers	0.549
	Low risk of ranchers for application GAP	0.594
Economical	No guarantees prices for safety products	0.632
	Ranchers accustomed to the use of chemical drugs and method because o Immediate effects and lesser cost	0.867

an eigenvalue of one as the cut-off value. That is, all factors in a particular solution must have eigenvalues greater than one. Rotation is used to reorient the factor loadings so that the factors are more interpretable. The Varimax rotation option, which tries to minimize the number of variables that load highly on a factor, was used Eigen values; variance percentage and the cumulative variance percentage of extracted determinants are presented in Table 5.

Accordingly, five factors were extracted (Table 6). Factors were examined and given a descriptive title that represented the characteristics of the constructs. The first factor was infrastructure barriers which explained 21.950 percent of variance. Other factors were informational - educational barriers, institutional-support barriers, personal barriers and economical barriers which explained 12.668 11.498, 10.425 and 10.425 percent of the total variance respectively. These factors explained 67.238 percent of the total variance. The five factors that were extracted are as follows:

Factor 1: The first factor accounted for 21.950 percent of the total variance and 7 variables loading significantly. These variables were "lack of equipments and facilities for using GAP technologies", 'The lack of Stalls health", "lack of using efficient decontamination in livestock sites", "lack of veterinary clinics and early detection of diseases", "lack of access to safety and health food", "lexture of old livestock buildings" and "lack of access to affordable and quality vaccines". So, this factor was termed "infrastructure barriers".

Factor 2: The second factor accounted for 12.668 percent of the total variance and 4 variables loading significantly. These variables were "lack of adequate information on GAP for livestock", "lack of sufficient knowledge and technique about organic and safety production" , "lack of familiarity with hygienic and health problem" , "low access or Lack of access to extension-education services". So, this factor was termed "informational-educational barriers".

Factor 3: The third factor accounted for 11.498 percent of the total variance and 3 variables loading significantly. These variables were "delay in payment of compensation by insurance", "lack of support for organic milk and dairy production", and "lack of avoiding laws on excessive use of antibiotics and hormones in livestock farming systems" So, this factor was termed "institutional-support barriers".

Factor 4: The forth factor accounted for 10.425 percent of the total variance and 3 variables loading significantly. These variables were "the high age of most ranchers", "the high age of most ranchers", and "low risk of ranchers for application GAP". So, this factor was termed "personal barriers".

Factor 5: The last factor accounted factor accounted for 10.425 percent of the total variance and 2 variables loading significantly. These variables were "no guarantees prices for safety products" and "ranchers accustomed to the use of chemical drugs and methods because of immediate effects and lesser cost". So, this factor was termed "economical barriers" (Table 6).

CONCLUSION AND RECOMMENDATION

The findings revealed that the major barriers to application of Good Agricultural Practices (GAPs) technologies in sustainability of livestock by of ranchers were infrastructure. Considering the percentage of variance explained by exploratory analysis factors "infrastructure barriers" with 21.950 percent of variance explained has the largest collection of major barriers to application of GAP technologies. In this context should be considered organization and integration of GAP of livestock programs in the region and infrastructure factors under supervision and manage of an organization (such as agricultural extension institution) and also to display FFS programs associated with early needs assessment of required infrastructure factors in the region for the implementation of GAP programs.

Information about being clear and explaining new technology causing more usable to farmers. Information reduces uncertainty and doubt about the application technology (Cawell *et al.*, 2001) Providing more information about the technology reduces negative attitude towards technology adoption, correct combination of information is required for the effectiveness and efficiency to adoption technology (Bonabana-Wabbi, 2002).

Recommended to implementation and encourage farmers in the villages to participation in group activities and active membership and participation in such as village councils and cooperative associations. Farmers with more interaction between the institutions and social organizations get more knowledge and their attitude is more favorable to GAP and application the higher levels of technology.

Obviously, opinion leader's impact on farmers in technology adoption, attention trusted people and local leaders to adoption new technology is important as an effective strategy. Therefore, more attention to identify the real needs, to make tangible of non-tangible needs and prioritizing them by their producers. With facilitative agricultural extension and research agents and personnel under FFS programs. Most effective solution in this case targeted training of ranchers is in connection with the capabilities and benefits of GAP and is necessity the importance of its goals for the implementation of better safety production programs in the region (accordant to the result of Razzaghi Borkhani *et al.*, 2013) the important role of agricultural extension agents which can affect farmers' perceptions and behaviors to adopt and apply new technologies with use of extension-participatory methods such as farmer field schools is proper strategy for creating positive attitude of farmers towards new technologies.

Since one of the barriers to adoption and application of technologies, (according to study done by Razzaghi Borkhani *et al.*, 2011) is farmers' low risk orientation, it is recommended to provide incentives such as loans and facilities for farmers who have low income, purchasing guaranteed products, Fixed prices policies. This will increase application level among farmers. Besides more coordination between the public and the private sectors for application GAP standard and to give certification of safety production of livestock farming system. In extension GAP standard in Iran is important.

A important factor to ranchers consume to use chemical drugs and hormones, because of immediate effects and lesser cost is. Therefore requires changing farmer's perception. Hence, agricultural extension agent can affect farmers' perceptions and behaviors, in order to increasing usage and application GAP practices. Commuter of the Internet as sources of information cannot be used by ranchers. Therefore the necessary infrastructure provided education and information through mobile messaging and commuter for dairy farmers. In this context, educational - extension programs are used local knowledge and experience in order to production a healthy crop and GAP. Agricultural extension agents presented indigenous knowledge and combining it with modern knowledge and technology transfer and information educational messages in minimum time for a family of ranchers.

Pay attention to major barriers to application of Good Agricultural Practices (GAPs) technologies among ranchers were informational-educational barriers. So for best application of GAP technology, it is recommended to establish extension workshops to increase farmers' knowledge toward GAP practices in livestock.

ACKNOWLEDGEMENT

The authors are grateful from ranchers of Meshkinshahr in order to collaboration with researchers this thesis and thanks to the their efforts in Iran economic development. '

REFERENCES

1- Banzon, A. T., Mojica, L.E., Angela, A., & Cielo, A.A. (2013). Adoption of Good Agricultural Practices (GAP) in the Philippines: Challenges, issues, and policy imperatives. *Journal Policy Brief Series - Southeast Asian Regional Center for Graduate Study and Research in Agriculture (SEARCA)* . (1). Available on: http://beta.searca.org/searca/ index.php/knowledge

2- Bonabana-Wabbi, J. (2002). *Assessing factors affecting adoption of agricultural technologies: the case of Integrated Pest Management (IPM) in Kumi District, Eastern Uganda.* Unpublished thesis, Virginia Polytechnic Institute and State University.

3- Caswell, M., Fuglie, K., Ingram, C., Jans, S., & Kascak. C. (2001). *Adoption of Agricultural Production Practices: Lessons Learned From the US. Department of Agriculture Area Studies Project. AER* (792), Resource Economics Division, Economic Research Service, Agriculture Economic Report, USDA, Washington, DC. 20036-5831.

4- Drost, D.T., Long, G., Wilson, D., Miller, B., & Campbell, W. (1996). Barriers to adopting sustainable agricultural practices. *Journal of Extension,* 34(6), 1-6.

5- Erbaugh, J. M., Donnermeyer, J., Amujal, M., & Kidoido, M. (2010). Assessing the impact of Farmer, Field and School participation on IPM adoption in Uganda. *Journal of International Agricultural and Extension Education,* 17(3), 5-17.

6- FAO (1994). *Livestock recognizing: Their role in sustainable agriculture*, FAO study. Rome.

7- GlobLGAP .(EurepGAP). (2010). Control Points and Compliance Criteria Integrated Farm Assurance. *All Livestock Base.* Code Ref.: IFA 3.0 CP, pp, 18, http://www.globalgap.org/uk_en/for-producers/crops/

8- Hobbs, J. E. (2003). Incentives for the adoption of Good Agriculture Practices (GAPs). Background paper for the FAO consultation on Good Agricultural Practices, FAO GAP Working series, 3, Rome, November 10-12

9- Karimi, A. (2009). *Factors affecting the development of organic farming experts' department of agriculture.* Unpublished thesis of Agricultural Extension and Education, Faculty of Agriculture, Tarbiat Modarres University (In Persian).

10- Kleinwechter, U., & Grethe , H. (2006). *The adoption of the Eurepgap standard by mango exporters in Piura, Peru.* Paper presented at the International Association of Agricultural Economists Conference, Gold Coast, Australia, August 12-18).

11- Koohsar, A. (2012). An overview of the safety production program, livestock. *Agro-Industry Magazine*, 154, 37-38.

12- Moradi, P., & Omidi Najafabadi, M. (2011). Barriers of application global good agriculture practices standard in Iran agriculture system. *Agricultural Extension and Education Research.* 4(1): 27-39 (In Persian).

13- National Bureau of Agricultural Commodity and Food Standards (2005). Good Agricultural Practices for Beef Cattle Farming. Bangkok, *Ministry of Agriculture and Cooperatives.*

14- Nowak, P. (1991). *Farmer adopt of production technologies.* Presented at the Soil and Water Conservation Society National Conference, August 8, Lexington, KY.

15- Nurul Islam, G Md., Mohamed Arshad, F., Radam, A., & Farha Alias, E. (2012). Good Agricultural Practices (GAP) of tomatoes in Malaysia: Evidences from Cameron Highlands. *African Journal of Business Management*, 6(27), 7969-7976.

16- Razzaghi Borkhani, F. Rezvanfar, A. Shabanalifami, H., & Pouratashi, Mahtab (2010). Investigating the major barriers to adoption of IPM technologies by Paddy Farmers. *American-Eurasian Journal of Toxicological Sciences*, 2(3), 146-152.

17- Razzaghi Borkhani, F., Shabanalifami, H., Rezvanfar, A., & Pouratashi, M. (2011). Application of IPM practices by Paddy farmers in Sari County of Mazandaran Province, Iran. *African Journal of Agricultural Research,* 6(21), 4884-4892.

18- Razzaghi Borkhani, F., Rezvanfar, A., Shabanalifami, H., & Pouratashi, M. (2013). Social factors influencing adoption of integrated pest management (IPM) technologies by Paddy farmer. *International Journal of Agricultural Management and Development* (IJAMAD), 3(3), 211-218.

19- Rasul, G. Thapa, G. B. (2004). Sustainability of ecological and conventional agricultural systems in Bangladesh. *An assessment based on environmental, economic and social perspectives Agricultural Systems*. 79, 327–351.

20- Rodriguez Baide, J.M. (2005). *Barriers to adoption of sustainable agriculture practices in the south: change agent's perspectives*. A Thesis Submitted to the Graduate Faculty of Auburn University in Partial Fulfillment of the Requirements for the Degree of Master of Science. In Salem Oregon in the USA.

21- Steinfeld, H., & Mack, S. (1995). Livestock development strategies. *World Animal Review*, 84(85), 18-24

22- Suppadit, T. (2003). The Recycle of broiler litter as a feed ingredient for cattle to reduce environmental pollution II. Nutrient values of broiler litter, *Thai Environmental Consultants Journal*, 7(1), 31-36.

23- Swinnen, J.F.M., & Maertens, M. (2007). Globalization, agri-food standards and development. Retrieved 4 August 2010 from *http://www.inea.it/public df-articoli*.

24- Suppadit, T., Phumkokrak, N., & Poungsuk, P. (2006). Adoption of Good Agricultural Practices for Beef Cattle Farming of Beef Cattle Raising Farmers in Tambon Hindard, Dankhunthod District, Nakhon Ratchasima Province, Thailand. *KMILT Science and Technology Journal,* 6(2).

25- Vanclay, F. (1992). Barriers to adoption: A general overview of the issues. *Rural Society Journal,* 2(2), 10-12.

4

Cropping Pattern and Comparative Advantage of Agricultural Products in Ilam Province

Roya Eshraghi Samani [1] and Alireza Poursaeed [2]

Abstract

In this study, the comparative advantage of main agricultural products and its relation to cropping pattern was studied in Ilam Province. For data analyzing the comparative advantage and government policies effects indices, Policy Analysis Matrix (PAM) were used. The obtained results for the comparative advantage indices of Domestic Resource Cost (DRC), Social Cost Benefit (SCB) and Net Social Profitability (NSP) show that production of irrigated wheat, dry-farming wheat and dry-farming barely has not comparative advantage in Ilam Province but the production of irrigated barely, corn, dry-farming pea, dry-farming lentil, watermelon, cucumber and tomato have comparative advantage in Ilam Province. In addition, the obtained results for NPIC index show the Nominal Government Protection of inputs in all products. Also, Nominal Protection Coefficient index indicated that nominal market protection of products like wheat (irrigated and dry-farming) and barely (irrigated and dry-farming) was positive and those of other products were negative and the Effective Protection Coefficient for wheat and barley had the highest values. The results showed that the existing cropping pattern was not allotted with comparative advantage but is related to the effective protection coefficient.

Keywords:
Comparative Advantage, Cropping Pattern, Policy Analysis Matrix, Ilam Province

[1] *Department of- Agricultural Management, Ilam Branch, Islamic Azad University, Ilam, Iran.*
[2] *Department of Agricultural Extension and Education, Ilam Branch, Islamic Azad University, Ilam, Iran.*
* *Corresponding author's email: r_eshraghi_s@yahoo.com*

INTRODUCTION

Using comparative advantage principle can determine a region pattern of crop and distribute resources optimally among different activities in order to optimizing use of resources, the production and exporting capabilities might be recognized and conditions for the effective investments might be provided as well. This applies to commercial integration process especially in joining the World Trade Organization is very useful and beneficial. Most of the researchers, on the growth and development of the agricultural sector and its role on the development of the national economy of the excess supply of surplus capital, supply food and industrial raw material, provision of foreign currencies; also on changes of agricultural structure in fundamental and comprehensive study of existing forces, participation, optimum utilization of existing facilities and resources, organization and conduction of planned and measured to the scientific lifestyle emphasis. Since the present world is a world of economic competition and every country has to plan economic designs with precision in order to maintain its political and economic supremacy and autonomy and due to the fact that the principle of comparative advantage is one of the most useful economic policing tools. Discovering and utilizing comparative advantages of production in different economic sectors not only improve the policies of country's allocation of resources and production pattern but also could determine export types and their combinations (Karbasi *et al.,* 2009).

The comparative advantage is the central point of trade and profit obtained through trade and its path. The growth and development of global trading in recent years, national, international organizations and institutions established for this sake, have made governments to contemplate seriously about competition and especially comparative advantage; since real and trouble-free competition is founded on the base of comparative advantage. The competition states that within a country or among several countries which production units (or generation of which products) could have a better activity in global markets on determined assumptions. So, paying attention to comparative advantage in production activities is one of the most important aspects of economic planning which is of considerable stability (Noori, 2003).

Salvatore (1998) describes free trade as the best policy for the world's countries believes that generating to the degree of complete independence of all products is to disadvantage of those countries. Smith argues that via free trade, each country could specialize in producing those products which have absolute advantage in their production. This method of trade causes the optimized allocation of production factors in countries and also increased production; consequently all countries will be at the same time benefited from trade. Since the absolute advantage theory operates weakly especially when a country has absolute advantage in generating all products, it has not been applied very often and limitations of this theory allowed for comparative advantage theory coming to stage. According to this theory, even if a country has not absolute advantage in producing of any product compared to another country, it still could be present in global market and be benefitted. This country should produce and export those goods which has comparative advantage in their production and import those goods which has not comparative advantage in their production. The theory of comparative advantage evolved so that since 1970 decade, it was identified as the basis for taking investment decisions and economic policies and eventually it was used as a tool for determining the pattern of the optimized production and trade in a country.

Fong and Fang (1999) studied the comparative advantage of main agricultural products of China for 1992 to 1998 using policy analysis matrix. Results of this study indicated that China had comparative advantage in labor-wanting agricultural products and lacked this advantage in land-wanting products. Yazdani and Eshraghi (2006) investigated the comparative advantage of almond production in Iran (Cheharmehal-o-Bakhtiari Province). According to the obtained results, almond production in the region of study had comparative advantage. Reig-Martínez *et al.* (2008) combined policy analysis matrix and data envelopment analysis techniques to model the analysis of profitability from farming. The main conclusion was that the usefulness of the policy analysis matrix might be substantially enhanced by simulating profitability after efficiency-improving managerial decisions have been adopted. Karbasi *et al.* (2009) in a study investigated the comparative advantage of the major agricultural products of Kerman using the indices of domestic resource cost, pure social profitability, economic advantage and advantage scale. According to the obtained results, irrigated wheat had comparative advantage in this province; in other words, generating this product was more economic than importing it. Pakravan *et al.* (2012) performed a computation of comparative advantage indices of agricultural products of the city of Sari. Results of this study showed that among the products under study, barely had not advantage; however, it had third rank in terms of acreage among the studied products. Also, wheat product had DRC index equal to 1 in the lowest value of exchange currency rate in IRR and this suggested that high social profitability of this product encouraged farmers even with the lowest social income to generate this product which also has the second rank of acreage in the present conditions. Rozane and Philippe (2013) evaluate the profitability and the effects of direct and indirect taxes on rice production in Brazil compared to other member countries of Mercosur, use the Policy Analysis Matrix. The results have shown that in 2010, rice production in Argentina and Uruguay had positive social and private profitability, while in Brazil and Uruguay there were negative private results. Secondly, a simulation of an alternative scenario for Brazil was performed, considering a reduction in the direct and indirect tax burden to a similar percentage between the countries compared. Kanaka and Chinnadurai (2013) used PAM to model the analysis of profitability from farming. The main conclusion is that the usefulness of the policy analysis matrix might be substantially enhanced by simulating profitability after efficiency-improving managerial decisions have been adopted.

Hasanpour *et al.* (2013) studied the comparative advantages of rainbow trout production through PAM. The results based on the DRC approved the comparative advantage of fish production. NPCO indicated that there is a direct subsidy on the producer; the amount of NPCI represented the indirect taxes on tradable inputs of fish, EPC showed that the government's policies support production process.

The agricultural sector of Ilam Province is considered an important and effective sector in the economy of this neglected province. Through proper planned investments in this sector, steps might be taken on economic growth of the province by increased production and offering products to inside and outside markets. Borderland geographical status of this province is a significant advantage in the development of exporting products. Because neighborhood with a country like Iraq and having more than 420 kilometers of international border with this country has provided a close potential market for internal surplus products of the province. Making use of this relatively appropriate market situation and the degree of success in important issue of economic boom of the province through the productive activities of the agricultural sector depends on productive potential and competitive ability of agricultural products generated in the province in global markets including the market of the stated foreign country. Due to restrictions of many production inputs and factors in agricultural sector and also different climatic and geographical characteristics of different regions, taking steps based on the principle of comparative advantage in each region is of high importance. Using comparative advantage principle, it is possible to determine regionalizing pattern of cropping and distribute the resources optimally among the different activities in order that in addition to optimized use of resources, producing and exporting capabilities might be recognized and context for effective investments might be prepared. Paying attention to this issue is very helpful in trade integration processes especially in joining to the world trade organization. Taking into account the comparative advantage of agricultural products in agricultural plans of this province could be viewed as an important step in generating the profitable products and providing the appropriate context for informed presence in the global markets which requires the recognition of capabilities and potentialities of each region. Since through using comparative advantage principle, it would be possible to determine the regionalizing pattern of cropping and also distribute resources in an optimized way between different activities so that beside optimal use of resources, production and exporting capabilities could be recognized and the necessary context for effective investing could be provided, the analysis of comparative advantage of the major agricultural products of Ilam Province and its relationship with the current cropping pattern seems necessary. Regarding the significance of the subject of this study, the comparative advantage of main products of the province was measured using domestic resource index and then cropping pattern based on comparative advantage was determined and the quality of government intervention was also determined via computing the effective protection coefficient. At last, the cropping pattern based on comparative advantage was compared against the current cropping pattern of major agricultural products of the province and their respective relationships with the protection coefficient were studied. The focus of this research was to investigate the comparative advantage of main agricultural products in Ilam Province and its relation to cropping pattern.

MATERIALS AND METHODES

Identifying the comparative advantage of generated products especially agricultural products is indeed moving toward the optimized allocation of potential resources and facilities of each specific region in order to develop producing and exporting these products. Since in many countries, government for protecting agriculture and increasing production amount carries out various supports in forms of inputs provision and distribution, guaranteed purchase, market adjusting, offering facilities, paying subsidies and so on, a reasonable growth has appeared in products of this sector in light of these protections. Since government intervention has brought about inevitable effects on total price and real cost of a product, those who determine the economic policies, are puzzled over real prices and costs of generating a product. On the other hand, unreal prices and costs governing on the product and input market deviates the economic evaluation of generating these products. In order to identify the amount of deviations and evaluating economically the generation of different products, a method called policy analysis matrix is usually used.

In this study, for investigating the comparative advantage of generating main agricultural products in Ilam Province, Policy Analysis Matrix is used. This method is considered one of the most comprehensive and applied methods of policy analysis and computing comparative advantage.

The Policy Analysis Matrix (PAM) is developed by Monke and Pearson (1989) and augmented by Masters and Winter-Nelson (1995), for measuring input use efficiency in production, comparative advantage among commodities, and the degree of government interventions. PAM is a product of two accounting identities, one defining profitability as the difference between revenues and costs and the other measuring the effects of divergences (distorting policies and market failures) as the difference between observed parameters and parameters that would exist if the divergences were removed. By filling in the elements of the PAM for an agricultural system, an analyst can measure both the extent of transfers occasioned by the set of policies acting on the system and the inherent economic ef-

Table 1: Policy Analysis Matrix.

	Revenues	Tradable Inputs	Domestic Factors	Profit
Private prices	A	B	C	D
Social prices	E	F	G	H
Divergences	I	J	K	L

Table Notes:
Private profits, D= A- (B + C) Input transfers, J= B- F
Social profits, H= E - (F + G) Factor transfers, K= C-G
Output transfers, I= A-E Net policy transfers, L= D-H
Source: Based on Monke and Pearson (1998)

Table 2: Indicators of comparative advantage and protection co-efficient for main crop in Ilam.

Indicator	tomato	cucumber	water-melon	dry-farming lentil	dry-farming pea	corn	dry-farming barely	irrigated barely	dry-farming wheat	irrigated wheat
DRC	0.52	0.21	0.19	0.39	0.75	0.29	1.46	0.93	1.76	1.12
SCB	0.58	0.24	0.25	0.56	0.80	0.46	1.29	0.97	1.37	1.08
NSP	97763	3823159	300709	181178	113460	886406	-76524	13786	-67333	-54719
NPCI	0.64	0.77	0.72	0.74	0.83	0.37	0.65	0.51	0.69	0.65
NPC	0.75	0.69	0.85	0.94	0.72	0.95	1.12	1.18	1.28	1.15
EPC	1.16	0.68	1.08	1.30	0.79	1.23	1.35	1.42	1.68	1.41
NPIR	36%	23%	28%	26%	17%	63%	35%	49%	31%	35%
NPR	25%	31%	15%	6%	28%	5%	12%	18%	28%	5%
EPR	16%	32%	8%	30%	21%	23%	35%	42%	68%	41%

ficiency of the system (Table1).

The data in the first row provide a measure of private profitability (N), defined as the difference between observed revenue (A) and costs (B+C). Private profitability demonstrates the competitiveness of the agricultural system, given current technologies, prices for inputs and outputs, and policy. The second row of the matrix calculates the social profit that reflects social opportunity costs. Social profits measure efficiency and provide a measure of comparative advantage. In addition, comparison of private and social profits provides a measure of efficiency.

A positive social profit indicates that the country uses scarce resources efficiently and has a static comparative advantage in the production of that commodity at the margin. Similarly, negative social profits suggest that the sector is wasting resources that could have been utilized more efficiently in some other sector. In other words, the cost of domestic production exceeds the cost of imports, which indicates the sector cannot survive without government support at the margin. The third row of the matrix estimates the difference between the first and second rows. The difference between private and social values of revenues, costs, and profits can be explained by policy interventions.

The PAM framework can also be used to calculate important indicators for policy analysis. The Nominal Protection Coefficient (NPC), a simple indicator of the incentives or disincentives in place, is defined as the ratio of domestic price to a comparable world (social) price. NPC can be calculated NPCI. The domestic price used in this computation could be either the procurement price or the farm gate price, while the world reference price is the international price adjusted for transportation, marketing and processing costs. The other two indicators that can be calculated from the PAM include the Effective Protection Coefficient (EPC) and Domestic Resource Cost (DRC). EPC is the ratio of value added in private prices (A-B) to value added in social prices (D-E). An EPC value of greater than one indicates that government policies provide positive incentives to producers, while values less than one indicate that producers are not protected through policy interventions. Domestic resource cost, the most useful indicator of the three, is used to compare the relative efficiency or comparative advantage between agricultural commodities, and is defined as the shadow value of no tradable factor inputs used in an activity per unit of tradable value added (F/(D-E)). The DRC indicates whether the use of domestic factors is socially profitable (DRC<1) or not (DRC >1). A region will have a comparative advantage in a given crop if the value of the DRC for that crop is lower than the DRC for other crops grown in that state. A good alternative to the DRC is the Social Cost/Benefit (SCB), which accounts for all costs (Fang and Beghin, 1999). The SCB is calculated as the ratio (E+F)/D. can compute the rates of protection by use below equation:

$$NPR = (NPC-1)*100 \quad (1)$$

$$NPIR = (1-NPI)*100 \quad (2)$$

$$EPR = (EPC-1)*100 \quad (3)$$

The shadow price

The shadow price is the real value of a product or an input in conditions of free competition and without interference of any factor outside the market resources. Providing these conditions inside a country is not a simple task especially in case of agricultural products. Because due to taking protection and taxing policies inside countries, the price of agricultural products has undergone deviation and in such conditions, domestic shadow prices could not reflect well the real values of products. Therefore, for computing shadow prices of products and inputs generated for export or replacement of imports inside country, the Fob border price of the product is multiplied by the shadow exchange rate to get its border shadow value in IRR. Then transfer cost of the product from the region of study to export borders subtracted from it. For obtaining shadow price of imported products or inputs, their CIF cost is multiplied by shadow exchange rate and the cost of transferring from the border to farm is added to it.

The tradable inputs include chemical fertilizers, toxins (herbicides, insecticides and fungicides) and a part of machinery cost. The shadow price of machinery, the use cost of agricultural products are mainly due to using sorts of tractors, sprayers, water engine, vehicles and thrasher. Among the machineries, operations of tractor are more frequent than others. Plowing, disk, troweling, seeding, spraying and other operations could be carried out by tractor. So, in this research, determination of the shadow price of tractor is dealt with as the shadow price of machinery. Machineries have ambivalent characters; some of their parts are considered among tradable inputs and some among non-tradable inputs. In this study, based on previous studies, 66% of this cost is considered as non-tradable cost and 34% belonging to tradable costs.

The shadow price of chemical fertilizer: chemical fertilizer (phosphate, urea, ammonium nitrate, ammonium sulfate, potassium sulfate, nitrogen and other fertilizers) is a fully-exchangeable product. The consumptive chemical fertilizer is provided partly from domestic production and partly through imports.

In which xi is the amount of type i fertilizer and Pi is the price for type i fertilizer and the shadow price of consumptive fertilizers is FOB price of imported fertilizers.

The shadow price of toxin: The most important consumptive toxins for agricultural products are herbicides, fungicides and insecticides.

In which xi is the amount of type i toxin and Pi is the price of type i toxin and the shadow price of consumptive toxins is FOB price of the imported toxin.

Non-tradable inputs include seed, human resource, land, water, organic fertilizer and internal transportation.

The shadow price of seed: Since seed is among the inputs usually produced and exchanged by farmers themselves and there is not usually any special dysfunction in its market, its market price is assumed its very shadow price.

The shadow price of human resource: In the present study, the cost of lost opportunity of the employed labor is considered as shadow price. For the computation of the cost of lost opportunity, the highest wage paid for different activities in generating the product of study is considered.

The shadow price of land: For measuring the shadow price of land, according to previous studies, there are different methods. In the studies of Gonzales *et al.* (1993) the mean rate of land's rent was used as the shadow price by applying the 85% coefficient. the applied coefficient is because the subsidies given to tradable inputs cause the land's renting price to be more than its real value; therefore, the land's renting rate mean was assumed as the shadow price of land by applying 85% coefficient in the major regions of cropping the selected products. In this study, for computing the shadow price of land, 85% of the highest price of land renting has been considered as the base price of land in regional tradition.

The shadow price of organic fertilizer: it's selling and buy price (market price) has been considered as the shadow price.

The shadow price of water: In this study, the shadow price of water has been computed on the basis of highest total cost of water harvesting. The cost of internal transportation is considered the mean transportation cost of every ton of products to external gates and vice versa. This has been computed based on inquiring the consented rate of transportation of products in the syndicate of garage-owners and drivers, truck-owners Corporation and also the transportation terminals organization.

The shadow exchange rate

Another variable needed for analyzing the comparative advantage is the shadow exchange rate.

The official exchange rate is in fact the domestic price of exchange currency. Influenced by different policies of government, this price also similar to the domestic prices of products deviates from its real value. The shadow rate of exchange currency has special importance in computation of the comparative advantage and determination of government protection rates. In fact, this rate is the base for an acceptable shadow price for tradable products and inputs (Karbasi *et al.*, 2009). There are various methods for computing the shadow rate of exchange currency; different assumptions have been considered in each method which results in different numeral quantities for the mentioned indices. In the present study, official exchange (dollar) rate is assumed as the shadow rate. The mean rate of dollar in the year 2010, according to the central bank of Islamic Republic of Iran report, is 10335 IRR.

RESULTS AND DISCUSSION

The results obtained through computing the indices of policy analysis matrix of 10 main agricultural products of Ilam province are reported in Table 2.

The obtained results from the computation of the comparative advantage indices of DRC, SCB and NSP show that Ilam Province has not comparative advantage in producing irrigated and dry-farming wheat and dry-farming barely but in producing corn, pea, lentil, cucumber, watermelon and tomato has comparative advantage and in other words social profitability. This is the case while the majority of the cultivated lands of Ilam Province are devoted to wheat and barley cropping.

Also, the results of computing NPIC index indicate the nominal protection by government of inputs in all products. The generators of this product have been given input subsidy; in other words, government intervention in inputs market has been to the advantage of generators of these products. The value of NPC index or nominal protection coefficient suggests that the nominal protection of the market of wheat (irrigated and dry) and barely (irrigated and dry) products has been positive and this index has been negative regarding other products of study. In other words, government policies has been beneficiary only to producers of wheat (irrigated and dry) and barely (irrigated and dry) and has had disadvantages for other products under study and toward not protecting internal generating of other products. It could be said that government has forced indirect taxes on generating other products. Effective protection coefficient EPC, which analyzes the effects of government intervention in inputs and product market at the same time, has highest values for wheat and barley.

The intervention of governments influences the cropping patterns and causes change in domestic superiorities (production in domestic prices). The protection amount of governments, short-term and long-term investments influence the private sector as well which may result in growth and blooming of these sectors. Since one aim of this study is the investigation of relationship between acreage and the comparative advantage of products in Ilam Province and the way government protection influences the cropping pattern in Ilam Province, the rank of acreage of each product is compared with the rank of effective protection rate and rank of comparative advantage of these products in the

Table 3: DRC, EPC and cultivated land.

Product	DRC	%EPC	Rank of EPC	Cultivated land(ha)	Rank of Cultivated land
Watermelon	0.19	8	10	4737	7
Cucumber	0.21	32	5	5936	5
Corn	0.29	23	7	10360	4
Dry-farming lentil	0.39	30	6	2109	8
Tomato	0.52	16	9	406	10
Dry-farming pea	0.75	21	8	5012	6
Irrigated barely	0.93	42	2	998	9
Irrigated wheat	1.12	41	3	49394	2
Dry-farming barely	1.46	35	4	47992	3
Dry-farming wheat	1.76	68	1	76952	1

region of study in Table 3.

According to the information of table 3, the cropping pattern for the province is not determined on the basis of comparative advantage index and those products with high acreage have not high comparative advantage. For instance, dry-farming wheat which lacks comparative advantage has the highest acreage and on the contrary products such as cucumber and watermelon that have the highest advantage, have not the highest acreage. Among reasons for the present situation, we may refer to: guaranteed purchase of products such as wheat and barley and also imbalance in the market of such products as cucumber and tomato and inconsistency in their price. As it can be seen in the table, it seems that acreage of agricultural products is ordered almost based on the government effective protection of the product ions.

CONCLUSION

Regarding the computed indices, Ilam Province has good comparative advantage in producing watermelon, cucumber and tomato; however, due to these products being high corruptible, it is necessary -beside programming for development of cropping these products- to pay attention to the peripheral issues such as processing industries, improvement in transportation, storage, packaging and marketing systems. Since the cropping pattern of agricultural products in Ilam Province is designed mainly on the basis of the effective protection of products, it is recommended that government protection be increased purposefully for those products which have higher comparative advantage in order that in addition to the development of cropping these products and consequently optimized use of production resources and factors, conditions might be provided in which producers have high competitive potential when faced with the trade integration process they could benefit mfro the trade integration process. Since the obtained values for protection indices indicate direct and indirect protections the agriculture's inputs and outputs market, attempts should be made to increase the range of these activities with the aim of developing agricultural exports of these products.

Despite the fact that Ilam Province has not comparative advantage in producing irrigated and dry-farming wheat and also dry-farming barely and that cropping these products has not social benefit, the majority of cultivated lands in Ilam are devoted to these products; easiness, fashion, low risk and old cropping methods have made farmers of this region not to be ready to crop any product other than wheat. So, it is not easy to remove the cropping of this product from the agricultural program. Increasing studies concerning the desired cropping pattern suited to the climatic and economic conditions of the region, increasing deep-seated studies about cropping new products and economic analysis of those products, educating farmers and comprehensive attempt to present practical solutions to improve yield and decrease costs for purchasing necessary inputs for farmers in order to arrive at a better state of comparative advantage are solutions for enhancing the profitability of wheat in the region.

The comparative advantage is a dynamic index and its value changes over time by changing factors. Therefore, aimed at permanent development of comparative advantage, the necessary investments must be carried out; structural and fundamental investing including permanent investment in researches, education and promotion is recommended.

ACKNOWLEDGEMENT

The authors are grateful for the funding supplied by the Ilam branch, Islamic Azad University.

REFERENCES

1- Fong, C. B. J., & Fang, C. (1999). *Measuring Small Holder Comparative Advantage and Agricultural Trade* (CARD – Working Paper. No: 99 19). Gardner, B. & G.

2- Fang. C., &, Beghin, J. (1999). *Food self-sufficiency, comparative advantage, and agricultural trade: a policy analysis matrix for Chinese agriculture* (Working Paper 99-WP 223). Center for Agricultural and Rural Development, Iowa State University, Ames, IA.

3- Gonzales, L. A., Kasrino, F., & Rsegrant, M. W. (1993). *Economic Incentive and Comparative Advantage in Indonesian Food Crop Production* (Report NO 93). International Food Policy Research Institute, Washington D. C.

4- Hassanpour, B., Asadi, E., & Biniaz, A. (2013).

Investigation of Policies Effects and Comparative Advantage of Rainbow Trout Farming in KB Province, Iran. *International Journal of Agriculture and Crop Sciences*, 6(1), 31-34.

5- Huang, J. J., song, Qiao, F., & Fuglio, O. (2003). *Sweet Potato in China: Economic Accept and Utilization in Fig Production. International potato Center* (IPC), Pogor, Bogor, Indonesia.

6- Kanaka, S., & Chinnadurai, M. (2013).The policy analysis matrix of rice cultivation in INDIA. *European Journal of Physical and Agricultural Sciences*, - 1 (1), 8-17.

7- Karbasi, A. R., Shamsodini, S., & Rastgaripour, F. (2009). Determining comparative advantage of main cultivated production in Kerman. *Agricultural Economic and Development*. 65, 1-15.

8- Master, W. A., & Winter– Nelson, A. (1995). Measuring the comparative advantage of agricultural activities: Domestic resource cost and the social cost-benefit Ratio. *American Journal of Agricultural Economics*, 77,243-250.

9- Noori, K. (2003). Defining the production comparative advantage in major rice groups in Guilan and Mazandaran Province. *Journal of Agricultural Economics and Development*, 10(4), 25- 46.

10- Monke, E., & Pearson, S. (1989). *The policy analysis matrix for agricultural development,* Ithaca. NY: Cornell University Press.

11- Pakravan, M. R., ZareMehrjerdi, M. R., & Kazemnejad, M. (2012). Comparative advantage investigation of cultivated production in Sari. *Agricultural Economic and Development,* 77, 1-28.

12- Reig-Martínez, E., Picazo-Tadeo, A. J., & Estruch, V. (2008).The Policy Analysis Matrix with profit-efficient data: evaluating profitability in Rice cultivation. *Spanish Journal of Agricultural Research,* 6(3), 309-319.

13- Rozane, Â.L.S., & Philippe, P.R.J. (2013). Rice production in Mercosur seen through. *Ravista de Politica Agricola*, 22(1), 55-71.

14- Salvatore, D. (1998). *Theory and Problem of International Economics*. Schaum' Outline Series, McGraw-Hill Book Company. Printed in Ney Publications, Tehran, Iran.

15- Yazdani. S., & R. EshraghiSamani (2006). Economic analysis of Almond production in Chahar Mahal and Bakhtiariprovince. *Iranian Journal of Agricultural Sciences,* 36(3), 537-545.

5

Determinants of Adoption of Mini-Tuber Seed Potato: A Case in Ardabil Province of Iran

Asghar Bagheri

Abstract

The aim of this study was to examine factors affecting adoption of mini-tuber seed potato. Survey research design was used in this study. A sample of 100 potato growers who were informed about mini-tuber was selected and divided into adopter and non-adopter groups. Questionnaire was used as the instrument for data collection and was validated by a panel of experts. A pilot study was conducted for reliability and Cronbach's alphas obtained 0.77- 0.87. Results revealed that other farmers were the main source of information for respondents. Higher yield, healthy seeds, marketability and high quality of potato produced from mini-tuber seeds were the main important reasons of adoption. Non-adopters mentioned high cost of mini-tuber seed as the main factor for rejection. The result of t-test showed that adopters had used more farm machineries and equipments. Result further showed that among farming unit characteristics, the number of pieces of owned farmland had negative effect on the adoption. However, the adoption was positively affected by the extent of owned farmlands and extent of owned potato acreage. Among personal characteristics, education level, number of literate household members had positive effect while farmers' age and farming experience had negative effect on the adoption, respectively.

Keywords:
Potato, Mini-tuber seed, Adoption, Ardabil

College of Agricultural Technologies and Natural Resources- University of Mohaghegh Ardabili.
** Corresponding author's email: bagheri_a2001@yahoo.com*

INTRODUCTION

Potato is one of the tuber crops grown in Iran and is used as third staple food after wheat and rice. It is grown by about 50 thousand farmers in 180 thousand hectares with average yield of 22 ton/hectares (FAO, 2008). Potato is regarded as a high-potential food security crop because of its ability to provide a high yield of high-quality product per unit input with a shorter crop cycle than major cereal crops like maize (Hirpa *et al.,* 2010). Potato farmers are faced with several problems. Shortage of healthy and good quality seed, high use of tuber seeds, lack of suitable storages and fridges, lack of improved, high yielding and disease-resistant/tolerant varieties are among the major constraints faced by potato farmers (Deffo and Demo, 2003; Lal *et al.*, 2011; Zarghami, 1999). Increase in potato acreage and yield calls for improvement of the quality of seed potatoes supplied to the ware potato production systems (Hirpa *et al.*, 2010).

Moreover, tuber seed production cost is the key problem in potato production. Depending on the region/country, it accounts for 30- 50 percent of production cost. To resolve this problem, mini-tuber seed production system became prominent and this is efficient and inexpensive because it produces low cost tuber seeds (Struik, 2007). Dissemination of mini-tuber seed technology can improve productivity of potato production systems. However, farmers' adoption of improved technologies is one of the most crucial factors affecting productivity of a cropping system which is affected by many factors (Neupane *et al.,* 2002; Rogers, 2003).

Keshavarz *et al.* (2010) found that price volatility, low quality and marketability, and low profitability of High Yielding rice Varieties (HYV) in comparison with local varieties, unavailability of inputs, long period of production cycle, more water and fertilizer needs, and susceptibility to pests/diseases had negative impact on the adoption of HYV in Guilan province of Iran. Kafle and Shah (2012) examined determinants of adoption of improved potato varieties in Bara district of Nepal. They found household size has negative influence on the adoption while land tenancy, contact with extension agents, age and level of education have positive influences. Lal *et al.* (2011) found that involvement of middle men, low sale price of potato, shortage of electricity, gluts, poor quality and adulterated fungicides, lack of cold storage facilities, low risk bearing ability of the potato farmers, lack of motivation from State Department of Horticulture and unavailability of good quality potato seed to the farmers were the most serious constraints perceived by potato growers in Bihar state of India. Ekwek and Onunka (2006) evaluated farmers' adoption of sweet potato production technologies in Abia State of Nigeria. They found that adoption of the technologies was relatively at a medium level. They found that accruable farm income; farm size cultivated as well as sweet potato output was positively related to farmers' adoption of the technologies. Mbanaso *et al.* (2012) assessed the adoption of potato production technology in Nigeria. Their findings showed 79.63% of the farmers were aware of the technology. Majority of the farmers had adopted most production practices. The constraints to increased adoption of the technology were scarcity of land, difficulty in integrating production technology into existing production system, low consumer preference associated with sweet potato products, lack of market, and unavailability of inorganic fertilizer. Deffo and Demo. (2003) characterized progress and constraints faced by potato farmers in adopting the new potato varieties in the main potato-producing regions of Cameroon. Results indicated that 98.6% of the potato farmers were aware of the new cultivars. Their main source of information was other farmers. These farmers reported that the resistance of Cipira to late blight was higher than that of local and European cultivars. The farmers' average yearly potato production increased during the period when these varieties were being adopted. The main constraint to the adoption was bacterial writ susceptibility. Another major problem was the lack of technical guidance for potato farmers from the agricultural extension service. Hirpa *et al.* (2010) analyzed seed potato systems in Ethiopia. They found informal, alternative and formal seed systems co-exist. The informal system, with low quality seed, is dominant. The formal system is too small to contribute significantly to improve that situation.

Despite numerous studies conducted on the adoption of agricultural technologies, there is dearth of information on the adoption of potato seed technologies and factors hindering or promoting their adoption in Ardabil province. This study was, therefore, conducted to assess adoption of mini-tuber seed potato technology and identify factors associated with its' adoption in Ardabil province of Iran.

MATERIALS AND METHODS

Study area

Ardabil province is located in North West of Iran (37°04' to 39°42' N. and 47° 02' to 48° 55' E.). There are 9 counties in this province. From which, potato is grown in 24000 hectares by 3500 farmers in Ardabil plain, i.e. Ardabil, Namin and Nir Counties. Figure 1 shows the study area. There are two mini-tuber producing cooperatives in this region. Some of their products are sold to local farmers but most of it goes to other potato producing regions of the country.

A sample consisting 100 potato growers who were informed about mini-tuber was selected randomly and the data collected using a questionnaire as the main instrument of study. It was validated by specialists and potato field experts. A pilot study was conducted using 30 farmers for reliability and Cronbach's alphas obtained 0.77- 0.87. Data was analyzed using SPSS and Excel software.

Adoption of mini-tuber seed potato was the dependent variable which measured by binary variable of adoption (1) and non-adoption (0). Adopters were potato growers who were using mini-tuber as seed potato and non-adopters were farmers who informed about mini-tuber seeds, but did not use it in their farms or rejected after using in one cultivation year. Independent variables include socio- economic and farming systems characteristics i.e., age (year), farming experience (year), number of household members, literate members in households, education level, marital status, potato owned farm lands (ha), potato rented farm lands (ha), irrigation water resources, main job, production target, extension activities, and farmers' reason for adoption, Logistic regression was used for analyzing factor affecting adoption of mini-tuber seed potato technology.

RESULTS AND DISCUSSION

Demographic results (Table 1) showed that mean age and potato farming experience of the respondents were 47.9 and 25.6 years, respec-

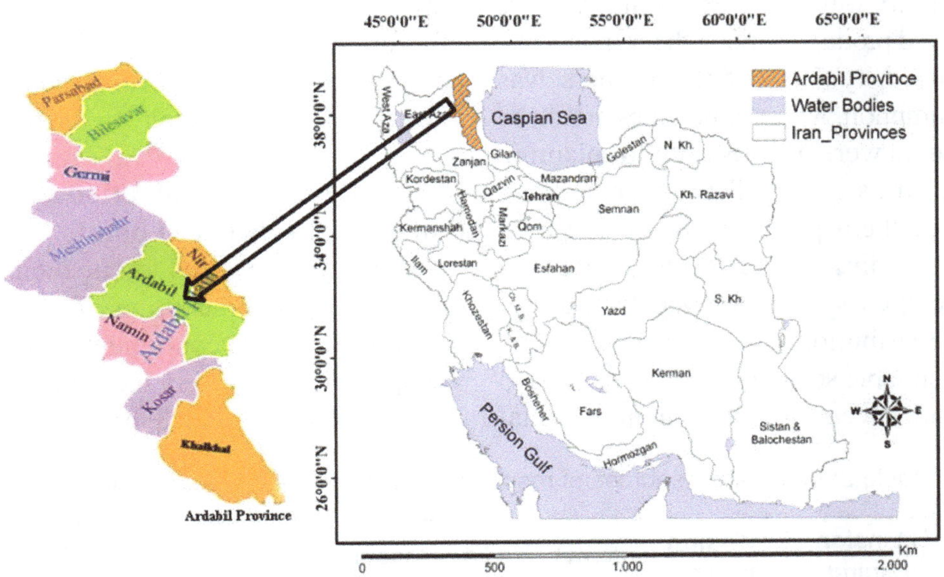

Figure 1. The area of study

Table 1: Socio-economic and farming system characteristics of respondents (n=100)

Variables	Means	SD	Variables	Means	SD
Age (year)	47.9	12.02	Potato owned farm lands (ha.)	8.45	7.02
Farming experience (year)	25.6	10.54	Potato Rented farm lands (ha.)	2.36	4.41
Number of household members	5.33	1.96	Irrigation water resources	Name of resource	Frequency
Literate members in households	2.24	1.69		Well	60
				Well and Aqueduct	10
	levels	Frequency		dam	13
Education level	Illiterate	5		Well and dam	17
	less than diploma	60	The main job	Agriculture	94
				Others	6
	diploma	23	Production target	Sale	94
	Above	12		Self-consumption	6
Marital status	single	97	Extension activities	Participation in Extension	49
	married	3			

tively. The main job of 94% was farming, 38% of them had not any off-farm income. All of them had owned farm lands. 33% of studied farmers had 1-9 hectares of cultivated area. Because of smallness of farm size, 45% of farmers were cultivating in rented lands in addition to their own farm operations. Education level of 67% of them was less than diploma certification.

Average potato growing area was 8.45 hectares in 9 pieces of farmlands. Regarding irrigation water resources, 60, 13, and 10 percent of them were using Well, Dam, Joint use of Well and Aqueduct, respectively. Other farmers were using two or more water resources jointly. There was not any seed producer among the respondents. Potato produced by 70% of the farmers was sold for consumption. A part of potatoes produced by other farmers were used as seed for planting in their own farms or to sell to other farmers. Six percent of them produced it for self-consumption. According to the results, there was not any developed system for management of production, distribution and plantation of certificated and suitable seed. As Hirpa *et al.* (2010) noted, there are several problems in non-formal and traditional systems of seed production such as high use of tuber seeds, yield decline and pests/diseases spread.

As depicted in Table 2, their main source of information for seed potato was other farmers. This result is in line with Deffo and Demo (2003). However, because of technical nature of mini-tuber seed technology, the other farmers had not enough experience of mini-tuber seed potato plantation and acted as a barrier to adoption. Information source of 31% of the studied farmers was mini-tuber producing cooperatives. Despite 49% of respondent who participated in extension programs, only 14% of them claimed that their main source of information was extension. It implies that extension programs did not aim at adoption and application of this technology in the region.

Farmers' reason for adoption and non-adoption of mini-tuber seed potato

Adopters were asked to rate their reasons for adoption and non-adoption of mini-tuber seed potato using a five continuum scale, from 0 for unimportant to 4 for very important, respectively. As depicted in Table 3 higher yield, healthy

Table 2: Information sources of respondents for mini-tuber seed potato (n=100)

Information sources	Extension training	Extension materials	Internet	Radio -TV	Researchers	Farm advisors	Other farmers	Seed cooperatives
Frequency	13	1	1	10	25	23	34	31

Table 3: Farmers' reasons for adoption of mini-tuber seed potato

Items	Mean	SD
Higher yield	3.62	0.681
Healthy seeds	3.35	0.900
Marketability and higher quality of potato	3.32	0.849
Durability compared with other varieties	3.20	0.947
Higher product prices compared with other varieties	3.09	0.945
Absorption of the more fertilizer for more yield	2.96	0.960
More profitable compared with other varieties	2.73	0.880
Priority in the use of farm advisors	2.22	1.058

Mean range: (0= Unimportant, 4= Very important)

Table 4: Farmers' reasons for non-adoption of mini-tuber seed potato

Items	M	SD	Rank
High costs of mini-tuber seed	3.50	0.820	1
Lack of information about mini-tuber seed cultivation	2.99	0.962	2
Lands fragmentation and dispersion	2.95	1.052	3
Technical nature of mini-tuber seed	2.94	0.979	4
Price uncertainty and volatility in potato market	2.94	1.058	5
Lack of timely supply of inputs and access to seed	2.93	0.991	6
Lack of machinery for mini-tuber seed cultivation	2.89	0.975	7
Lack of price guarantee	2.73	1.159	8
Lack of product market	2.66	1.096	9
Susceptibility of mini-tuber seed to pests and diseases	2.45	1.003	10
Higher water requirement	2.22	1.089	11
More fertilizer Need	2.17	0.966	12
Long period of mini-tuber seed cultivation	1.85	0.877	13

Mean range: (0= Unimportant, 4= Very important)

seeds, marketability and high quality of potato produced from mini-tuber seed were placed in higher importance. Priority of mini-tuber seed growers in the use of farm advisors was placed in the last importance.

In contrast, non-adopters mentioned the high costs of mini-tuber seed as the most important reason for rejection, followed by technical nature of mini-tuber seed and lack of information, land fragmentation and dispersion. Other reasons such as more water and fertilizer needs and cultivation duration problem were perceived as lower important reasons (Table 4).

The results also showed that infrastructural reforms of lands and credit facilities and the development of extension education programs can facilitate the adoption and application of mini-tuber seed potato by farmers.

The result of t test showed that there was significant difference between the two groups regarding farm mechanization (Table 5) and adopters had used more farm machineries and equipment in their farms. This result implies that poor mechanization is a barrier to adoption.

Logistic regression was used to examine the effect of selected variables on adoption of mini-tuber seed potato (Table 6). For this reason, respondents were divided into technology adopter (1) and non-adopter (rejecters and abandoners) (0) groups

Table 5: Comparing the two groups of farmers on the use of farm mechanization

Variable	Adoption	n	M	SD	t	p-value
Farm Mechanization	Yes	46	8.46	2.66	2.94	0.004
	No	54	6.50	3.78		

Table 6: Logistic regression analysis of factors affecting the adoption of mini-tuber seed potato

Variables	B	S.E.	Wald	Exp (B)
Age	-0.136**	0.048	7.980	1.146
Farming experience	-0.134**	0.048	7.776	0.875
The number of household members	-0.07	0.159	0.194	0.932
The number of literate members in households	0.191**	0.180	1.132	1.210
The main job	0.254	0.461	0.303	1.289
Education level	0.343**	0.221	2.407	1.409
Total area of owned farmland	0.041*	0.026	2.476	1.041
Total area of rented farmland	0.043	0.080	0.289	1.044
Extent of owned potato acreage	0.048*	0.058	0.674	0.953
Extent of rented potato acreage	0.060	0.100	0.358	1.062
Number of pieces of owned farmland	-0.035*	0.036	0.938	1.036
Number of pieces of rented farmland	-0.060	0.139	0.189	0.941
-2 Log likelihood		107.894		
-2 Log likelihood		75.645		

**p<0.01 *p<0.05

As depicted in Table 6 the model correctly predicted 86.4 percent of the observations. It implies that high percentage of the dependent variable was correctly separated by the model and the adoption was explained by variables in the equation. According to the model, the extent of rented land and rented potato farms, household number and farmers' main job had no significantly effect on the adoption of mini-tuber seed. -2 Log likelihood acts as goodness of fit and decreases when model is well matched with data. As shown in Table 6, it decreased from 107.894 to 75.645 which confirm the goodness of fit of the model.

Among farming units characteristics, number of pieces of owned land negatively impacted on the adoption. The adoption was also positively affected by the extent of owned farmlands and extent of owned potato acreage. Among the personal characteristics, education level and number of literate household members positively impacted while farmers' ages and farming experience negatively impacted on the adoption, respectively.

CONCLUSION

Although potato is the fourth staple crop in Iran, its ability to provide high yield of product per unit input with a shorter crop cycle than major cereal crops like maize gives it a high potential for food security (Hirpa *et al.*, 2010). However, in the conventional farming systems, farmers utilize high amount of seed tuber that threatens food security. For this reason mini-tuber seed potato get more importance.

According to the result, respondents perceive higher yield per unit of mini-tuber seeds in comparison with other seed tubers as the most important factor affecting adoption and use of the technology. Higher quality and marketability of potato produced from mini-tuber, production of healthy seed, and more durability of mini-tuber compared with other varieties were placed in the next ranks. As average yield per hectare of potato in Iran is 22 tons (FAO, 2008), there is high potential for enhancement of potato production and its' cost reduction. This result is in line with Hirpa *et al.* (2010) and Struik (2007).

According to logistic regression analysis, age and farming experiences had negative effect on adoption of mini-tuber technology. In this line, Iqbal *et al.* (1999) found age as an affecting factor on adoption. Furthermore, Faraji and Mirdamadi (2006), Ommani and Chizari (2006), Tabarraei and Hasannejad (2009), reported that there is negative and significant relationship between age and technology adoption. In contrary, farmers' education and literacy levels of family members were important factors affecting the adoption. Evidence has shown education level as a positive affecting factor on the adoption of innovations Joshi and Pandi (2005), Pezeshki and Masaeli (2002), Rajaei and Najafloo (2011). Negative effect of land fragmentation and dispersion on adoption of mini-tuber technology was the other finding of logistic regression. Many other studies conducted in Iran Bagheri and Malek-mohammadi, 2005; (Kohansal *et al.*, 2009; Pezeshki-rad and Masaeli, 2002) revealed the negative effect of land fragmentation and dispersion on the adoption of agricultural technologies. Based on the findings of the study, the following recommendations are presented:

High participation of studied farmers in extension programs and their satisfaction from farm advisors' performance despite no effect of these programs on adoption and use of mini-tuber technology imply that this technology overlooked by extension program planner that should be reconsidered. Adoption was negatively affected by lack of technical information of respondents. Other farmers were the main information source of respondent. Because of technical nature of the technology, special extension program is needed to promote potato growers' technical knowledge. Considering the results related to the effect of land fragmentation and dispersion on adoption as well as sensitivity of mini-tuber seed potato to water tension, improvement of land infrastructure, i.e. land consolidation and irrigation technology is necessary.

ACKNOWLEDGEMENT

The research reported here was financially supported by the University of Mohaghegh Ardabili (UMA). I would like to express my gratitude to the UMA and the anonymous reviewers.

REFERENCES

1- Bagheri, A., & Malek-mohammadi, I. (2005). Adoption behavior of Sprinkler irrigation in Ardabil province. *Iranian journal of Agricultural sciences,* 36(6), 1479-1488.

2- Deffo, V., & Demo, P. (2003). Adoption of two new potato varieties in Cameroon: progress and constraints. *American Journal of Potato Research*, 80, 263-270.

3- Ekwek, C., & Onunka, B.N. (2006). Adoption of sweet Potato production technologies in Abia State, Nigeria. *Journal of Agriculture and Social Research*, 6(20), 92-100.

4- FAO. (2008). International year of the Potato. Retrieved from http://www.fao.org/potato-2008/en/world/asia.html

5- Faraji, A., & Mirdamadi, (2006). Investigation of extension rule in the adoption of the apple crop insurance by growers of Damavand County. *Journal of Agricultural Science*, 12 (3), 500-489.

6- Hirpa, A., Meuwissen, M.P.M., Tesfaye, A., Lommen, W.J.M.-, Oude Lansink, A., Tsegaye, A. & Struik, P.C. (2010). Analysis of seed Potato systems in Ethiopia. *American Journal of Potato Research,* 87, 537–552.

7- Iqbal, M., Bashir, A., & Farooq, U. (1999). Factors affecting the adoption of hybrid maize varieties in the irrigated Punjab. *International Journal of Agriculture & Biology*, 1(3), 149-151.

8- Joshi, G., & Pandi, S. (2005). *Effects of farmer's perception on the adoption of modern rice varieties in Nepal.* Conference on International Agricultural Research for Development, October 11-13, Stuttgart-Hohenheim.

9- Kafle, B., &., Shah, P. (2012). Adoption of improved Potato varieties in Nepal: a case of Bara district. *The Journal of Agricultural Sciences*, 7 (1), 14-22.

10- Keshavarz, F., Allahyari, M., Azarmi Sesari, Z., & Khayati, M. (2010). Factors affecting non adaption of high-yielding rice varieties in Guilan province. *Journal of Agricultural Extension and Education*, 3 (4), 109-112.

11- Kohansal, M., Ghorbani, M., & Rafiee, H. (2009). Factors affecting of environmental and non-environmental on the adoption of Sprinkler irrigation. A case study in Khorasan Razavi Province. *Journal of Agricultural Economics and Development*, 17 (65), 112-97.

12- Lal, B., Sinha, T.K., Kumar, A., Pandit, A., & Pandey, N.K. (2011). Constraints perceived by the farmers in adoption of potato technology, *Potato Journal,* 38 (1), 73-77.

13- Mbanaso, E.O., Agwu, A.A.C., &., Asumugha, G. N. (2012). Assessment of the extent of adoption of sweet potato production technology by farmers in the southeast agro-cological zone of Nigeria. *Journal of Agriculture and Social Research,* 12(1), 124-136.

14- Neupane, R.P., Sharma, K.R., & Thapa, G.B. (2002). Adoption of agroforestry in the hills of Nepal: A logistic regression analysis. *Journal of Agricultural Systems,* 72, 177-196.

15- Ommani, A. F., & Chizari, M. (2006). Determination of wheat farmers' the social, economic and farming characteristics in Counties of Ahvaz, Dezful and Behbahan with respect to the adoption of low input sustainable agriculture. *Journal of Agricultural and Natural Resources Science and Technologies*, 10 (1), 23-13.

16- Pzeshki-Rad, G., & Masaeli, M. (2002). Economic factors affecting the adoption of IPM in the control of rice stem borer in Esfahan province. *Journal of Agricultural and Natural Resources Science and Technologies,* 6(4), 53-64.

17- Rajaee, Y., & Najafloo, B. (2011). Investigation of dryland wheat farmers' attitude in Mianeh county on the adoption of sustainable agriculture practices (case study: the outskirt of Mianeh county), the first national congress on science and new technologies in agriculture, September 10-12, Zanjan, Zanjan University.

18- Rogers, E. M. (2003). *Diffusion of innovations* (4th ed.). New York: Free Press.

19- Struik, P.C. (2007). The canon of Potato science: 25- mini-tubers. Potato Research 50,305–308.

20- Tabarraei, M., & Hasannejad, M. (2009). Investigation of performance and factors influencing the adoption of agricultural extension programs implemented in the course of the development process, a case study of wheat farmers in Mashhad County. *Journal of Agricultural Economics and Development* (Agricultural Science and Technology), 23 (1), 68-59.

21- Zarghami, R. (1999). Effect of culture medium and hormones used on rooting and the single nodes growth of Potato meristem culture. Proceedings of the First International Conference on Biotechnology of the Islamic Republic of Iran, March. 14-15, Tehran.

6

Energy Efficiency Analysis and Modeling the Relationship between Energy Inputs and Wheat Yield in Iran

*Fakher Kardoni [1], Majid Jami-Al Ahmadi [2] and Mohammad Reza Bakhshi [3]**

Abstract

Wheat is the dominant cereal crop constituting the first staple food in Iran. This paper studies the energy consumption patterns and the relationship between energy inputs and yield for Wheat production in Iranian agriculture during the period 1986 – 2008. The results indicated that total energy inputs in irrigated and dryland wheat production increased from 29.01 and 9.81 GJ ha^{-1} in 1986 to 44.67 and 12.35 GJ ha^{-1} in 2008, respectively. Similarly, total output energy rose from 28.87 and 10.43 GJ ha^{-1} in 1986 to 58.53 and 15.77 GJ ha^{-1} in 2008, in the same period. Energy efficiency indicators, input–output ratio, energy productivity, and net energy have improved over the examined period. The results also revealed that non-renewable, direct, and indirect energy forms had a positive impact on the output level. Moreover, the regression results showed the significant effect of irrigation water and seed energies in irrigated wheat and human labor and fertilizer in dryland wheat on crop yield. Results of this study indicated that improvement of fertilizer efficiency and reduction of fuel consumption by modifying tillage, harvest method, and other agronomic operations can significantly affect the energy efficiency of wheat production in Iran.

Keywords:
Wheat, Energy efficiency, Input-output, Econometric analysis

[1] *Former Graduate Student of Agro ecology, Department of Agronomy and Plant Breeding, Faculty of Agriculture, University of Birjand, Birjand, Iran.*
[2] *Associate Professor, Department of Agronomy and Plant Breeding, Faculty of Agriculture, University of Birjand, Birjand, Iran.*
[3] *Assistant Professor, Department of Agronomy and Plant Breeding, Faculty of Agriculture, University of Birjand, Birjand, Iran.*
* *Corresponding author's email: mbakhshi@birjand.ac.ir*

INTRODUCTION

Agriculture is an important sector in Iran's economy wherein wheat and barley are the main crops cultivated. Wheat constitutes the primary staple food of the people of Iran after being adopted since ages due to local wisdom, its immense nutritional value as well as low price. Therefore, wheat is the dominant cereal crop including almost 70 % of the aggregate cereal production. Based on statistics of Ministry of Jihad-e-Agriculture of Iran (Anonymous, 2010), wheat production in Iran was about 13.5 million tons in 2010, out of -which 8 million tons was from irrigated and 5.5 million tons was from dryland lands, respectively. According to the statistics of FAO (2013), Iran is the 12th leading producer of wheat in the world, with an average production of 14 million tons in 2011.

Wheat production is a direct function of high-yielding varieties, chemicals, fertilizers, mechanization and other energy inputs. It is produced using energy resources ranging from human and animal power to power of heavy machinery (Singh *et al.*, 2007). Energy requirements in agriculture are divided into four groups: direct and indirect, non-renewable and renewable. Direct energy is required to perform various tasks related to crop production processes such as land preparation, irrigation, planting, different crop management operations, harvesting and transportation of agricultural inputs and farm products (Singh, 2000). Direct energy consumption in Iranian agriculture amounts to around 204.37×1015 J.yr^{1} that constitutes 3.5% of the national consumption of fuel and electricity (MOE, 2006). However, a large part of the energy consumption in agriculture is indirect. Indirect energy consists of the energy used in manufacturing, packaging and transporting agricultural inputs (fertilizers, herbicides, etc.) and farm machinery. Non-renewable energy includes diesel, chemicals, fertilizers and machinery, and renewable energy consists of human labor, seed and manure (Mohammadi *et al.*, 2008). Energy consumption in wheat production has been increasing in response to increasing population, limited supply of arable land and most importantly, by supporting government's policies. Intensive use of energy causes problems threatening public health and environment. Efficient use of energy in agriculture will minimize environmental problems, prevent destruction of natural resources, and promote sustainable agriculture as an economical production system (Kizilaslan, 2009). The development of energy- efficient agricultural systems with low input energy compared with the output of food could help reduce the greenhouse gasses emissions in agricultural production (Dalgaard *et al.*, 2001).

Given the importance of energy consumption in agriculture sector, numerous works have been conducted on energy efficiency in agricultural systems and showed that, in most countries, Agricultural sector was highly dependent on different sources of energy and intensive use of energy causes some environmental problems such as an increase in global warming,CO_2 emissions, and non-sustainability. These studies revealed that the output of production could increase by ensuring optimal energy inputs in agricultural systems.

In case of Turkey, for example, the study by Canakci *et al.* (2005), Esengun *et al.* (2007), Hatirli *et al.* (2005), Hatirli *et al.* (2006), Kizilaslan (2009), Ozkan *et al.* (2004) showed that the inputs used in agricultural production were not used efficiently and led to many environmental problems. Hence, they suggested that sustainable agriculture should be extended and conscious farming should be provided. Results of Beheshti Tabar *et al.* (2010), Mohammadi and Omid (2010), Shahan *et al.* (2008) revealed that the contribution of energy inputs to crops yields in Iranian agriculture was significant. Also energy use in Iranian agriculture showed an increasing trend during the period 1990-2006 as the total energy input increased from 32.4 GJ ha^{-1} in 1990 to 37.2 GJ ha^{-1} in 2006. At the same period, total output energy increased from 30.85 to 43.68 GJ ha^{-1}. Also, Singh *et al.* (2007) showed that western Rajasthan had the maximum energy consumption in wheat production of India because textured soil was light which required intensive irrigation. The maximum energy ratio was observed at Punjab (5.2) and Ultra Pradesh (4.2) areas. The work of Martin *et al.* (2006) in

USA is another example of the important energy inputs in agricultural systems with different scales and management. According to this study, food production systems with large yields are more dependent on renewable energies.

Regarding the energy scarcity and wheat importance in Iran, this study was carried out to examine energy use pattern for irrigated and dryland wheat production for the period of 1986–2008. Furthermore, this study aims to explore the relationship between energy inputs and wheat yield using various functional forms. In this regard, the relationship is examined for different energy sources as renewable and non-renewable, direct and indirect energy. The results can be used by policy makers or other relevant agents in order to ensure sustainability and more efficient energy use in wheat production in agricultural economics of Iran. Some studies have been conducted on energy use in wheat production (Beheshti Tabar *et al.*, 2010; Shahan *et al.*, 2008; Taki *et al.*, 2012). However, most of these studies have used cross-sectional data rather than time series data. Besides, none of them explored functional relationship between energy inputs and wheat yield. In fact, little attention has been paid to the relationships between input energy and wheat yield using functional forms in these research studies where energy use in agriculture was examined. In this study, we have used time series data and focused on the relationship between energy inputs and wheat yield which differ from the previous researches.

MATERIALS AND METHODS

The data used in this study is based on yearly statistics for the period of 1986 to 2008. Information about energy consumption was obtained from the Ministry of Jihad-e Agriculture of Iran and FAO. The study has also benefited from the previous researches and studies conducted on energy analysis in agriculture. The study starts in 1986 because confident statistics are available only for this period of time. To calculate energy ratio, five major inputs in wheat production were considered including human labor, fertilizer, pesticide, irrigation, seed, and diesel Fuel and yield value. Energy content embodied in the crop was calculated based on review of literature and ASAE standards (Kitani. 1999). Energy equivalents of the inputs and outputs are given in Table 1.

In order to do an energy analysis, it is necessary to consider the use of human and animal power in agricultural processes. The amounts of agricultural labor work required for all of the operations for each crop were collected from the statistical Yearbooks on production cost (Anonymous. 2007). These data were given as labor-day ha^{-1}. Assuming 8 h of work a day and considering 1.96 MJ h^{-1} as human labor energy equivalent (Beheshti Tabar *et al.*, 2010; Mohammadi *et al.*, 2008) , the labor-day ha^{-1} was converted into GJ ha^{-1}. Considering the negligible share in the total energy input, animal power was omitted. Pesticides (insecticide, fungicide, and herbicide) and fertilizer (N, P, and K) consumption inputs data for irrigated and dryland wheat in each

Table 1. Energy equivalent of inputs and outputs in agricultural production

Item	Unit	Energy equivalent (MJ/unit)	Reference
A. Inputs			
Human labour	h	1.96	[3, 5, 10]
Diesel Fuel	L	47.3	[10]
Chemical fertilizers			
Nitrogen (N)	kg	78.1	[3, 8]
Phosphate (P_2O_5)	kg	17.4	[3, 8]
Potassium (K_2O)	kg	13.7	[3, 8]
Irrigation	m3	1.02	[11]
Insecticides	kg	58	[12, 8]
Fungicides	kg	115	[12, 8]
Herbicides	kg	295	[12, 8]
Seeds	kg	25	[13]
B. Outputs			
Wheat grain yield	kg	14.7	[14]

year were collected. Another essential input for agriculture production is seed. Its amount was obtained based on the data for wheat crop (Anonymous. 2007).

Since there is no data available for diesel fuel consumption for machinery used in agriculture, the total diesel energy input for the last year of the investigation was collected through field investigation using 75 hp tractors, taking into account the differences in the field operations of different crops. The fuel consumption for all the field operations for every year of the study period was estimated based on the differences in mechanization levels of each operation at different years compared with the mechanization level of 2008 (Beheshti Tabar *et al.*, 2010).

Data on electricity consumption in agriculture is not available for wheat crop. Therefore, to calculate the amount of electricity consumed in crop production, the energy required to pump water for irrigating crop was estimated. Average depth of wells used for agricultural purposes in Iran is reported to be 80 m (MOE. 2006). A weighted average of the water requirement of wheat crop sown in different parts of the country was obtained and the energy required to pump the required amount of water was calculated using the following formula:

$$DE = \gamma gHQ/\varepsilon q \qquad (1)$$

Where DE: direct energy (GJ ha^{-1}); g, water density (1000 kg m^{-3}); g, gravity (9.81 m s^{-2}); Q: net water requirement (m^3 ha-1); H: the total head (m); eq: overall efficiency, taken to be 0.18 (Ercolia *et al.*, 1999). Net water requirement was calculated by the CropWat. Based on the energy equivalents of the inputs and output (Table 1), the energy ratio (energy use efficiency), energy productivity and the specific energy were calculated (Demircan *et al.*, 2006; Sartori *et al.*, 2005).

$$\text{Energy use efficiency} = \text{Energy Output (MJ ha}^{-1}\text{)/Energy Input (MJ ha}^{-1}\text{)} \qquad (2)$$

$$\text{Energy productivity} = \text{Grain Output (MJ ha}^{-1}\text{)/Energy Input (MJ ha}^{-1}\text{)} \qquad (3)$$

$$\text{Specific energy} = \text{Energy Input (MJ ha}^{-1}\text{)/Grain Output (MJ ha}^{-1}\text{)} \qquad (4)$$

$$\text{Net energy} = \text{Energy Output (MJ ha}^{-1}\text{)} - \text{Energy Input (MJ ha}^{-1}\text{)} \qquad (5)$$

In order to analyze the relationship between energy inputs and yield, several mathematical functions were tried. The Cobb–Douglas function was selected as the suitable function pattern. Several authors used Cobb–Douglas function to evaluate the relationship between energy inputs and production (Hatirli *et al.*, 2005; Hatirli *et al.*, 2006, Mohammadi and Omid, 2010). Cobb–Douglas function is expressed as follows:

$$Y = f(x)\exp(u) \qquad (6)$$

Eq. (6) can be linearized and be further re-written as:

$$\ln Yi = a + \sum_{j=1}^{n} \alpha_j \ln(x_{ij}) + e_i \qquad i = 1, 2, \ldots, n$$

(irrigation wheat) (7)

$$\ln Yi = a + \sum_{j=1}^{n} \alpha_j \ln(x_{ij}) + e_i \qquad i = 1, 2, \ldots, n$$

(dryland wheat) (8)

Where Yi denotes the yield of the i'th farmer, Xij is the vector of inputs used in the production process, a is a constant, aj represents coefficients of inputs which are estimated from the model and ei is the error term. Eq. (7) and (8) can be expressed as follows;

$$\ln Y_i = \alpha_0 + \alpha_1 \ln x_1 + \alpha_2 \ln x_2 + \alpha_3 \ln x_3 + \alpha_4 \ln x_4 + \alpha_5 \ln x_5 + \alpha_6 \ln x_6 + e_i \qquad (9)$$

Where human labour energy is (X_1), diesel fuel energy is (X_2), fertilizer energy is (X_3), pesticide energy is (X_4), water of irrigation energy is (X_5), seed energy is (X_6) in irrigation wheat production.

$$\ln Y_i = \alpha_0 + \alpha_1 \ln x_1 + \alpha_2 \ln x_2 + \alpha_3 \ln x_3 + \alpha_4 \ln x_4 + \alpha_5 \ln x_5 + e_i \qquad (10)$$

Where human labour energy is (X_1), diesel fuel energy is (X_2), fertilizer energy is (X_3), pesticide energy is (X_4), seed energy is (X_5) in dryland wheat production.

With respect to this pattern, the impact of the energy of each input on the output energy was studied using (9) – (10). Then, the impact of DE and IDE energies, and RE and NRE energies on the output energy was studied. For latter purposes, Cobb–Douglas function was determined as (11) and (12), respectively, in irrigation and dryland wheat production:

$$\ln Y_i = \beta_0 + \beta_1 inDE + \beta_2 \ln IDE + e_i$$

(11)Where Yi is the i'th farm's yield, DE and IDE are direct and indirect energies used for wheat production, respectively, and β_i is the co-efficient of exogenous variables.

$$\ln Y_i = \gamma_0 + \gamma_1 inRE + \gamma_2 \ln NRE + e_i \qquad (12)$$

Where Y_i is the i'th farm's yield, RE and NRE are renewable and nonrenewable energies used for wheat production, respectively, and γ_i is the coefficient of exogenous variables.

Eqs. (9)– (12) were estimated using ordinary least square (OLS) technique. Basic information on energy inputs and cucumber yields was entered in Excel's spreadsheet and SHAZAM 9.0.

RESULTS AND DISCUSSION

Energy inputs and output values used in irrigated and dryland wheat production for the period of 1986-2008 are shown in Table 2. The energy amount of various inputs during the studied period showed an increasing trend in wheat production. Among all the various inputs of energy in irrigated wheat production, fertilizers were the highest energy consuming input (33.2%), followed by irrigation (31.1%) and diesel fuel (18.7%). In dryland farms, diesel fuel (41.3%) consisted of the highest energy share of total energy input. followed by fertilizer (30.92%) and seed (24.9%). Among the fertilizer energy inputs, irrigated and dryland wheat production increased by 135% and 68%, respectively over the past 23 years. The percent of N fertilizer of the total fertilizer energy was about 93% in irrigated and 87% in dryland wheat during the same period. Due to use of improper methods for application of fertilizers and lack of sufficient knowledge among farmers, fertilizer energy had an increasing trend during the studied period.

The results in Table 2 show that irrigation energy in irrigated wheat has increased from 9.74 GJ ha^{-1} in 1986 to 13.04 GJ ha^{-1} in 2008. Irrigation energy accounts for 31.1% of the total energy inputs per ha^{-1}. The energy value of irrigation in irrigated wheat production of Iran is higher than that of wheat production systems in Australia (Khan *et al.*, 2010) and Pakistan (Hussain *et al.*, 2010), because irrigation is commonly performed with surface irrigation methods such as furrow irrigation in Iran. The average diesel fuel energy in the study period was 6.91 GJ ha^{-1} for irrigated and 4.53 GJ ha^{-1} for dryland wheat production. Increase of the diesel fuel consumption is due to increase of agricultural mechanization index from 0.28 kW ha^{-1} in 1986 to 0.53 kW ha^{-1} in 2008. Agricultural mechanization index is a ratio of the available

Table 2: Energy input and output values in wheat production for the period 1986-2008.

	Years					Average of Total	% [a]
	1986	1991	1996	2001	2008		
Irrigated wheat							
A. input							
human labor (GJ ha^{-1})	0.61	0.69	0.51	0.40	0.69	0.61	1.65
diesel fuel (GJ ha^{-1})	6.56	6.95	6.90	7.03	7.10	6.91	18.70
fertilizers (GJ ha^{-1})	7.24	10.06	11.21	10.6	17.03	12.26	33.20
pesticide (GJ ha^{-1})	0.06	0.22	0.21	0.21	0.39	0.24	0.67
irrigation (GJ ha^{-1})	9.74	10.88	11.91	11.52	13.04	11.47	31.10
seed (GJ ha^{-1})	4.68	4.72	5.20	5.96	6.23	5.37	14.56
B. outputs							
grain yield (GJ ha-1)	28.86	42.64	46.24	52.75	58.53	45.57	
Dry land wheat							
A. input							
human labor (GJ ha^{-1})	0.20	0.18	0.22	0.39	0.25	0.24	2.23
diesel fuel (GJ ha^{-1})	4.44	4.67	4.63	4.64	4.69	4.53	41.32
fertilizers (GJ ha^{-1})	2.44	3.11	3.03	4.76	4.1	3.39	30.92
pesticide (GJ ha^{-1})	0.02	0.09	0.05	0.06	0.06	0.06	0.62
irrigation (GJ ha^{-1})	-	-	-	-	-	-	-
seed (GJ ha^{-1})	2.7	2.65	2.73	2.51	3.22	2.73	24.9
B. Outputs							
grain yield (GJ ha^{-1})	10.4	11.75	10.59	15.7	15.77	13.06	

[a] Indicate percentage of total energy input.

Table 3: Energy parameters in wheat production for the period 1986-2008

	1986	1991	1996	2001	2008	Average of Total
Irrigated wheat						
Energy input (GJ ha^{-1})	29.01	33.71	36.14	36.01	44.67	36.80
Energy outputs (GJ ha^{-1})	28.87	42.64	46.24	52.76	58.53	45.57
Energy use efficiency	0.99	1.26	1.27	1.46	1.31	1.22
Net energy (GJ ha^{-1})	-0.13	8.93	10.1	16.75	12.9	8.46
Energy productivity(Kg MJ^{-1})	0.06	0.08	0.08	0.09	0.09	0.08
Specific energy (M J Kg^{-1})	14.77	11.62	11.49	10.03	11.22	12.21
Dryland wheat						
Energy input (GJ ha^{-1})	9.81	10.73	10.67	12.12	12.35	10.96
Energy outputs (GJ ha^{-1})	10.43	11.75	11.35	15.7	15.77	13.06
Energy use efficiency	1.06	1.09	0.99	1.29	1.27	1.16
Net energy (GJ ha^{-1})	0.62	1.02	0.67	3.58	3.42	2.05
Energy productivity(Kg MJ^{-1})	0.07	0.07	0.06	0.08	0.08	0.08
Specific energy (MJ Kg^{-1})	13.81	13.42	14.82	11.34	11.51	13.03

mechanical power to the total agricultural land. The inconsistency observed in the increase of diesel energy stems from differences in agricultural practices and the type of machinery used.

The seed energy consumed in irrigated and dryland production increased from 4.68 and 2.7 GJ ha^{-1} in 1986 to 6.23 and 3.22 GJ ha^{-1} in 2008, respectively. In recent years, more applications of board cast seeders instead of grain drills have led to use of more seeds per hectare during this period. Human energy in irrigated and dryland wheat production increased from 0.611 and 0.20 GJ ha^{-1} in 1986 to 0.696 and 0.25 GJ ha^{-1} in 2008, respectively. The shares of these inputs in total energy use per hectare for irrigation and dryland wheat were 1.65% and 2.23%, respectively. Use of cheap foreign labors in the agricultural sector especially in irrigation and harvesting wheat has been reason for the increase of labor energy.

It is clear from Tables 2 that pesticides were the least demanding energy input in wheat production with 0.249 (0.67% of the total energy input) and 0.06 GJ ha^{-1} (only 0.62% of the total energy input) in irrigated and dryland farms, respectively. Pesticide energy consumption in irrigated and dryland wheat production had a growing trend in the study period and increased from 0.066 and 0.02 GJ ha^{-1} in 1986 to 0.392 and 0.06 GJ ha^{-1}, in 2008, respectively. Energy input and output, energy ratio, net energy, productivity energy and specific energy of wheat production for different years of the study period are shown in Table 3. Total input energy in irrigated wheat production increased by approximately 53.9% and in dryland wheat increased by about 25.8% from 1986 to 2008, whereas total output energy in irrigated and dryland farms increased by about 102.7% and 51.1% from 1986 to 2008, respectively. (Table 3). Improvement of seed varieties and application of new technologies have affected the energy output.

Mean energy ratio was calculated as 1.22 for irrigated and 1.16 for dryland wheat production. Energy ratio in irrigated wheat production increased from 0.99 to 1.31 and in dryland wheat from 1.06 to 1.27 during the study period. Thus, energy ratio in wheat production shows an increasing trend from 1986 to 2008. Beheshti Tabar *et al.* (2010) reported an increasing trend in the energy ratio of Iran agriculture from 0.95 in 1990 to 1.17 in 2006. However, in Turkey, Ozkan *et al.* (2004) reported a decreasing energy ratio from 2.23 to 1.18 for the periods of 1975 – 2000. Higher growth rate in energy output (yield) than the input of energy is the main reason for rising trend of energy ratio in this review period.

Energy productivity is the term used to estimate the yield of marketable product per unit of energy consumption. Average energy productivity of irrigation and dryland wheat was 0.08 kg MJ^{-1}. As can be seen in Table 4, energy productivity of irrigated and dryland wheat has shown a significant increase from 0.06 and 0.07 kg MJ^{-1} in 1986 to 0.09 and 0.08 kg MJ^{-1} in 2008,

Table 4: Total energy input in the form of direct, indirect, renewable, Non-renewable energy for wheat (Average of period 1986-2008)

Types of energy	Irrigated wheat		Dryland wheat	
	(GJ ha^{-1})	%[a]	(GJ ha^{-1})	%[a]
Direct energy	18.99	51.6	4.77	43.55
Indirect energy	17.81	48.39	6.18	56.44
Renewable energy	17.38	47.24	2.97	27.13
Non-renewable energy	19.42	52.76	7.98	72.87
Total energy input	36.80		10.96	

[a] Indicate percentage of total energy input.

respectively, generally reflecting improvement of efficiency in wheat production over the years.

During the 23- year period, the net energy in irrigated and dryland wheat in 2008 compared to 1986 has a 93 fold and 5 fold increases, respectively. Also, specific energy showed an increasing trend during this period. Specific energy indicated the amount of energy spent to produce a unit of marketable product. The average amount of energy required to product 1 kg of irrigated and dryland wheat was calculated as 12.21 and 13.03 MJ.

Also, the distribution of inputs used for the production of wheat based on the direct, indirect, renewable and non-renewable energy group is given in Table 4. As can be seen from the Table, wheat production has mainly depended on non-renewable energy sources. Similar results have been reported in several studies (Hatirli *et al.*, 2005; Beheshti Tabar *et al.*, 2010).

One of the main objectives of this study was to explore the relationship between total output and energy inputs in some detail. For this purpose, Cobb–Douglas production function was specified and estimated using ordinary least square estimation technique. One of the features of this production function is that the estimated coefficients represent elasticity. Furthermore, Cobb–Douglas production function imposes a priori restrictions on patterns of substitution among inputs. In particular, elasticity of substitution among all inputs must be equal to unity.

Since time series data were used in this study, autocorrelation might be a potential concern, and therefore should be tested using the Durbin–Watson test. The computed Durbin–Watson values were calculated as 2.03, 2.15, 2.05, 1.9, 2.01 and 2.08 for Eqs. (9) – (12), showing that there was no autocorrelation at the significance level of 5% in the estimated models.

The impact of energy inputs on yield was also investigated by estimating Eq. (9)-(10). Regression result for this model was shown in Tables 5 and 6. Irrigation water in irrigated wheat and

Table 5: Econometric estimation results of inputs for irrigated wheat production

Endogenous variable: yield / Exogenous variables	Coefficient	t-Ratio
Model1: $\ln Y_i = \alpha_0 + \alpha_1 \ln x_1 + \alpha_2 \ln x_2 + \alpha_3 \ln x_3 + \alpha_4 \ln x_4 + \alpha_5 \ln x_5 + \alpha_6 \ln x_6 + e_i$		
Constant	3.36	0.56
Human labour	-0.24	-1.97
Diesel fuel	-0.87	-0.97
Fertilizer	0.21	1.15
Pesticide	0.14	1.81
Water for irrigation	0.83	2.26**
Seeds	0.71	2.11**
Durbin-Watson	2.03	
R^2	0.89	

*p<0.01, **p<0.05

Table 6: Econometric estimation results of inputs for dryland wheat production

Endogenous variable: yield Exogenous variables	Coefficient	t-Ratio
Model2: $\ln Y_i = \alpha_0 + \alpha_1 \ln x_1 + \alpha_2 \ln x_2 + \alpha_3 \ln x_3 + \alpha_4 \ln x_4 + \alpha_5 \ln x_5 + e_i$		
Constant	- 9.14	-0.69
Human labour	0.41	2.46**
Diesel fuel	0.38	0.81
Fertilizer	0.58	2.64*
Pesticide	-0.055	-0.51
Water for irrigation	0.11	0.26
Seeds		
Durbin-Watson	2.15	
R^2	0.52	

*p<0.01, **p<0.05

Table 7: Econometric estimation results of direct, indirect, renewable, and non-renewable energies for irrigated wheat production

Endogenous variable: yield Exogenous variables	Coefficient	t-Ratio
Model 3: $\ln Y_i = \beta_0 + \beta_1 \ln DE + \beta_2 \ln IDE + e_i$		
Constant	9.05	3.41*
Direct energy	- 0.25	- 0.89
Indirect energy	0.43	2.52**
Durbin–Watson	2.05	
R^2	0.86	
Model 4: $\ln Y_i = \gamma_0 + \gamma_1 \ln RE + \gamma_2 \ln NRE + e_i$		
Constant	- 7.2	- 3.47*
Renewable energy	1.32	4.07*
Non-renewable energy	0.51	2.71*
Durbin–Watson	1.9	
R^2	0.83	

*p<0.01, **p<0.05

fertilizer in dryland wheat had the highest impact among the other inputs and with elasticity of 0.83 and 0.58, respectively.

This indicates that additional use of irrigation water by 1% in irrigation wheat and fertilizer in dryland wheat would lead to increase of yield by 0.85% and 0.58%, respectively. Hatrili *et al.* (2006) developed an econometric model for greenhouse tomato production in Antalya Province of Turkey and showed that human labor, chemical fertilizers, biocides, machinery, and water energies were important inputs, which significantly contributed to yield.

The regression coefficients of direct and indirect energies (Models 3 and 5) as well as renewable and non-renewable energies (Models 4 and 6) on yield in irrigated and dryland production were investigated through Eqs. (11) and (12), respectively. The results are given in Tables 7 and 8. As shown, the regression coefficients of indirect, renewable and non-renewable energies were all statistically significant at level of 1%. The impacts of direct, indirect, renewable, and non-renewable energies on irrigation wheat were estimated as 0.25, 0.43, 1.32 and 0.51, respectively, whereas the regression coefficient of direct, indirect, renewable and non-renewable energies in dryland wheat were estimated as 3.38, 1.08, 0.98 and 1.38, respectively. In the literature, similar results have been in line with reported results of irrigation

Table 8: Econometric estimation results of direct, indirect, renewable, and non-renewable energies for dryland wheat production.

Endogenous variable: yield / **Exogenous variables**	**Coefficient**	**t-Ratio**
Model 5: $\ln Y_i = \beta_0 + \beta_1 \ln DE + \beta_2 \ln IDE + e_i$		
Constant	- 28.7	3.41*
Direct energy	3.38	- 0.89
Indirect energy	1.08	2.52**
Durbin–Watson	2.01	
R^2	0. 53	
Model 6: $\ln Y_i = \gamma_0 + \gamma_1 \ln RE + \gamma_2 \ln NRE + e_i$		
Constant	- 10.88	- 2.73*
Renewable energy	0.98	1.81
Non-renewable energy	1.38	2.64*
Durbin–Watson	2.08	
R^2	0.51	

*$p<0.01$, **$p<0.05$

wheat. For example, the impact of indirect energy was more significant than the impact of direct energy on yield (Hatrili *et al.*, 2006).

CONCLUSIONS

This study analyzed the energy consumption and investigated influences of energy inputs and energy forms on output levels in wheat production of Iran during the period of 1986–2008. For this purpose, total human labor, diesel fuel, fertilizer, pesticide, irrigation, seed, and output energies were estimated and the energy consumption pattern was examined as direct, indirect, renewable, and non-renewable energy classifications.

The results revealed that both input and output energy had grown over the study period. The total input energy value for irrigated and dryland wheat increased from 29.01 and 9.81GJ ha^{-1} in 1986 to 44.67 and 12.35 GJ ha^{-1} in 2008, respectively. Similarly, output energy in irrigated and dryland wheat increased from 28.87 and 10.43 GJ ha^{-1} in 1986 to 58.53 and 15.77 GJ ha^{-1} in 2008, respectively. Fertilizers in irrigated wheat production are high-energy resources and supply 33.2% of total energy input, while diesel fuel (41.32%) in dryland wheat consumed the most energy of total energy input. From an environmental perspective, excessive usage of chemical fertilizers and pesticides has negative effects on sustainable agricultural production. In addition, mechanization that depends on fossil fuel usage is another negative factor. This is significant because use of farm manure, biological control, integrated pest management and saving diesel fuel by improving tillage can be effective on sustainable agricultural production.

The input-output ratio in irrigated and dryland wheat increased from 0.99, 1.06 to 1.31, and 1.27, respectively in the examined period. Similarly, energy productivity and net energy indicators have also increased during this period. Improvement of energy indicators indicates the enhancement of wheat productivity in Iran during past two decades. It seems that application of optimal irrigation methods and policymaking toward proper distribution of fertilizer can lead to an increase in the energy ratio of wheat in Iran. Irrigation water and fertilizer had the highest impact on irrigation and dryland wheat production, respectively among the other inputs. The impacts of direct, indirect, renewable, and non-renewable energies in irrigation wheat were estimated as 0.25, 0.43, 1.32, and 0.51, respectively, whereas the regression coefficients of direct, indirect, renewable, and non-renewable energies in dryland wheat were estimated as 3.38, 1.08, 0.98, and 1.38, respectively.

Future researches should apply energy analysis in various low and high input systems along with long -term economic, environmental, and societal analysis, which further explain the suitability and compatibility of production system for establishing sustainable development.

ACKNOWELEDGMENTS

The authors wish to thank Mr Iman Beheshti Tabar for his guidance.

REFERENCES

1- Anonymous. (2007). Annual Agricultural Statistics (vol. 1). Agronomy and Horticultural Products; Tehran, Iran (In Persian).

2- Anonymous. (2010). Annual Agricultural Statistics (vol. 1). Agronomy and Horticultural Products; Tehran, Iran (In Persian).

3- Beheshti Tabar, I., Keyhani, A., & Rafiee, S. (2010). Energy balance in Iran's agronomy (1990–2006). *Renewable and Sustainable Energy Reviews*, 14, 849–55.

4- Canakci, M., Topakci, M., Akinci, I., & Ozmerzi, A. (2005). Energy use pattern of some field crops and vegetable production: case study for Antalya region, Turkey. *Energy Conversion Management,* 46, 655–66.

5- Dalgaard, T., Halberg .N., & Porter ,JR. (2001). A model for fossil energy use in Danish agriculture used to compare organic and conventional farming. *Agriculture, Ecosystems and Environment.* 87,51-65.

6- Demircan, V., Ekinci, K., Keener, H. M, Akbolat, D., & Ekinci, C. (2006). Energy and economic analysis of sweet cherry production in Turkey: A case study from Isparta province. *Energy Conversion Management*, 47,1761-1769.

7- Ercolia, L., Mariottib, M., Masonib, A., & Bonaria, E. (1999). Effect of irrigation and nitrogen fertilization on biomass yield and efficiency of energy use in crop production of Miscanthus. *Field Crops Research,* 63, 68–81.

8- Esengun, K., Gunduz, O., & Erdal, G. (2007). Input-output energy analysis in dry apricot production of Turkey. *Energy Conversion and Management.* 48,592-8.

9- Food and Agricultural Organization (FAO)(2013). Crop prospects and food situation. Retrieved from : http:/www.fao.org/giews/

10- Hatirli, S. A., Ozkan, B., & Fert, C. (2005). An econometric analysis of energy input–output in Turkish agriculture. *Renewable and Sustainable Energy Reviews*, 9, 608–623.

11- Hatirli, S. A., Ozkan , B., & Fert, C. (2006). Energy inputs and crop yield relationship in greenhouse tomato production. *Renewable Energy,* 31, 427–38.

12- Hussain, Z., Azam Khan, M., & Irfan, M. (2010). Water energy and economic analysis of wheat production under raised bed and conventional irrigation systems: A case study from a semi-arid area of Pakistan. *Soil Tillage research,* 109, 61-67.

13- Khan, S., Khan, M. A., & Ltife, N. (2010). Energy requirements and economic analysis of wheat, rice and barley production in Australia. *Soil and Environment,* 29(1), 61-68.

14- Kitani, O. 1999. *CIGR handbook of agricultural engineering,* (Vol. 5). American Society of Agricultural Engineers Publication.USA.

15- Kizilaslan, H. (2009). Input-output energy analysis of cherries production in Tokat province of Turkey. *Applied Energy*, 86,1354-8.

16- Martin, J.F., Diemont, S.A.W., Powell, E., Stanton, M., & Tacher, S.L. (2006). Energy evaluation of the performance and sustainability of three agricultural systems with different scales and management. *Agriculture, Ecosystems and Environment,* 115, 128–140.

17- Ministry of Energy of Iran (MOE). (2006). *Electrified wells report.* Water Resources Management Institute; Tehran, Iran (In Persian).

18- Mohammadi, A., Tabatabaeefar, A., Shahin, S., Rafiee, S., & Keyhani, A. (2008). Energy use and economical analysis of potato production in Iran, a case study: Ardabil province. *Energy Conversion Management,* 49, 3566–70.

19- Mohammadi, A., & Omid, M. (2010). Economical analysis and relation between energy inputs and yield of greenhouse cucumber production in Iran. *Applied Energy*, 87, 191-196.

20- Ozkan, B., Akcaoz, H., & Fert, C. (2004). Energy input–output analysis in Turkish agriculture. *Renewable Energy*, 29, 39–51.

21- Sartori, L., Basso, B., Bertocco, M., & Oliviero, G. (2005). Energy use and economic evaluation of a three year crop rotation for conservation and organic farming in NE Italy. *Biosystems Engineering*, 91(2), 245-256.

22- Shahan, S., Jafari, A., Mobli, H., Rafiee, S., & Karimi, M. (2008). Energy use and economical analysis of wheat production in Iran: A case study from Ardabil province. *Journal of Agricultural Technology,* 4(1), 77-88.

23- Singh, H., Singh, A. K., Kushwaha, HL., & Singh, A. (2007). Energy consumption pattern of wheat production in India. *Journal Energy*, 32,1848-1854.

24- Singh, J. M.(2000). *On farm energy use pattern in different cropping systems in Haryana*, India. Unpublished thesis International- Institute of Management, University of Flensburg, Germany-.

25- Taki, M., Ajabshirchi,Y., & Mahmoudi, A. (2012). Prediction of output energy for wheat production using artificial neural networks in Esfahan province of Iran. *Journal of Agricultural Technology*, 8(4). 1229-1242.

7

Analysis of Agro-Chemical Inputs Use in Maize Production among Small-Scale Farmers in Iwo Local Government, Osun State, Nigeria

Kemi Funmilayo Omotesho [1], Azeez Muhammad-Lawal [2] and Gafar Olawumi Iyiowu [2]*

Abstract

In spite of the development of various improved cultivars of maize, yield in Nigeria has continued to be hampered by the over cropped nature of most farmlands as well as a wide range of pests and diseases. This study therefore carried out an analysis of agro-chemical inputs' use in maize production among small scale farmers in Iwo Local Government Area of Osun State. Specifically, the study investigated the relationship between agro-chemical inputs' use and maize yield in the study area. It determined the factors affecting agro-chemical inputs' use and the major constraints to its usage. Data was collected using a structured questionnaire from 105 respondents obtained through a two stage random sampling procedure. Analytical tools employed for the study included Pearson Moment correlation and linear regression analysis. The study revealed a 78.5% positive and significant relationship between agro-chemical inputs' use and maize yield ($p< 0.01$). High cost and irregularity of supplies, insufficient farm income and fear of health hazards were the major constraints to the use of agro-chemicals. The determinants of agro-chemical inputs' use among the farmers were; farm size, income at $p<0.01$ and educational status and household size at $p<0.1$. The study concluded that a strong positive relationship existed between agro-chemical inputs' use and maize yield in the study area. It is recommended that there should be guaranteed minimum price for agricultural produce and that farmers should be empowered to increase their farm size through agricultural mechanization.

Keywords:
Agro-chemicals, Maize, Iwo Local Government Area, Pearson moment correlation, linear regression

[1] *Department of Agricultural Extension and Rural Development, University of Ilorin, Ilorin.*
[2] *Department of Agricultural Economics and Farm Management, University of Ilorin, Ilorin.*
* *Corresponding author's email: kfomotesho@yahoo.com*

INTRODUCTION

In terms of area under cultivation and total production, Maize (*Zea mays*. L.) is the world's third most important crop coming after wheat and rice (Purseglove, 1992). With average yield of 4.47tons per hectare as compared to 2.67 and 3.84 of wheat and rice respectively in 2003 (FAO, 2004), it can be concluded that aside from being one of the oldest crops in the world, maize is undoubtedly the highest yielding grain crop (Monlruzzaman *et al.*, 2009). It is a basic staple food for the teeming population of developing countries and hence significant to world food security (FAO and ILO, 1987). Africa contributes 6.5% of worldwide maize production of 785million tons while Nigeria; the largest African producer contributes about 8 million tons (IITA, 2013b). Land area cultivated with maize in Nigeria is about 3% of the world's total of 158million hectares (FAO, 2007). In Nigeria, Maize is becoming the miracle seed for the country's agricultural and economic development. Being an integral component of most cropping patterns, it provides food for man, feed for livestock and raw materials for some agro-based industries (Ahmed, 1996; Iken and Amusa, 2004; Onuk *et al.*, 2010). The dependence of many Nigerian rural families on income derived from the cultivation, processing and marketing of Maize (in various forms) has been observed to be significant and hence its contribution to rural livelihood (Degrande and Duguma, 2000).

According to Bello *et al.* (2010), agro-chemical inputs such as inorganic fertilizers, pesticides and herbicides are of immense importance to agricultural production because they eliminate drudgery, excessive operation cost, problem of scarcity of labour and improve both quality and quantity of output. Fertilizers are inorganic chemicals that are applied to the soil by crop farmers to improve production. Biggs and Courtney (1989) opined that farmers generally are favorably inclined to use fertilizers because of its wide range of benefits which include; their cheapness, cleanliness and ease of handling, storage and transport. The guaranteed composition of inorganic fertilizers makes it easier for users to determine the rate of application and to predict the effect upon yield. Iken and Amusa, (2004) emphasized the need for the application of the right form, combination and quantity of fertilizer and stressed the importance of nitrogen, phosphorus and potassium alongside micro nutrients such as sulphur, zinc and magnesium to good growth and yield in maize production. For West African savannas, a combination of NPK 15:15:15 at 50kg/ha and 50kgN/ha in the form of urea has been found to be most adequate (IITA, 2013a).

In spite of the existence of over 50 species, several of which are improved cultivars developed to resist known pests and diseases of the crop, Maize yields have continued to suffer from a range of diseases which have varied over the years in intensity (IITA, 2013; IITA 2013; Oladipo *et al.*, 1993) listed downy mildew, rust, stalk and ear rots, maize streak virus, leaf spot and leaf blight as major diseases of maize in Nigeria. Stem borers, grain borers, beetles, moths, rootworms, white grubs, army worms, termites, grasshoppers and weevils have been reported as major pests causing losses to maize production (Iken and Amusa, 2004). Though a number of grassy and broadleaf weeds have been associated with the maize plant, striga is about the most devastating weed of the plant which has been known to cause significant losses by attacking maize plants underground (Carsky *et al.*, 2000). According to Dugje *et al.* (2006), lamdacyhalothrin, cypermethrin, maricozeh, glyphosphate, alachor, oxidiaxone, propanil, altrazine and paraquat are some of the most common pesticides used in maize production in Nigeria.

While an increase of over 44% has been recorded in Nigerian maize production over the immediate past decade, more of this has been attributed to increase in area cultivated rather than yield (US Department of Agriculture, 2013). Demand for the plant has however continued to increase with increase in population and its increasing status as a commercial crop in high demands by agro-allied industries. Inadequacy of agro-chemical input use in maize production can significantly reduce yield thereby worsening the already existing gap between demand and supply of the crop (Crawford *et al.*, 2006; Daramola, 1985; Iken and Amusa, 2004).

Knowledge of existing relationship between agro-chemical input usage and maize production will provide a basis for favorable policies on agro-chemical input promotion so as to enhance its accessibility to smallholder farmers. It is also possible that certain socio-economic factors affect agro-chemical input usage among the farmers. Scientific information on the possible constraint militating against the use of agro-chemicals by maize farmers will provide insight for decision making on best ways of enhancing its use among farmers. It is against this background that the study is designed to answer that following research questions;

1. What are the socio- economic characteristics of maize farmers in Iwo Local Government, Osun State?
2. What is the relationship between agro-chemical input use by the farmers and their maize Yieldproduction?
3. What are the challenges militating against agro-chemical input use among maize farmers in the study area?
4. What factors affect agro-chemical input use among maize farmers in Iwo Local Government, Osun State?

Objectives of the study

The general objective of the study is to carry out an analysis of agro-chemical inputs use in maize production among small scale farmers in Iwo Local Government, Osun State, Nigeria. Specifically, the study;

(1) Highlights the socio-economic characteristics of maize farmers in Iwo local Government of Osun state.

(2) Assessed the relationship between agro-chemical input usage and maize yield.

(3) Examined the constraint militating against the use of agro-chemical inputs among maize farmers in the study area

(4) Identify the factors affecting agro- chemical input use in maize production in the study area.

MATERIALS AND METHODS

Study area

The study was carried out in Iwo Local Government Area (LGA) of Osun State, Nigeria. The local Government is one of the 30 LGAs in Osun State. Iwo LGA consists of 15 wards namely; Oke adan 1, Oke adan 2, Oke adan 3, Gidigbo 1,Gidigbo 2, Gidigbo 3, Molete 1, Molete 2, Molete 3, Isale oba 1, Isale oba 2, Isale oba 3, Isale oba 4, Oke oba 1, Oke oba 2. According to the 2006 Census report, Iwo is the most populous Local Government in Osun State with a population of 191,346. Iwo is situated north east of Osun State on latitude 7.58°N and longitude 4.28° E. The primary economic activity in Iwo LGA is agriculture as the soil and climatic condition in the area is favourable to plant growth. Iwo LGA is well known for maize cultivation given its vast savanna land area. Other crops produced in the area include; cocoa, Palm oil, cassava, vegetables, and yam.

The population for the study comprised of all maize farmers in Iwo LGA. The sampling frame for the study was prepared in conjunction with all the wards associations. A two-stage random sampling technique was used to select a total of 105 respondents used for the study. The first stage comprises of random selection of one community from each of the 15 wards that constitute Iwo LGA. The second stage was random selection of seven farmers from each of the selected communities based on the list of maize farmers.. The instrument for data collection was a structured questionnaire.

A combination of descriptive and inferential statistical tools was employed to analyze data collected. Frequency distribution, percentages and mean were used to present the socio-economic characteristics of the respondents. The Pearson moment Correlation analysis (PMCC) was used to test the relationship between usage of agro-chemical by the respondents and their maize production. A four point Likert-type scale was used to elicit information on the extent to which the respondents considered specific challenges to constitute constraints to their usage of agro-chemicals. A list of possible constraints drawn from literature was presented against which the farmers responded as to their severity using the Likert-type scale graduated thus;

Not serious=1, moderately serious=2, very

serious=3, extremely serious=4

A mean score was obtained for each constraints and this was used to rank the constraints in order of severity. The regression analysis was used to determine the factors affecting agro-chemical input use in maize production among the respondents. The regression model in its implicit form is given as;

$Y= f(X_1, X_2, X_3, X_4, X_5, X_6, X_7, U)$..........1

Where:

Y= Investment in agro-chemical input measured in Naira

X_1 = Farm Size measured in hectares

X_2 = Total Income measured in naira as the addition of farm income, non farm income and available income from other household members

X_3 = Age measured in years

X_4 = Highest Educational Attainment measured as a dummy variable 1 for the possession of formal education and 0 otherwise

X_5 =Gender measured as a dummy variable. 1 for male and 0 for female

X_6 = Farming Experience measured as number of years spent in farming

X_7 = Household size measured as the number of persons living under the same roof and eating from the same pot

U = Error term

RESULTS AND DISCUSSION

Socio-economic characteristics of Maize farmers

This study examined selected socio-economic characteristics of maize farmers. This was with a view of determining gender, age, educational status, farming experience, income, farm size and labour use among the farmers. Table 1 presents selected socio-economic characteristics of the respondents. Table 1 reveals the dominance of maize cultivation in Iwo Local Government by male as over 97% of the respondents were

Table 1: Socio-economic characteristics of the respondents

Socio-economic Characteris	Frequency	Percentage
Gender		
Male	102	97.1
Female	3	2.9
Total	**105**	**100**
Age		
<30	28	26.7
30-50	60	57.1
>50	17	16.0
Total	**105**	**100**
Educational status		
No Formal Education	8	7.6
Quoranic Education	4	3.8
Adult Education	7	6.7
Primary Education	71	67.6
Secondary Education	12	11.4
Tertiary Education	3	2.9
Total	**105**	**100.0**
Farming experience		
≤10years	47	44.8
>10years	58	55.2
Total	**105**	100.0
Farm Size		
<1ha	56	53.3
1-2ha	45	42.9
>2ha	4	3.8
Total	**105**	**100.0**
Household size		
≤6	24	22.9
>6	81	77.1
Total	**105**	**100.0**

Table 2: Result of Pearson correlation between agro-chemical usage and Maize yield

		Agro-chemical use	Maize Yield
Chemical used	r	1	0.785***
Significant			0.00
N		105	105

*** p<0.01

found to be male. This may be a pointer to the fact that women are more into its processing and possibly marketing of the crop. With a mean age of 38years and only 16% of the respondents being of over 50 years in age, it can be inferred that there was evidence of active participation of the younger generation in the production of maize in the study area. This is particularly remarkable because of its positive implications for the sustenance of the interest of the youth in agriculture given the effect of rural urban migration on agriculture in Nigeria. Table 1 also reveals that only 7.6% of the respondents had no formal education while about 14% had a minimum of secondary school education. Majority of the respondents (67.6%) had primary school education. More than half of the respondents (55.2%) had been farming for an upward of ten years. This goes to show that majority of the respondents were experienced farmers who were likely to be conversant with agrochemicals. The mean farm size of the respondent was 1.2ha and only few (3.8%) had more than 2 ha of farmland. On the average, the household size of the respondents was considerably high with a mean of 8. Over 77% of the respondents had household sizes of above 6persons with positive implications for availability of labour to carry out farming activities including the application of agro-chemicals.

Relationship between farmers' Agro-chemical input use and Maize production

In order to determine the relationship between the agro-chemical input use and maize production, Pearson correlation was used to test the correlation between the amounts spent on agro-chemical input and maize yield.

As shown in Table 2, the correlation coefficient (r) was estimated at 0.785 which was significant at 1% level. This implies that there was a relationship of about 78.5% between amount spent on agro-chemical inputs (fertilizer, herbicide and pesticide) used by the farmers and their maize yield. This relationship was positive and therefore as the amount of chemical used by the farmers increased, the maize yield also increased.

Constraints to Agro-chemical input use among respondents

This sub-section analyzed the constraints to the use of agro-chemical inputs among maize farmers in the study area. Table 3 presents the major constraints to its use as well as ranked the constraints in order of severity. Table 3 shows some variations in the respondents' aggregate ranking of the severity of the various constraints as it affects fertilizers, herbicides and pesticides. The major constraints to the use of fertilizers among the respondents were high

Table 3: Major constraints to Agro-chemical use among respondents

S/N	Constraint	Fertilizer		Herbicide		Pesticide	
		M	Rank	M	Rank	M	Rank
1	Irregular Supplies	2.10	2	1.71	7	1.72	7
2	High Cost of Supplies	2.16	1	3.16	1	2.15	2
3	Poor Knowledge of Application	1.42	7	2.96	2	2.42	1
4	Health Hazard	1.71	6	2.10	4	2.09	4
5	Large Farm Size	1.90	4	1.91	5	1.93	5
6	Insufficient Farm Income	2.10	2	2.90	3	2.10	3
7	Long Distance from Source	1.77	5	1.76	6	1.73	6
8	Poor Quality	1.36	8	1.10	8	1.43	8

Table 4: Determinants of Agro-chemical Input Use in Maize Production

Variables	Co-efficient	p-value
Constant	14413.952	.423
Farm Size	7978.215***	.000
Income	.050***	.000
Age	.991	.865
Education	2.690*	.075
Gender	1.079	.813
Farming Experience	-18.895	.787
Household Size	-810.053*	.084

Dependent Variable: Investment in Chemical Use. *** ($p<0.01$), * ($p<0.1$)
$R^2 = 0.759$, Adjusted $R^2 = 0.742$

cost of supply with the highest mean score of 2.16, irregular supplies and insufficient farm incomes with mean scores of 2.10 each. High cost of supplies with a mean score of 3.16 was also the most sever constraints to the use of herbicides among the respondents. This was followed in order of severity by poor knowledge of application, insufficient farm income and health hazards with mean scores of 2.96, 2.90 and 2.10 respectively. However, poor knowledge of application was the most severe constraint to the use of pesticides among the respondents with a mean score of 2.42. Other constraints considered to be severe included high cost of supplies, insufficient farm income and fear of health hazards.

Factors affecting Agro-chemical input use among respondents

This sub-section determines the factors affecting agro-chemical input use among maize farmers in the study area. Table 4 presents the result of linear regression analysis carried out to establish the determinants of agro-chemical input use among the respondents.

To determine the factors affecting chemical inputs use in maize production, regression model was fitted to determine the effect of some independent variable on the investment in chemical use. The Durbin-Watson statistics (2.226) shows that the data fitted well with the model. The coefficient of determination (R square adjusted) with value 0.745 shows that the explanatory variables explained about 74.5% of the variations in the use of chemical inputs on the maize farms leaving about 24.5% unexplained. The F statistic ($df_{9,97}$, 43.743) was significant at α=1.0%. This shows that the variables have joint significant effect on the investment on chemical use.

Table 4 shows the effect of each independent variable on the dependent variable (investment on chemical use). The farm size (ha) has a coefficient of 7,978.215. This was significant at α= 1.0%, this implies that if the farm size is increased by 1 ha, the investment on chemical input use will increase by #7,978.28. Another factor with significant effect on the investment of chemical input use was income of the farmers. The variable had a coefficient with value of 0.050. This was also significant at α=1.0%. This implies that if the farmers' income increased by #1.00, the investment on chemical use by the farmers will increase by #0.050. The coefficient of highest educational attainment was also found to be significant at α=5% meaning that the more educated the farmers were, the more their chances of investing in the use of agro-chemicals. This is likely due to the established relationship between education and the rate of adoption of innovation. The household size with coefficient of -810.053 was however found to be though significant, inversely related to investment in agro-chemical input use. An increase in household size therefore will result in a reduction in agro-chemical input use. It is possible that high household size presents a drain on the family income and hence a reduction in the available income to procure agro-chemical. The large family size could also provide the necessary manpower to carry out manual weeding as opposed to the use of herbicides.

CONCLUSION AND RECOMMENDATIONS

The study concluded that a strong positive relationship existed between agro-chemical input use and maize yield in the study area. It identified high cost of supplies, poor knowledge of application, insufficient farm income, and irregular supplies as the major constraints faced by farmers in the use of agro-chemicals in maize production. Farm size, income, educational attainment and household size were found to be the determinants of usage of agro-chemical inputs in Iwo Local Government Area of Osun State, Nigeria. The study therefore recommends as follows;

- There is a need to monitor and ensure that the farmers actually are the beneficiaries of government's subsidies to agriculture either directly in form of providing agro-chemicals at reduced prices or by removing import duties on them.
- Manufacturers of the various agro-chemicals should be encouraged to adopt packaging with user friendly labels which could be self explanatory to farmers.
- Farmers in the study area should take advantage of adult literacy classes to improve upon their levels of education, awareness and adoption of innovation.
- Agricultural extension units within the study area should sensitize farmers on preventive measures against the health hazards of agro-chemicals.
- There should be guaranteed minimum price for agricultural produce and that farmers should be empowered to increase their farm size through agricultural mechanization

ACKNOWLEDGEMENT

The Authors acknowledge the contributions of the management and staff of the Kwara State Agricultural Development Project for their role in facilitating the data collection stage of the study.

REFERENCES

1- Ahmed, B. (1996). *Economic analysis of fertilizer used on Maize production in the Northern Guinea Savannah of Nigeria,* Unpublished dissertation, Department of Agricultural Economics And Rural Sociology, Ahmadu Bello University, Zaria, Nigeria

2- Bello, M., Ibrahim, H.I., & Salau, E.S. (2010). Factors influencing the adoption of agro-chemical technology by small-scale farmers in Kwali Area Council of Abuja, FCT, Nigeria. *Journal of Environmental Issues and Agriculture in Developing Countries,* 2 (1), 176-190

3- Briggs, D., & Courtney, F. (1989). *Agriculture and environment.* Longman: London

4- Carsky, R.J., Berner, D.K. Oyewole, B.D. Dashiell, K., & Schulz, S. (2000). Reduction of striga hermonthica parasitism on Maize Using Soybean Rotation. *International Journal of Pest Management,* 46,115–120

5- Crawford, E.W, Jayne, T.S., & Kelly, V.A. (2006). *Alternative approaches for promoting fertilizer use in Africa. Agriculture and rural development discussion* - Washington, DC: World Bank. pp. 22

6- Daramola, A.M. (1985). Comparative effectiveness of carbofuran and carbaryl for the control of maize stem borers. *Nigerian Journal of Plant Protection,* 9, 54-59.

7- Degrande, A., & Duguma, B. (2000) Adoption potential of rotational hedge row intercropping. *Humid and Extension Network,* 103, 1-7

8- Dugje, I.Y., Kamara, A.Y., & Omoigui, L.O. (2006). Infestation of Crop Fields by Striga species in the Savanna Zones of North-East Nigeria. *Agriculture, Ecosystems and Environment,* 116, 251–254.

9- FAOSTAT (2004). Food and Agriculture Organization, Retrieved at 2005.

10- FAOSTAT (2007). Food and Agriculture Organization, Retrieved at 2008.

11- FAO and ILO (1987). *Maize in human nutrition intermediate level handbook.* FAO and ILO publication, Rome, Italy

12- IITA (2013a). Maize. Retrieved on 16th June, 2013 from www.iita.org/maize

13- IITA (2013b) Best Practices for Maize Production in the West African Savannas, IITA R4D Retrieved on 16th June, 2013 from www.iita.org/c/document_library/get_file?uuid=8330df34-cb4f...

14- Iken, J.E. & Amusa, N.A (2004), Maize research and production in Nigeria. *African Journal of Biotechnology,* 3, 301-307.

15- Monlruzzaman, M. S., Rahman, M., Karim, K. & Alam, Q. M (2009). Agro-economic analysis of Maize production in Bangladesh: A farm level study. *Bangladesh Journal of Agricultural, Resources,* 34(1), 15-24

16- Oladipo, J.L., Fajemisin, J.M., & Olanya, O. (1993). Diseases of maize damage and control in Nigeria. Maize improvements, production and Utilization in Nigeria, pp. 1181-1188

17- Onuk, E.G., Ogara, I. M., Yahaya, H., & Nannim, N. (2010), Economic Analysis of Maize Production in Mangu Local Government Area of Plateau State, Nigeria. *Production, Agriculture and Technology,* 6 (1), 1-11

18- Purseglove, J.W. (1992). *Tropical crops: Monocotyledons*. Longman Scientific and Technical, New York. Pp. 300 – 305

19- US Department of Agriculture (2013). Nigerian corn production by year. Retrieved on 16th June, 2013 from http://www.index mundi.com/ agriculture?country=ng&com

8

Effective Factors on Application of Sustainable Agricultural Practices by Paddy Farmers (Case of Rural Production Cooperatives Members)

Amir Ahmadpour

Abstract

One of the central duties of rural production cooperatives is to promote the adoption and application of sustainable agricultural practices by farmers, which hinges on several factors. Therefore, the study intends to evaluate the factors influencing the application of sustainable agricultural practices by paddy farmers. The research population included from all includes all paddy farmers as members of the production cooperatives in rural areas of Sari County, Iran (N=4160). A total of 162 farmers were selected through stratified sampling technique. The data was gathered through questionnaires, that its face validity was verified by a panel of experts, and its reliability was obtained through a pilot test (Alpha Cronbach coefficient, 0.93). The results showed that the application of sustainable agricultural practices was generally good. Moreover, the findings showed that there was a positive and significant relationship between the age, using information sources, family members, cultivating area, production amount, social status, access to inputs, participating in extension training courses and economic factors with the application of sustainable agricultural practices. The stepwise linear regression analysis results showed that the variables of cultivating area/hectare, economic facilities, participating in extension training courses and age can determine 61.8% variation of the application of sustainable agricultural practices by paddy farmers as members of rural production cooperatives.

Keywords:
Sustainable agriculture, Agricultural practices, Rural cooperative

Department of Agricultural Extension and Education, Sari Branch, Islamic Azad University, Sari, Iran.
** Corresponding author's email: Ahmadpour@iausari.ac.ir*

INTRODUCTION

The evaluation of population growth trends shows that the world population of six billion people in 1998 has grasped an upward trend of about to 1.8% per year and it is expected to reach 8 and 9.4 billion people in 2025 and 2050, respectively (Ahlander, 1994). Regarding the population growth and the needs for lots of food products, basic and traditional farming methods with low efficiency no longer meets these requirements. In recent years, deforestation and destruction of grasslands have significantly increased the cultivating areas and these requirements have been partly overcome by the application of industrial technologies and new ways of farming, also applying modified seeds, fertilizers, and chemical pesticides. However, lack of knowledge and technical expertise among farmers led to inefficient application of these inputs as the existing evidence suggests that the overuse of agricultural inputs not only failed to increase production, but also reduced the production in the long run (Omani, 2001). Moreover, it brought about loss in soil organisms, low quality of products, and reduced soil productivity, (Saad, 2007); also it caused the destruction of the natural environment and endangered the agricultural commodities consumers' health. Excessive consumption of pesticides not only made pests resistant and caused the emergence of new pest's generations, but also endangered the sustainability of the environment and human health (Oskou *et al.*, 2007). Thus, although the application of mechanical and chemical methods flourished agriculture, many disadvantages were obtained regarding issues on sustainable agriculture (Pishro and Azizi, 2009). Agriculture becomes sustainable when natural resources management such as the lack of pollution and land degradation, climate and giving priority to agricultural and rural communities to encourage existing communities for agricultural activities as part of the agreement on productive life are improved along with the economic efficiency and profitability. In fact, sustainable agriculture puts emphasis on the economical, ecological and rural culture balance with each other (Alipour *et al.*, 2008). Thus, the concept of sustainable agriculture features such factors of long-term protection of natural resources, optimal production with minimum production inputs, engendering sufficient income from each operation unit, and meeting all the demands of the rural population and other inevitabilities (Brown, 1987). One of the strategies of different countries, both developed and undeveloped, to increase the quantity and quality of production and achieve food security and environmental sustainability in overcoming problems, such as small units of agricultural production, environmental degradation and unsustainable production, is establishing agricultural cooperatives in different production fields because the agricultural production cooperatives can settle and resolve the abnormalities caused by the presence of small land area, scattered and heterogeneous arable land and many other problems of smallholder farmers and play an important role in rural development and sustainable development (Alizadeh and Omani, 2011). Achieving the objectives of increasing the quality and quantity of production, sustainable agriculture and environmental protection, the production cooperatives were established in Iran. The presence of such cooperatives is considered as one of the manifestations of a common corporate culture of partnership and assistance in meeting the needs of community members and resolves their social problems and human activities in removing the barriers to sustainable agriculture in society. Hence, one of the aspects that determines the achievement and attainment of sustainable agriculture in cooperatives is the evaluation of the application of sustainable agricultural practices by rice farmers as members of rural production cooperatives. Accordingly, the current study intends to answer the question on the status of the application of sustainable agricultural practices and to determine the factors influencing such status. What follows is the research conducted in this area. In their study on the role of ICT in developing sustainable agricultural practices, Afrukhteh and Saraie (2013) demonstrated that there was a significant relationship between rice famers' practices of sustainable agriculture and their access to ICT. Shiri *et al.* (2013), using correlation analysis in their study, showed that there was a significant positive correlation between the variables of agricultural land size, the annual income, farmers' educational level, total participation in the participating in extension training courses, using resources and communication- information channels, having knowledge on conservation practices and attitudes towards soil conservation practices and adopting soil conservation practices. Moreover, the results of the stepwise multiple regression analysis showed that three variables of attitude towards soil conservation practices, total participation in the educational extension classes, and the agricultural land size explained 56.2% of the dependent variable variance of soil conservation practices adoption by farmers. Using regression analysis, Sharifi *et al.* (2011) in their study on factors affecting the greenhouse cultivation system sustainability in Jiroft and Kahnuj demonstrated that 5 variables of sustainable agriculture knowledge, education level, attitude towards sustainable agriculture, participating in extension training courses, and greenhouse work experience explained 53.9% of sustainability in greenhouse cultivation system. Motieilangeroudi *et al.* (2010) in their path analysis to explain the causal factors affecting the sustainability of operation system showed that the 6 variables of capital investment, farmers' age, participation level, the agricultural land size, access to the inputs, and machineries have direct effects and variables of using agricultural resources and farmers' literacy level have indirect affect on the sustainability of farming operation system. In a study on evaluating and explaining the sustainable activities of corn farmers in Khuzestan conducted by the Enayatirad *et al.* (2009), the results showed that the majority of farmers used crop rotation and micronutrient fertilizers to fertilize their farm. Using green fertilizers, animal manures, using legumes in crop rotation when cultivating corn and using a combination of approaches to fight weeds were low and insignificant. The results showed that age, number of children, the amount of pesticides used and the farmers' ownership type had a significant negative relationship with their sustainability activities. Multivariate regression analysis showed that the use of pesticides, having access to extension services, the use of animal manure and corn yield explained 28% of the variation in sustainable activities. Shahrudi *et al.* (2009) in a study on beet farmers' knowledge, attitudes and skills in Khorasan Razavi province towards sustainable soil management practices showed that education, information resources, the use of technology in soil management, agricultural soil quality, extension contact, land area, age, and social status as the main distinguishing factors of farmers' behavioral classes have properly classified a total of 66.9% for all respondents on the basis of diagnostic functions. Chaharsooghiamin and Mirdamadi (2008) in their study showed that there was a positive and significant correlation between the age, history of rice growing, rice production, the amount of cultivated land (paddy field), social participation, the annual income from the sale of rice and ecological awareness with sustainable agriculture. Furthermore, they added that there was a significant negative correlation between the number of household members and participating in rice cultivation with sustainable agriculture. The results of the study by Lashgarara and Asadi (2008) showed that the adoption of sustainable agricultural practices among wheat farmers in Lorestan Province was at a relatively low level and there was a significant positive correlation between education, social participation, market access, using communication tools, participation level in courses and farmers' attitude and knowledge towards the adoption of sustainable agriculture. In their study on factors affecting the sustainable agriculture adoption in wheat farmers in Semnan, Korkehabadi (2005) came to the conclusion that sustainable agriculture adoption was at a relatively low level and there was a significant positive relationship between the education level, wheat farmers' green manure application, social participation, educational activities, market access, the means of communication and participating in extension training courses, recommendations compatibility given on sustainable agriculture and farmers' attitudes towards agriculture with the adoption of sustainable agriculture. On the other hand, there was a significant negative relationship between the wheat farmers' age, use of fertilizers and chemical pesticides with sustainable agriculture. Shariatzadeh-Joneidi (2012), in their study on factors affecting the sustainability of agricultural production systems in the country, indicated that 66.78% of production cooperatives were classified as relatively sustainable. They also showed that there was a significant positive relationship between age, farming experience, social status, using communication resources, participation in extension training courses, training activities compatibility, positive attitude towards sustainable agriculture, product type, being satisfied with the input prices, the area of the land, the total area of agricultural land, sales income, performance, using farm workers, adequate farm loans, sustainable agriculture knowledge with sustainable agricultural production systems.

Reviewing the previous studies shows that the majority of studies are associated with the adoption of sustainable agriculture and there is

Table 1: Variables of extracted from studies

Authors \ Variables	Age	Education Level	Number of family members	Area under cultivation	Work experience	Production Production amount	Economic factors	Accessing the inputs	Using information sources	Social involvement	Social status	Participating in extension training courses	Attitude towards sustainable agriculture	Knowledge towards sustainable agriculture
Shiri *et al.*, 2013.	•	•		•	•		•		•			•	•	•
Afrukhteh and Saraie, 2013									•					
Shariatzadeh-Joneydi, 2012.	•			•	•	•	•		•		•	•	•	•
Korkehabadi, 2004.	•	•							•	•		•	•	
Charsoughiamin and Mirdamadi, 2008.	•		•	•	•	•	•			•		•		•
Shahrudi *et al.*, 2009.	•	•		•					•		•	•		
Enayatirad *et al.*, 2009.	•		•			•						•		
Motieilangrudi *et al.*, 2009.	•	•		•				•	•	•				
Sharifi *et al.*, 2010.		•			•							•	•	•
Lashgarara and Asadi, 2008.		•							•	•		•	•	•

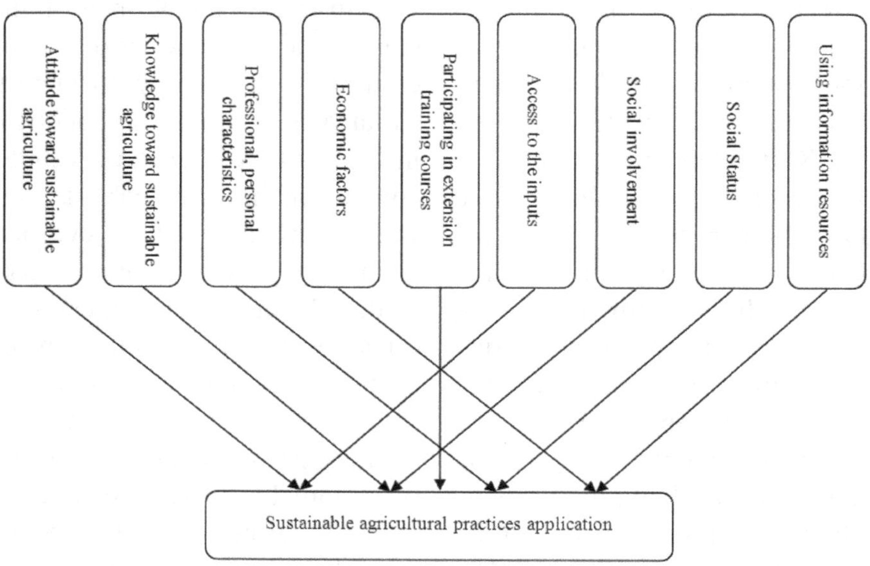

Figure 1: Research theoretical framework

no research about the application and implication of sustainable agricultural practices by farmers. Because of the importance of this subject, current study intends to answer the question on the status of applying sustainable agricultural practices and to determine the factors influencing such status by farmers as members of rural production cooperatives in Sari, Iran.

We conducted a detailed study of the research and extracted the significant relevant variables as shown in Table 1.

Based on data extracted from the literature review, the theoretical framework is given in Figure 1.

Regarding the research question and the literature, the main goal of the present study was to investigate the factors influencing the application of sustainable agricultural practices by rice farmers as members of rural production cooperatives in Sari County, Iran. The research objectives were as follow:

1. Evaluating the personal and professional characteristics of paddy farmers
2. Evaluating the application of sustainable agricultural practices by paddy farmers

MATERIALS AND METHODS

This applied study adopted a descriptive-causative field approach as the study procedure. The research population includes all rice farmers as members of the production cooperatives in rural areas of Sari, Iran (N=4160). 162 farmers were selected through stratified sampling technique in Cochran formula. The size of the target population for each cooperative is shown in Table 2.

To measure the dependent variable (application of sustainable agricultural practices), 12 open-ended questions were used (Table 5). Independent variables included demographic and professional characteristics of rice farmers (the open-ended and closed-end questions), using information resources (16 items), social status (5 items), social participation status (6 items), having access to inputs (5 items), the participation in extension training courses (5 items), the economic factors (8 questions), knowledge of sustainable agriculture (21 items), and attitude towards sustainable agriculture (18 items). It should be noted that the Likert scale was used as variables measurement from the very low=1to very high=5 and (strongly disagree=1 to strongly agree=5 as the attitude scale). In order to describe the independent and dependent variables of the study qualitatively, the Interval of Standard Deviation from the Mean (ISDM) was used (Sadighi and Ahmadpour Kakhak, 2005). Using such approach, the acquired score conversion into four levels were estimated as follows:

A = Poor: A <Mean- Sd
B = Average: Mean- Sd <B <Mean
C = High (Good): Mean <C <Mean + Sd
D = Very high: Mean + Sd <D

The data was gathered through questionnaires, which face validity was verified by a panel of experts, and its reliability was obtained through a pilot test (Alpha Cronbach coefficient: 0.93). Then, the data was analyzed through SPSS software (Version 16). Finally, correlation analysis and regression model were applied to understand the cause and effect relationship between the dependent variable (application of sustainable agricultural practices) and independent variables.

Table 2: The separate population and sample size for rural production cooperatives

Rural production cooperatives	Production cooperatives population	Statistical sample size	Returned questionnaires
Sahele Sabz Seyed Mahaleh	520	22	22
Bahar Gostar Chukela	573	24	22
Sarsabz Miarkola	322	13	13
Falahat Asram	750	31	26
Tallaie Sefid	1200	50	46
Chatre Sabz Surak	700	29	29
Darabkola	95	4	4
Total	4160	173	162

Table 3: The frequency distribution of rice farmers' general characteristics

Variables	Group	Frequency	%	Mean	SD
Age (year)				43.90	11.71
Education level	Illiterate	12	7.5	-	-
	Ability to read and write	18	11.3		
	Junior degree	24	15.1		
	Diploma	43	27.0		
	Bachelor degree	54	34.0		
	Post graduate	8	5.0		
Gender	Female	12	7.5	-	-
	Male	149	92.5		
Marital status	Married	144	89.4	-	-
	Single	17	10.6		
Rice farming experience (year)	-	-	-	26.65	13.05
Land under cultivation (hectare)	-	-	-	1.65	1.04
Rice production (ton)	-	-	-	4.70	2.49
Membership in cooperatives (year)	-	-	-	8.85	5.18
Family member (person)	-	-	-	4	-

RESULTS

Rice farmers' general characteristics

Table 3 shows the general characteristics of rice farmers. According to Table 1, the average age of farmers' was 43.9 years old. Most of the farmers had Bachelor's degrees (34%). About 93% of the respondents were men and just 7% of them were women, and 89.4% were married. The agricultural work experience of the farmers was about 21 years on average. The average rice cultivation was 1.65 hectare. The mean years of membership in the production cooperative and the number of family members were 9 years and 4 people, respectively.

Qualitative description of independent variables

Table 4 illustrates the frequency distribution of independent variables. According to Table 4, 38.9% of the paddy farmers are at the good level of information sources use, and 29% are at the average level, 39.5% of rice farmers are at the good level of social status, and 30.2% are at the average level. 35.8% of the participants had an average social participation and only 16.7% are at the highest level. Considering the access to agricultural inputs, 38.8% were at the average level. Pointing to the participation in extension training courses, 38.3% of the rice farmers considered it at highest level. Regarding the economic

Table 4: The frequency distribution of independent variables

Variables	Variable level	Poor	Average	Good (high)	Very good
Using information resources	Frequency	32	47	63	20
	%	19.8	29.0	38.9	12.3
Social status	Frequency	27	49	64	22
	%	16.7	30.2	36.5	13.6
Social participation	Frequency	29	58	48	27
	%	17.9	35.8	29.6	16.7
Access to agricultural inputs	Frequency	22	58	49	33
	%	13.6	38.8	30.2	20.4
Participating in extension training courses	Frequency	32	51	62	17
	%	19.8	31.5	38.3	10.5
Economic situation	Frequency	37	61	26	38
	%	22.8	37.7	16.0	23.5
Knowledge (towards sustainable agriculture)	Frequency	34	53	37	38
	%	21.0	32.7	22.8	23.5
Attitude (towards sustainable agriculture)	Frequency	26	56	52	28
	%	16.0	34.6	32.1	17.3

Table 5: The frequency distribution of the questions related to the sustainable agricultural practices application

Variables	Mean	SD	Minimum	Maximum
Organic Fertilizer consumption (kg/ha)	96.22	258.28	0	2000
Compost and manure application rate (kg/ha)	498.15	2527.05	0	30000
Herbicide consumption rate (liter/ha)	3.327	1.53	0	7
Using tillage and plowing to remove weeds (ha)	1.20	1.28	0	6
Using grazing for weed control (ha)	0.78	1.28	0	6
Using mechanical operations to remove weeds instead of herbicides (ha)	0.96	1.24	0	6
Burnt remains of plants after harvest (ha)	0.74	1.13	0	4
Area of crop rotation to control pests and diseases (ha)	0.84	1.25	0	6
Area of crop rotation to control weeds (ha)	0.81	1.23	0	6
Area of crop rotation for soil fertility (ha)	0.84	1.24	0	6
Conservation tillage area (ha)	0.75	1.10	0	4
Changing the date of planting crops for pest control (ha)	0.83	1.26	0	6

aspects, 37.7% considered it at the average and 23.5% at the highest level. Considering rice farmers' knowledge towards sustainable agriculture, 32.7% are at the average. On rice farmers' attitude towards sustainable agriculture, 34.6% considered it at the average and 32.1% at the highest level.

Evaluating the application of sustainable agricultural practices by paddy farmers

Based on table 5, the average consumption of fertilizers among rice farmers was 96.22 Kg/ha (SD=258.28). The average use of compost and manure by rice farmers was 498.15 kg/ha (SD=2725.08). The average consumption of herbicides was 3.327 liters per hectare (SD=1.53). Using the plowing and tillage to remove weeds by rice farmers was 1.20 hectare with the standard deviation of 1.28. The average area of grazing for weed control was 0.78 hectare (SD=1.28). Average area of using mechanical operations to remove weeds instead of herbicides was 0.96 ha (SD=1.24). The mean and the standard deviation of the burnt remains of plants after harvest, the area of crop rotation to control pests and diseases, and the area of crop rotation to control weeds, the area of crop rotation for soil fertility, conservation tillage area and changing the date of planting crops for pest control had the respective values as showed in table 5.

Considering the fact that the application of sustainable agricultural practices questions were open-ended and had different scales and directions, and in order to classify and evaluate the implementation of sustainable agricultural practices, the following steps will be followed:

1. Homogenization: due to the fact that various items may be positive or negative and in order to align the items, attempts should be taken to homogenize them. Since the research questions on herbicide consumption liters/ha (greater use stands on negative performance in sustainable agriculture) and an area of burnt remains of plants after harvest (ha) (more area means negative performance in sustainable agriculture) were not homogeneous with other questions (questions or more area meaning taking positive in sustainable agriculture) and in order to homogenize the given questions, a division was conducted on the questions.

2. Standardization: after conducting the homogenization and as the items are not aligned in terms of measurement units (liters/hectare, kg/ha, ha), the questions were standardized to conduct the same unit of measurement.

3. After homogenization and standardization, the variables were computed. Then, the rate of applying sustainable agricultural practices was classified through the Interval of Standard Deviation from the Mean.

As Table 6 illustrates, the application of sustainable agricultural practices by rice farmers' are as follow, 61.7% is at average level, 21.6% is at high levels, 13.6% is at very high-level, and 3.1% at poor level.

Evaluating the correlation between independent variables and application of sustainable agricultural practices

Based on Table 7, the results showed that

Table 6: The frequency distribution of the questions related to the sustainable agricultural practices application

Sustainable agricultural practices application	Frequency	Valid Percentage	Cumulative percent
Poor	5	3.1	3.1
Average	100	61.7	64.8
High (good)	35	21.6	86.4
Very high	22	13.6	100.0

Table 7: Correlation between independent variables and sustainable agricultural practices application

Independent variables	Correlation test	r	p-value
The extent of paddy field	Pearson	0.674**	0.000
Rice production	Pearson	0.615**	0.000
Participating in participating in extension training courses	Pearson	0.450**	0.000
Social status	Pearson	0.365**	0.000
Number of family members	Pearson	0.358**	0.000
Economic factors	Pearson	0.356**	0.000
Knowledge (towards sustainable agriculture)	Pearson	-0.253**	0.001
Accessing agricultural inputs	Pearson	0.227**	0.004
Age	Pearson	0.172*	0.035
The use of information sources	Pearson	0.165*	0.036
Attitude (towards sustainable agriculture)	Pearson	0.120	0.128
Membership in the production cooperative	Pearson	0.118	0.151
Education Level	Spearman	-0.109	0.171
Social participation	Pearson	0.058	0.460
Work experience	Pearson	-0.048	0.549

*p<0.05 **p<0.01

Table 8: Regression model (dependent variable: Application of sustainable agricultural practices)

Independent variables	B	Beta	t	p-value
Constant value	-15.980	-	-	-
Cultivating area /ha (x_1)	2.419	0.372	5.293	0.000
Economic factors (x_2)	1.145	0.409	5.918	0.000
Participating in extension training courses (x_3)	0.276	0.158	2.554	0.012
Age (year) (x_4)	0.077	0.126	2.059	0.042

R= 0.786 R^2= 0.618 F= 51.276

there was a significant positive correlation with 95% confidence level between age and the use of information sources and the application of sustainable agricultural practices. There was a significant positive relationship between the number of family members, the rice cultivation area, production amount, social status, accessing the inputs, economic facilities and participation in extension training courses with the application of sustainable agricultural practices at 1% level. Accordingly, the results showed that there was a significant negative correlation between rice farmers' knowledge towards sustainable agriculture and sustainable agricultural practices application. But there was no relationship between the other variables and the application of sustainable agricultural practices.

Factors influencing the application of sustainable agricultural practices by rice farmers

Regression model was used to investigate cause-and-effect relationship between the dependent and independent variables. The stepwise linear regression analysis was conducted to investigate the factors influencing the application of sustainable agricultural practices by rice farmers in production cooperatives. Having entered all of the independent variables with significant correlation, variables of the land area/ha,

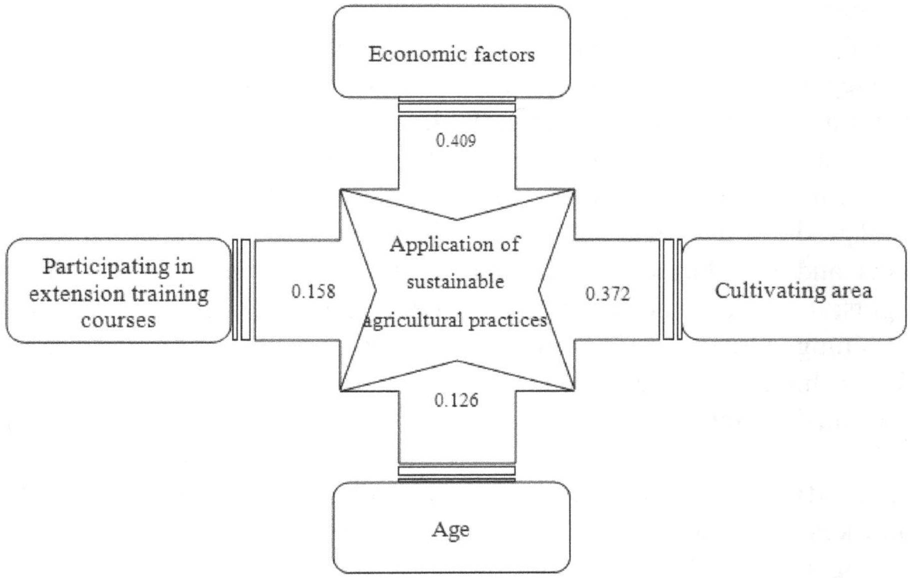

Figure 2: Factors influencing the application of sustainable agricultural practices

economic factors, participating in extension training courses and age remained in the model, these variables were able to determine 61.8% of the variation in the application of sustainable agricultural practices (Table 8).

CONCLUSION

Sustainable agriculture involves the fertility and productivity of soil, fertilizer, and pesticides control, management strategies, human needs, the potential economic viability, social acceptance, ecological capacity, time potentiality and philosophical ethics (implying moral satisfaction and goals). One of the institutions that could play a major role in the development of sustainable agriculture is production cooperatives working in rural areas. Such public institutions provide rice farmers services like sustainable agriculture, which can influence the behavior of rice farmers using sustainable practices. Therefore, evaluating rice farmers' application of sustainable agricultural practices in production cooperatives, and assessing the factors in the development of sustainable agriculture are critically important. Based on the data collected, there was a significant positive relationship between the application of sustainable agricultural practices and age (being consistent with the studies by (Chaharsooghiamin and Mirdamadi, 2008; Motieilangeroudi *et al.*, 2010; Shahrudi *et al.*, 2009; Shariatzadeh-Joneydi, 2012; but inconsistent with researches by Enaiatirad *et al.*, 2009; Korkehabadi, 2005) and using information sources (based on the studies by Afrukhteh and Saraie, 2013; Korkehabadi, 2005; Motieilangeroudi *et al.*, 2010; Shahrudi *et al.*, 2009; Shariatzadeh-Joneydi, 2012; Shiri *et al.*, 2013) at 5% at the confidence level of 5%. In other words, the older rice farmers had acceptable behavior in terms of sustainable agriculture. And the more rice farmers use the information resources, the more acceptable their behavior would get towards sustainable agriculture. There was a significant positive relationship between the number of family members (inconsistent with research by Chaharsooghiamin and Mirdamadi, 2008), the extent of rice growing land (consistent with Chaharsooghiamin and Mirdamadi, 2008; Motielangeroudi *et al.*, 2010; Shiri *et al.*, 2013; Shahrudi *et al.*, 2009; Shariatzadeh-Joneydi, 2012) production amount (compatible with the research by Chaharsooghiamin and Mirdamadi, 2008; Shariatzadeh-Joneydi, 2012), social status (according to the studies by Shahrudi *et al.*, 2009; Shariatzadeh-Joneydi, 2012), the access to agriculture inputs (in line with the research by Motieilangeroudi *et al.*, 2010), the participation in extension training courses (consistent with the research by Enaiatirad *et al.*, 2009; Korkehabadi, 2005; Shahrudi *et al.*, 2009; Sharifi *et al.*, 2011; Shariatzadeh-

Joneydi, 2012; Shiri *et al.,* 2013) and economic factors (compatible with the research by Shariatzadeh-Joneydi, 2012; Shiri *et al*., 2013) and the application of sustainable agricultural practices at 1%. In other words, increase in number of rice farmers' family member; rice growing land and production amount; improving the social status and rice farmers' access to agricultural inputs; increase in the participation of extension training courses, and economic factors have led to higher application and implication of sustainable agricultural practices by rice farmers.

There was a significant negative correlation between rice farmers' knowledge in sustainable agriculture (as opposed to related research by Chaharsooghiamin and Mirdamadi, 2008; Lashgrara and Asadi, 2008; Shiri *et al.,* 2013) and the application of sustainable agricultural practices. In other words, knowledge is not effective in the application of sustainable agricultural practices. The results of the stepwise linear regression analysis showed that the four variables of paddy field area / hectare, economic factors, participating in extension training courses and age could explain 61.8% of the total changes in the application of sustainable agricultural practices by rice farmers as members of rural production cooperatives. Regarding the standardized coefficients, it was determined that economic factors (40.9%) had greater contribution than other factors in explaining the application of sustainable agricultural practices.

Considering the results of the study, the following suggestions are offered:

1- Considering the relative contribution of economic factors as the most important factor in explaining the application of sustainable agricultural practices, it is suggested that the produced rice should be favorably priced by practitioners and should be purchased from rice farmers in a simple process with no complicated paperwork to achieve relative satisfaction of farmers. Also, farmers should be helped to use rice production inputs and their related tools by providing them the necessary and required mechanism. Considering such issues, a long step can be taken towards the application of more sustainable agricultural practices.

2- Considering the relative contribution of land area of paddy field/hectare (37.2%) as the second most important factor in explaining the application of sustainable agricultural practices, it is recommended that in order to implement sustainable practices and sustainable agricultural development, experts and especially agriculture extension agents initially focus on rice farmers with more agricultural land to consider more application of sustainable agricultural practices and such desired behavior (using sustainable agricultural practices) will be extended to other rice farmers.

3- Another important factor in the application of sustainable agricultural practices, is the participation in extension training practices which had an important role. Accordingly, it is suggested that more favorable conditions for farmers' participation in extension training activities should be provided by the authorities.

ACKNOWLEDGEMENT

The authors would like to thank Islamic Azad University, Sari Branch for financial support of this project.

REFERENCES

1- Afrukhteh, H., & Saraie, S. (2013). The role of ICT in the development of sustainable agriculture practices: a case study among rice farmers in Rasht. *Rural Development,* 16 (2), 155-178.

2- Ahlander, A. M. (1994). Environmental problems in the shortage economy. Cheltenham Edward Elgar Publishing Ltd. *Journal of Environmental Economics and Management*, 30, 381-395.

3- Alipour, H., Fallah, R., & Moghadas farimani, N. (2008). Researchers' knowledge and tendency in extension, education and agricultural research organization towards sustainable agriculture. *Research and Development,* 81, 110-119.

4- Alizadeh, M., & Omani, F. (2011). *Evaluating the farmers' attitudes towards the role of agricultural production cooperatives in sustainable agricultural development*. Paper presented National Conference on Sustainable Agriculture, Islamic Azad University of Varamin-Pishva, and December.

5- Brown, B., Hanson, M., Liverman, D., & Merideth, R. (1987). Global sustainability: toward definition.

Environmental Management, 11 (6), 713-719.

6- Chaharsooghiamin, H., & Mirdamadi, S. (2008). Identifying and analyzing the factors affecting sustainable agriculture among women farmers in Bandar Anzali, with emphasis on indigenous knowledge. *Journal of Agricultural Science*, 5 (1), 61-83.

7- Cooperative Office of Mazandaran Province. (2014). Production cooperatives Statistics. Sari County, Mazandaran Privince.

8- Enayatirad, M., Ajili, A., & Rezaeimoghaddam, K. (2009). Investigating and explaining the sustainability activities among corn farmers in Khuzestan Province. *Journal of Agricultural Economics and Development Research*, 40-2 (7), 99-107.

9- Korkehabadi, N. (2005). *Factors affecting the adoption of sustainable agriculture among wheat farmers in Semnan*. Unpublished thesis of Agricultural Extension and Education, Faculty of Agriculture, Islamic Azad University, Science and Research Branch, Tehran.

10- Lashgarara, F., & Asadi, A. (2008). Analyzing factors affecting the adoption of sustainable agriculture among wheat farmers in Lorestan province. *Iranian Journal of Agriculture Science*, 39-2 (1), 97-104.

11- Motieilangeroudi, H., Rezvani, M. R., Farajisabokbar, H., & Khajehshokuhi, A. (2010). Analyzing the agricultural stability of a family farm operation and rural production cooperatives (Case study: Aghghala). *Iranian Journal of Agricultural Economics and Development*, 41-2 (3), 323-333.

12- Omani, A. (2001). *Evaluating the social, economic and agricultural features effective wheat farmers on the adoption of sustainable agriculture regarding the low input crops in Khuzestan Province*. Unpublished thesis in Agricultural Extension and Education, Faculty of Agriculture, Tarbiat Modarres University.

13- Oskou, T., Chizari, M., & Rasouli, S.f. (2007). Evaluating the effect of Farmer Field School (FFS) participatory approach on rice farmers' knowledge and attitudes on biological control of rice stem borer. *Journal of Agricultural Sciences*, 2-1 (38), 109-119.

14- Pishro, H., & Azizi, P. (2009). Sustainable agricultural development through agricultural income stabilization. *Human Geography*, (4) 1, 1-20.

15- Saad, O. (2007). On stability of proper efficient solutions in multiobjective fractional programming problems under fuzziness. *Mathematical and Computer Modelling*, 45, 221- 231.

16- Sadighi, H., & Ahmadpour Kakhak, A. (2005). Assessing farmers' attitude toward cultivation and development of saffron production and investigating their difficulties and problems (A case study In - Farmers Attitude Toward Cultivation and Develop ment of Saffron Production and Investigating Their Difficulties and Problems (A Case Study In Gonabad, Khorasan, Iran). *Iranian Journal of Agriculture Science*, 36 (3), 689-699.

17- Shahrudi, A., Chizari, M., & Pezeshkirad, Gh. (2009). Beet farmers' knowledge, attitudes and skills in Khorasan Razavi Province toward sustainable soil management practices. *Knowledge of Modern Agriculture*, 5 (16), 43-53.

18- Shariatzadeh-Joneydi, M. (2012). Factors affecting in sustainability of agricultural production systems in Iran. *Annals of Biological Research*, 3 (9),4578-4583.

19- Sharifi, A., Rezaei, R., & Borumand, N. (2011). Evaluating factors affecting the stability of greenhouse cultivation in Jiroft and Kahnuj. *Iranian Journal of Agricultural Economics and Development*, 42-2 (1), 143-152.

20- Shiri, N., Hashemi, S., Mirekzade, A., & Eshaghi, S. (2013). Factors affecting the adoption of soil conservation practices by farmers in Ilam Province. *Agriculture Economic Research and Development*, 44 (2), 308-297.

9

Ranking and Level of Development According to the Agricultural Indices, Case Study: Sistan Region

Ali Sardar Shahraki [1]***, Javad Shahraki*** [2*] ***and Seyed Arman Hashemi Monfared*** [3]

Keywords:
Levels of Agricultural Development, Numerical Taxonomy, Fuzzy Hierarchical Analysis (FAHP), Sistan

Abstract

Sistan region is one of the most important agricultural areas in the province of Sistan and Baluchistan. Therefore, given the heterogeneity in agriculture and recognizing these differences, the aim of this study was to obtain the level of development of agriculture in the Sistan region. To obtain this purpose Fuzzy Analytical Hierarchy Process (FAHP) and the numerical taxonomy were used in a view of 20 indicators in the agricultural sector in the region. The required data were achieved by filling out the questionnaire certified experts and statistical yearbooks in the agricultural sector. Data analysis was used by Matlab and SPSS softwares. Results of numerical taxonomy showed that Markazi, Shibab and Poshteab sectors component parts were less developed. Also, Jazinak and Miyankangi are in the category sections were undeveloped. The results of Fuzzy Analytical Hierarchy Process (FAHP) model indicated that Markazi, Shibab and Poshteab sectors are in the first rank of development, in terms of agricultural indices in the region. Jazinak and Miyankangi are in the fourth and fifth ranking. Therefore, in general, it is clear that the level of development of agricultural in Sistan region isn't in good condition. In this regard it is suggested that appropriate planning to promote agricultural development is on the agenda should be applied.

[1] *Ph.D. Candidate of Agricultural Economics, University of Sistan and Baluchestan, Iran*
[2] *Associate Professor, Agricultural Economics, University of Sistan and Baluchestan, Iran*
[3] *Assistant Professor, Civil Engineering, University of Sistan and Baluchestan, Iran*
* *Corresponding author's email: j.shahraki@eco.usb.ac.ir*

INTRODUCTION

During the past decade, agricultural development has been considered as one of the main objectives of development policies in many countries. Development policies, such as agricultural development, specifically in developing countries, are considered as the center of development projects. Considering the fact that agricultural development is a highway for improving the quality of rural life, welfare, food security and etc., balancing them with respect to levels of agricultural development necessitates categorizing and rating of districts (Moradi *et al.*, 2015).

Agricultural development denotes the quality of agricultural system of a region; it is a multidimensional concept which mainly includes development in a real strength of cropped land, improvement in farm practices/system, improved farm implements, irrigation system and irrigated area, high yielding improved varieties of seeds, chemical fertilizers, insecticides and pesticides, intensity of cropping and specialization and commercialization of agriculture (Mohammed, 1980).

Many economic activities depend on agriculture, including marketing, processing, and export of agricultural products. Because of this fact, development of the regional economy heavily depends upon the speed with which agricultural growth is achieved (SardarShahraki *et al.*, 2014). Accordingly, the different development programs gave due to the emphasis of development of the agricultural sector with a view to stimulating rapid growth in total production and improvement in productivity (Badri *et al.*, 2007). Nonetheless, development agents have been found to differ in perception and performance of their roles and satisfaction on their job (Peet, 1999). In Sistan region majority of its population depend upon agriculture. So a vast rural mass tries to earn their livelihood from agricultural land. With fast increasing pressure of population on agricultural land, old methods and techniques of production cannot cope with growing demand. As a result, new technologies and commercial crops are adopted to develop agro-economy. For these reason emphases on the diffusion of agricultural innovation are stressed.

MCDM approaches were developed in 1960s to assist decision-makers to incorporate many options, reflecting the opinions of the actors concerned, into a potential or retrospective framework. They were designed to define the relationship between the data input and the data output. MCDM can be separated into main two main groups of methods; multi objective and multi attribute (Malczewski, 1999).

Shafiee *et al.* (2012) in research of optimum selection of integrated marketing communication tools with FAHP approach, this result is achieved that advertisement has the highest rank among marketing communications channels in "product-market" of interest. Thus, it is the main channel in marketing communications of that "product-market". Shafaiyanfard *et al.* (2015) were examined superior options exploitation of water resources by using WEAP model and analysis and multiple attribute decision making. According to results of this research, the scenario of further development in summer planting was selected. Sasikumar and Mujomdar (1998) proposed a multi objective fuzzy model for quality management Systems River. In this research, qualitative goals of the organizations responsible for river water quality protection and dump various contaminants considered in the form of models fuzzy multi attribute decision making.

Chuntian (1999) used fuzzy multi-criteria optimization model to manage water resources in times of flood. Fu (2008) used Fuzzy multi-criteria decision-making method and their application in management and flood control in the agricultural sector. SardarShahraki *et al.* (2014) in determining the level of development of agriculture and economy in rural of Iran by using 68 economic indicators and 46 indicators of agriculture in rural areas using two numerical taxonomy and factor analysis were studied. The results of their study showed that thatprovincialIsfahan,Tehran,Mazandaran,Fars,Iran,Yazd, Qomprovincescategoriesdevelopedintermsofagricultureandruralsector,ofwhichonlythreeprovincesofTehran,FarsandYazdeconomicdevelo pmentsaid. The results of their study showed that Isfahan, Tehran, Mazandaran, Fars, Golestan, Yazd and Qom in the developed provinces in terms of agriculture in the rural sector, of which, only three provinces of Tehran, Fars and Yazd have economic development. Khakpoor and BavanPoori (2009) analyzed inequity in development level in the regions of mashed city using 32 general indices, the obtained results showed that 25% urban regards of Mashhad are in very ownership class, 25% are in ownership class, 25% regions are in deprived class and 16.7% are in very deprived class. Ebrahimzadeh and Eskandarisani (2011) explain spatial pattern of urban-region development level in Iran during 1996 and 2006 using analytical- comparative method. In this study 35 different indices of development have been used. Curies of country class if 4 developed (32 cities), half to high developed cities (46 cities), less to low developed (126 cities), deprived (106 cities). Ramatu and Xinshen in 2007 with the application of factor analysis to examine regional disparities Ghana paid during the period 2000-1990. The results showed that the gap between the more developed northern regions of the South of Ghana.

Sharama in 2004 to investigate regional disparities in the state of Indian coefficient of variation using principal component analysis is paid.

Sistan region with abundant natural resources, skilled labor and manufacturing capability of diverse products has not been able to fit their abilities and opportunities relative to other regions earn appropriate development and development level of the region is very improper. Therefore, evaluating the level of development of this region in terms of having agricultural development indicators for suitable regional planning is important.

Purpose of study

Studyof development situation in the Sistan region with factor on agricultural development by numerical taxonomy and FAHP methods.

Present suitable solutions and strategies to eliminate shortage in developed and less developed sectors.

MATERIALS AND METHODS

Numerical Taxonomy

Data matrix includes table that its columns consist of indices and its lines consist of all economical sections or field of economic activities. Correlation matrix is used to have internal communication of indices, the values of the main diameter of matrix are 1 and numbers below its diameter are the: repetition of the numbers upper diameter, because correlation of each index is 1 and correlation of index 2 to 1 is always index 1 to 2.Extracting factors obtained using correlation matrix between indices. Using factor matrix, comma on factors of each index is known. Then specific vectors computed nunzero for all specific values. In fact, specific vectors are elated to loading value corresponds to each index that are called factor loads. Interpretation of factors will be simple, if each index, curries on one factor and or loaded values in factor are big, positive or close to zero, interpretation of factors will be hard, if the loads values of each index are average values on multi-factor and are given to obtain simple structure. In this research, varimax method has been used to ratite factors. According to the correlation amount of each index, we can select suitable titles or names for each one of them (SardarShahraki *et al.*, 2014).

Numerical taxonomy is one of the best ranking methods of various activities to use some indices. This method is also suitable to rank and compare various areas regarding development degree. This method can divide one set to more or less homogenous set and present suitable scale to recognize enjoying considered indices and social and economic development degree that has been used in the analysis of activities. Taxonomy analysis is performed in several phases that are as follows:

1. To from data matrix
2. To standardize data matrix
3. To from distance matrix
4. To determine homogenous distances
5. To rank homogenous activities for the studied indices.
6. To calculate ownership degree of homogenous activates (SardarShahraki *et al.*, 2014).

In this phase, we designed matrix for all series of activities according to the studied indices, So that matrix dimensions have been n×m namely this matrix has line for the number of activity series and column for indices number. As an example, $X_{n\times m}$ element indicates index m of n activity series:

$$\text{A ij} = \begin{bmatrix} x_{11} & \cdots & x_{1m} \\ & \cdots & \\ x_{n1} & \cdots & x_{nm} \end{bmatrix} \quad (1)$$

I=1, …., n; j=1 , …., m

In this matrix, columns indicate the used indices and lines showed the studied sub-section. As an example X_{ij} shows the related value to sub-section and indicate.

Considering that indices are measured with various units, thus, to delete the effects of these units and substitute unit scale and also to delete the effect of origin, first mean Z_{ij} is calculated, And in first step, the mean of columns is obtained:

$$\bar{x}_1 = \frac{1}{n}\sum_1^n x_{ij} \tag{2}$$

At the next step, standard derivation is obtained for each column for matrix X_{ij}.

$$S_j = \sqrt{\frac{1}{n}\sum_{i=1}^{n}(X_{ij} - \bar{X}_j)^2} \tag{3}$$

The third step is that to standardized members of matrix A_{ij} from a new matrix framework called standard matrix to be calculated (Sardar-Shahraki *et al.*, 2014).

$$z_{ij} = \frac{\bar{x}_{ij} - x_j}{s_j} \tag{4}$$

Matrix Z has n×m dimension and is a standard matrix, different matrices of indices have been converted to unit scale, when variable changes. It is clear that statistically average each column of standardized matrix Z is equal to zero and its standard deviation is equal to 1:

$$\text{z ij} = \begin{bmatrix} z_{11} & \cdots & z_{1m} \\ & \cdots & \\ z_{n1} & \cdots & z_{nm} \end{bmatrix} \tag{5}$$

i=1… , n ; j=1,…, m

By standard matrix, the next step is to give difference value or distance value of two points from each other (1, 2, ,…, n) for each m variable or index that formed distance matrix. In this step combined distances between different activities is obtained as n for m indices and follows according to standardized numbers in standard matrix z. these distances are same Euclidean distance generalization that presented as follows:

$$c_{ab} = \sqrt{\sum_{k=1}^{m}(z_{ak} - z_{bk})^2} \tag{6}$$

a,b = 1, 2, ….,n

Which Cab is the distance between two activity a and b. If the distance of activity series in Paris be obtained then the combined distance matrix shows as follows:

$$C_{ab} = \begin{bmatrix} c_{11}c_{12} & \cdots & c_{1n} \\ c_{21}c_{22} & \cdots & c_{2n} \\ & \cdots & \\ c_{n1}c_{2n} & \cdots & c_{nm} \end{bmatrix} \tag{7}$$

Because distances matrix is a symmetric matrix, we can conclude distance of activity series is equal to b to from a to b is equal to b to a and the distance of each activity series of itself is also equal to zero:

$$C_{aa} = C_{bb} = \ldots = C_{nm} = 0\ ,\ C_{ab} = C_{ba} \tag{8}$$

$$C_{ab} = \begin{bmatrix} 0\ c_{12} & \cdots & c_{1n} \\ c_{21}\ 0 & \cdots & c_{2n} \\ & \cdots & \\ c_{n1}c_{n2} & \cdots & 0 \end{bmatrix} \tag{9}$$

As it is identified this matrix has been symmetrical and its diameter is zero. Also it is square matrix by n×n dimension. Each element of matrix Cab indicates the distance between each two activity strings in the considered index. In this matrix, it is the least distance between two activity strings in each line and it is written in the separate column (for example column) then we compute mean and standard deviation of the least distances of each line namely same column.

$$d_j = \begin{bmatrix} d_1 \\ d_2 \\ \vdots \\ \vdots \\ d_n \end{bmatrix} \tag{10}$$

$$\bar{d} = \frac{1}{n}\sum_{i=1}^{n} d_j \tag{11}$$

$$S_d = \sqrt{\frac{1}{n}\sum_{i=1}^{n}(d_j = \bar{d})^2} \tag{12}$$

Distances of high limit d (+) and low limit d (-) is calculated according below relation to identify homogenous activity series. It is necessary to mention that number 2 is the Z of normal distribution here that has been determined in 95 percent level:

$$\begin{aligned} d(+) &= \bar{d} + 2S_d \\ d(-) &= \bar{d} - 2S_d \end{aligned} \tag{13}$$

In this step, activity series that are the least distances between two high and low limit, have been homogenous and are in one group. If the least differences between two activates strings are higher than high limit and or less then low limit, then the series of abode activates must be deleted as inhomogeneous.

Ranking homogenous activity series from the studied standers. If all activity series aren't in one homogenous group, then it forms data matrix for homogenous activity strings, then it standardizes and considers ideal case in standardized indices matrix for each one of indices and the desired ownership for each activity series for is computed to find ideal values:

$$C_{io} = \sqrt{\sum_{i=1}^{m}(Z_{ik} - Z_{ok}} \quad (14)$$

i= 1, 2, …, n

Z_{ok} is ideal quantity for k standardized index, Z_{ok} is the standardized index for I activity and C_{io} is desired ownership for i activity. We consider as ideal the highest number in each column and if index direction is negative, the bigger number shows lack of ownership that we select as ideal value the smallest. Value in this phase, combination index in traduced as ownership degree that it had limited range and is between zero and one values. If ownership degree of I choice shows with f_i , then we have:

$$f_i = \frac{C_{io}}{C_o} \quad (15)$$

C_o is called high limit of the desired ownership and obtain from relation:

$$C_o = C_{io} + 2\ S_{cio} \quad (16)$$

So that C_{io} is average desired ownership for all I activity and S_{cio} is their standard deviation:

$$S_{cio} = \sqrt{\frac{1}{n}\sum_{i=1}^{n}(C_{io} - \bar{C}_{io})} \quad (17)$$

Whatever f_i is near to zero, the considered activity string is more ownership it is near to one, it indicates lack of ownership of related activity string (SardarShahraki *et al.*, 2014).

Fuzzy Analytical Hierarchy Process (FAHP)

Fuzzy AHP method is a popular approach for multiple criteria decision-making. In this study the extent fuzzy AHP is utilized, which was originally introduced by Chang (1996). Let X = {x_1, x_2, x_3,..., x_n} an object set, and G = {g_1, g_2, g_3,....., g_n} be a goal set. Then, each object is taken and extent analysis for each goal is performed, respectively. Therefore, *m* extent analysis values for each object can be obtained, with the following signs:

$$\tilde{M}^1_{g_i}, \tilde{M}^2_{g_i}, ..., \tilde{M}^m_{g_i} \quad (18)$$

Where $\tilde{M}^j_{g_i}$ *(j= 1,2,3,..., m)* are all triangular fuzzy numbers. The steps of the Chang's (1996) extent analysis can be summarized as follows:

Step 1: The value of fuzzy synthetic extent with respect to the it objects is defined as:

$$S_i = \sum_{j=1}^{m} \tilde{M}^j_{g_i} \otimes [\sum_{i=1}^{n}\sum_{j=1}^{m} \tilde{M}^j_{g_i}]^{-1} \quad (19)$$

Where ⊗ denotes the extended multiplication of two fuzzy numbers. In order to obtain $\sum_{j=1}^{m} \tilde{M}^j_{g_i}$, we perform the addition of m extent analysis values for a particular matrix such that,

$$\sum_{j=1}^{m} \tilde{M}^j_{g_i} = (\sum_{j=1}^{m} l_j, \sum_{j=1}^{m} m_j, \sum_{j=1}^{m} u_j) \quad (20)$$

And to obtain $[\sum_{i=1}^{n}\sum_{j=1}^{m} \tilde{M}^j_{g_i}]^{-1}$ we perform the fuzzy addition operation of $\tilde{M}^j_{g_i}$ *(j= 1,2,3,..., m)* values such that,

$$\sum_{i=1}^{n}\sum_{j=1}^{m} \tilde{M}^j_{g_i} = (\sum_{j=1}^{n} l_j, \sum_{j=1}^{n} m_j, \sum_{j=1}^{n} u_j) \quad (21)$$

Then, the inverse of the vector is computed as,

$$[\sum_{i=1}^{n}\sum_{j=1}^{m} \tilde{M}^j_{g_i}]^{-1} = \left(\frac{1}{\sum_{i=1}^{n} u_i}, \frac{1}{\sum_{i=1}^{n} m_i}, \frac{1}{\sum_{i=1}^{n} l_i} \right) \quad (22)$$

Where $u_i, m_i, l_i \succ 0(22)$

Finally, to obtain the *Sj*, we perform the following multiplication:

$$S_i = \sum_{j=1}^{m} \tilde{M}^j_{g_i} \otimes [\sum_{i=1}^{n}\sum_{j=1}^{m} \tilde{M}^j_{g_i}]^{-1} = (\sum_{j=1}^{m} l_j \otimes \sum_{i=1}^{n} l_i, \sum_{j=1}^{m} m_j \otimes \sum_{i=1}^{n} m_i, \sum_{j=1}^{m} u_j \otimes \sum_{i=1}^{n} u_i) \quad (23)$$

Step 2: The degree of possibility of $\tilde{M}_2 = (l_2, m_2, u_2) \geq \tilde{M}_1 = (l_1, m_1, u_1)$ is defined as:

$$V(\tilde{M}_2 \geq \tilde{M}_1) = Sup[Min(\tilde{M}_1(x), \tilde{M}_2(y))] \quad (24)$$

This can be equivalently expressed as,

$$V(\tilde{M}_2 \geq \tilde{M}_1) = hgt(\tilde{M}_1 \cap \tilde{M}_2) = \tilde{M}_2(d) = \begin{cases} 1 & If \quad m_2 \geq m_1 \\ 0 & If \quad l_1 \geq u_2 \\ {l_1 - u_2}/{(m_2 - u_2) - (m_1 - l_1)} & Otherwise \end{cases} \tag{25}$$

Step 3: The degree of possibility for a convex fuzzy number to be greater than k convex fuzzy numbers M_i *(i=1, 2, ..., k)* is defined as:

$$V(\tilde{M} \geq \tilde{M}_1, \tilde{M}_2, ..., \tilde{M}_k) = Min\ V(\tilde{M} \geq \tilde{M}_i)$$

i= 1, 2, ..., k (26)

Step 4: Finally, $W = (Min\ V(S_1 \geq S_k), Min\ V(S_2 \geq S_k), .., Min\ (S_n \geq S_k))^T$, is the weight vector for K=1,…, N (Rafiee *et al.*, 2013).

Indicators used

InTable1shown the index used by the Department of Agriculture in Sistan region.

RESULTS AND DISCUSSION

The results of numerical taxonomy and Fuzzy Analytical Hierarchy Process method are presented in this section. In numerical taxonomy, first, it was determined homogeneous sectors and then, they were classified. In this regard, the lower and upper limits to determine the homogeneity sectors in the Matrixes intervals were calculated. Section that minimum distances from other sectors outside the contour was 1.87222<d<5.1521, Heterogeneous sectors deemed and from ratings are removed. According to the results in Table 3, heterogeneous section did not exist. The first column in table 3 is section name, second column is cumulative relative frequency, third column is degree of development and fourth columnist level of development. The results of the ranking in terms of development of agriculture in the Sistan region shown in Table3.

According to the results in Table 3, parts that cumulative frequency there is from 0- 0.25. In batches developed part is that, no part was not in this category. Also, Parts that their Cumulative Frequency is at intervals of 0.25-0.5, 0.5-0.75, 0.75-1, in batches relatively developed, less developed and underdeveloped placed, Respectively. Hence, Poshteab, Shibab and Markazi sectors are in the category of less developed. Jazinak and Miyankangi sectors are in the category of underdeveloped. In Figure 1, the results of FAHP technique shown.

According to the results of FAHP medal, Matkazi sector is the first rank. Poshteab and Shibab sector are in second and third ranking. Jazinak and Miyankangi are in Fourth and fifth ranking.

Table 1: List of indicators of development in agriculture

	Indicators
1	Yieldof barley (kg/ha)
2	Yieldof forage plants (kg/ha)
3	Yield of wheat (kg/ha)
4	Yield of Watermelon (kg/ha)
5	Yieldofrapeseed (kg/ha)
6	Yield of Melon (kg/ha)
7	Yield of tomato (kg/ha)
8	Yieldofpotatoes (kg/ha)
9	Yield of grain (kg/ha)
10	Yield of herb (kg/ha)
11	Yield of bean (kg/ha)
12	Yield of pea (kg/ha)
13	Yield of Grapes (kg/ha)
14	Yield of corn (kg/ha)
15	Percapita production of banana seach operation (kg)
16	Cultivation of flowers and ornamental plants (ha)
17	Average production of Beehive
18	Yield of fish of water cold (kg/ha)
19	Proportion of agricultural lands and horticultural cooperatives covering a area under cultivation (ha)
20	The employment rate per hundred thousand rural industries (person)

Table 2: Section of Sistan

No	Sector
1	Poshteab
2	Shibab
3	Markazi
4	Miyankangi
5	Jazinak

Table 3: Ranking of the Sistan region of agricultural development

Section Name	Cumulative relative frequency	Degree of development	Level of development
Poshteab	0.5425	0.8797	Less Developed
Shibab	0.5874	0.8897	Less Developed
Markazi	0.5412	0.9254	Less Developed
Jazinak	0.7547	0.9745	Underdeveloped
Miyankangi	0.8745	0.9547	Underdeveloped

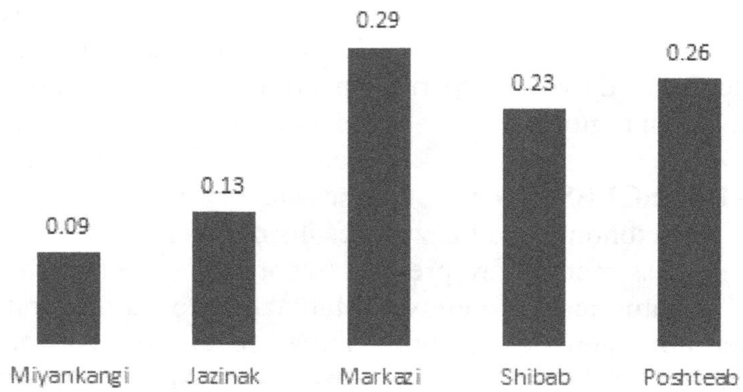

Figure 1: Ranking of the Sistan region in terms of agricultural development

CONCLUSION AND RECOMMENDATIONS

Sistan region is very complicated in terms of agricultural development. The present study reveals that the spatial distribution of variables and agricultural development is not uniform in this region and it provides very significant information about the level of agricultural development.In the present study, Parts of Sistan region were ranked according to 20 indicators agriculture regarding development using fuzzy hierarchical analysis and numerical taxonomy. According to the results, the following suggestions are offered:

Adequate planning for improve the status of agricultural development to be performed areas such as Jazinak and Miyankangi. One of the objectives of the regional plan, a balanced distribution of development in different areas of a region; Look for in a systematic, spatial inequalities in deprived areas and back ward areas of the country led to the creation of infrastructure facilities and services, employment, income, welfare. In other words, the uneven distribution of development, causing transmission problems in disadvantaged areas is rich areas. Therefore, the plan to eliminate the gap between different areas of the development is considered as a necessity. Regional planning, agricultural development, in coordination with other economic sectors (industry, services) as well as future plans are. However, before anything else, an urgent need for regional planning of agricultural gains knowledge of the state to draw favorable conditions.

ACKNOWLEDGEMENT

The authors would like to thank the anonymous reviewers for their constructive comments. The second author thanks the regional water experts of Sistan and Baluchestan Province for provide information and data relating and cooperation in this study.

REFERENCES

1- Badri, A., Akbarian, S., &Javaheri, H. (2007). Demining the level of development in rural areas of Kamyaran city. *Journal of Geographical Research*, 2(82), 17-29.

2- Chuntian, C. (1999). Fuzzy optimal model for the flood control system of the upper and middle reaches of the Yangtze River. *Journal Hydrological Sciences*, 44(4), 573-582.

3- Ebrahimzadeh, A., & Eskandarisani, M. (2011). The location of factor analysis to explain the development environment and development. *Journal of Geography and Development,* 8(17), 7-28.

4- Fu, G. (2008). A fuzzy optimization method for multi-criteria decision-making: An application to reservoir flood control operation. *Expert Systems with Applications*, 34(1), 145-149.

5- General Censusof Population and Housing Censusof Iran, (2012).

6- Khakpoor, B., & BavanPoori, A. (2009). Analysis of Mashhad city inequality in levels of development. *Journal of Knowledge and Development*, 16(27), 182-202.

7- Malczewski, J. (1999). *GIS and multi criteria decision analysis.* USA and Canada, John Wiley & Sons.

8- Mohammed, A. (1980). Regional Imbalances in Levels and Growth of Agricultural Productivity -A Case Study of Assam. *The Geographer, Aligarh Geographical Society*, Aligarh.

9- Moradi, ZH., Mirakzadeh, A., Rostami, F., Karimi, F. (2015). Measuring of agricultural development levels in villages of Qaratureh Dehestan using TOPSIS technique. *Journal of Research and Rural Planning,* 4(2), 21-23.

10- Peet, R. (1999). *Theories of Development*, Guilford, New York and London.

11- Rafiee, R., Ataei, M., & Jalali, S.M.E. (2013). The optimum support selection by using fuzzy analytical hierarchy process method for Beheshtabad water transporting tunnel in Naien. *Iranian Journal of Fuzzy Systems,* 10(6), 39-51.

12- Ramatu, M., & Xinshen, D. (2007). *Regional disparities in Ghana: Policy option and public investment implications*. International Food Policy Research Institute (IFPRI), Washington, discussion paper series, No. 693, 1-55.

13- SardarShahraki, A., Karim, M.H., & Sheikh Tabar, M. (2014). Determine the level of economic development of agriculture and the rural sector in Iran. *Journal of Rural Development*, 16(1), 21-36.

14- Sasikumar, K., & Mujumdar, P.P. (1998). Fuzzy optimization model for water quality management of a river system. *Journal Water Resource Planning and Management*, 124 (2), 79-80.

15- Shafaiyan Fard, D., Kohiyan Afzal, F., & Yakhkeshi, M. (2015). Determine Superior options of water resources with the use WEAP model and Multi-criteria decision analysis (case study: Zaringhol basin). *Journal of Watershed Management*, 5(9), 29-45.

16- Shafiee, M., Ketabi, S., & Shaker Ardakani, M. (2012). Optimum selection of integrated marketing communication tools with FAHP approach (a home appliances group case study). *Journal of Operational Research and Its Applications (Journal of Applied Mathematics)*, 9(3), 13-26.

17- Sharama, B. (2004). Regional Disparities in agricultural labor productivity in the Brahmaputra valley. *International Journal of Interdisciplinary Social Sciences,* 4(1), 43-56.

10

Land Suitability Classification of East Azerbaijan Research Station for Tomato, Potato, Onion and Bean

Parisa Alamdari * and Setareh Amanifar

Abstract

In this study, qualitative land suitability evaluation was made for research station of East Azarbaijan, for tomato, potato, onion and bean. The study was done using Simple Limitation Method (SLM), Limitation Method regarding Number and Intensity (LMNI) and Parametric Methods (PM) such as the square root and the Storie methods. The results from different methods indicated that the most important limiting factors are Organic Matter (OM), salinity and sodicity and calcium carbonates, either alone or in combination. Evaluation indicated that the (SLM) and (LMNI) mainly show similar suitability classes, but Parametric methods especially square root method which it's results revealed to be more realistic show different suitability classes in many cases. According to the square root method, the area can be recommended as marginally suitable for cultivation of tomato and potato and expected to yield about 40-65% of optimal production.

Keywords:
SLM, LMNI, PM, Storie, Square root methods

Assistant Professors of Soil Science, University of Zanjan, Iran.
** Corresponding author's email: esfand1970@yahoo.com*

INTRODUCTION

With respect to increasing world population and scarce of arable lands, for food security, land evaluation accuracy should be considered. In this regard, land suitability evaluation can play key role in getting the best decides by developers and agriculturists in order to match the land optimum use, because it has long been found that land suitability is a part of a "rational" cropping system (FAO, 1976). Land evaluation assesses the performance of land based on a more or less systematic analysis of the physical land conditions. Land evaluation is a tool or a technique to compare the various use potentials and benefits that can be obtained from the land. The land evaluation in its simplest form selects the best lands for specific purpose. Main objective of land evaluation is achieving optimum usage from each land and soil with assessment physical, social and economic phases. In Iran 32 million ha of agricultural soils are suitable for crop cultivation; either rain fed or irrigated so to indicate suitability of these soils for particular use under particular climate, soil studies are carried out (Sys and Verheye,1974). Onion, potato and been are important crops, that mainly produced in major parts of Iran and also in the East and West Azerbaijan provinces, where their production depends on soil and climate. In different parts of Iran, evaluation of land suitability was done for some of crops by Jafarzadeh *et al.* (2005), Jafarzadeh *et al.* (2008), Kamkar *et al.* (2014), Safari *et al.* (2013), in order to find an optimum use for each land unit. Therefore this study not only confirm some obtained results for previous researchers about the different methods but also presents a land suitability based on SLM, LMNI and PM methods and evaluates the capabilities of the study area for the above named crops.

MATERIALS AND METHODS

This research station with about 200 ha located between 47°18' and 47°30' East latitudes and between 35°11' and 35°19 North longitudes. The altitude of the region is about 1230 m above sea level. The climate characteristics necessary for land suitability estimation (temperature, rainfall and relative humidity) were collected from the Ahar meteorological station was shown in Table 1, where the average annual total rainfall is about 302.8 mm, with the mean and maximum annual temperatures being 1.8°C and 19.5°C, respectively.

In order to obtain reliable soil data, the available soil survey reports were inspected and, based on this; seven representative soil profiles were chosen investigation within different land units with different average slope steepness. Soil profile descriptions, samplings and analyses were made using standard terminology and procedures (Soil Survey Staff, 1993). The soils were classified according to USDA classification system (Soil Survey Staff, 2014). As shown in Table 2, the soils were classified as an Aridisols order, Sodic Haplocambids and Typic Haplocambids suborders. According to the map of soil temperature and moisture regimes of Iran (Banaei, 1998), the soil temperature and moisture regimes of the area were identified as mesic and aridic, respectively. In the process of qualitative land suitability evaluation, it is the physical soil characteristics (texture, structure, stones, profile depth, $CaCO_3$ status and gypsum status), the fertility characteristics not easy to correct (Apparent Cation Exchange Capacity (ACEC)), sum of exchangeable base cations, pH in H_2O, organic matter) and the salinity and alkalinity that play an important role. The land qualities such as the moisture and oxygen availability and the foothold for root development depend to a large extent on the soil texture. The content of coarse fragments, influence the capacity of soil to retain nutrients and water and tillage conditions. Investigations have indicated that most crops produce excellent yields with an effective root zone depth of 90 to 100 cm. Therefore, for annual crops, the dense root system is usually assumed to occur within the upper 100 cm, while most tree crops have a dense to moderate root system up to 150 cm depth. The textural classes to be used for land suitability evaluation were recalculated using depth weighting factors up to the depth 1 m for annual crops and up to 1.5 m or up to an impermeable layer for perennial crops. When the content of gypsum in the root zone is higher than 25% and the

Table1: Climatic characteristics of the meteorological station of Tabriz

Characteristics	Jan	Feb	Mar	Apr	May	Jun	Jul	Aug	Sep	Oct	Nov	Dec
Mean max t(°C)	2.6	3	8.8	16.3	20	25.5	17.7	27.1	24.6	17.8	11.5	5.5
Mean min t(°C)	-5.5	-5	-0.8	2.3	8	11.8	15.2	14.9	10.8	6.3	1.6	-2.3
Absol max t(°C)	15.6	17.4	21	26.5	30.2	34.2	36.4	36.4	35	29	22.5	17
Mean month t(°C)	-1.5	-0.6	4	10.3	14.5	18.7	21.5	21	17.7	12	6.6	1.6
Rainfall (mm)	18.44	19.2	38.9	42.7	14.4	29.4	5.3	9.1	9	33	31.1	22.2
Relative humidity (%)	68.7	69.7	67.4	60.7	59.8	56.3	51.4	55.8	56.1	62.1	61.5	67.5
Sunshine (hours)	4.74	5.15	5.13	6.29	7.57	9.1	9.59	8.82	8.08	6.16	5.36	4.36
Calculated ETp(mm)	13.35	17.8	27.9	47.8	65.5	85.9	95.7	89.8	69.1	42.8	25.3	17.1

mean lime and gypsum content decreases with depth within the top 30 cm layer, then the lime and gypsum content in the soil was evaluated for this upper 30 cm only. In the other cases, the recalculated lime and gypsum content, using depth-weighting factors, was taken. The apparent CEC (ACEC) of the B horizon, or at 50 cm depth for A–C profiles, or just at the lithic or paralithic contact if this was present within 50 cm from the surface, was calculated as the weighted average of the sum of the exchangeable Ca, Mg and K, taking into account pH and Organic Matter (OM) in the upper 25 cm of the soil. In the irrigated land, salinity and alkalinity evaluation was made for the 100 cm depth from the soil surface, while the salinity evaluation for annual crops with shallow root systems was calculated as a weighted average of the upper 50 cm only.

The crop requirements with respect to climate, landscape and soil were summarized in separate tables according to Sys *et al.* (1993). With use of these tables, the qualitative land suitability evaluation was done for onion (*Allium cepa* L.), potato (*Solanum tuberosum* L.), tomato (*Solanum lycopersicum* esculentum L.) and bean (*Phaseolus vulgaris* L.) by comparing the actual soil characteristics and qualities with the crop requirements. The Simple Limitation Method (SLM), the Limitation Method regarding Number and Intensity (LMNI) and two Parametric Methods (PM), namely, the square-root and the Storie methods (Sys *et al.*, 1991a, b) were used to lands classification.

Simple Limitation Method (SLM): The simple limitation method implies that the crop requirement tables are made for each land utilization type. For each characteristic, the tables define the class-level criteria. The methodology suggests that, in the first place, an evaluation of the climatic characteristics is made, with an aim to determine a climate class level to be used in the following evaluation. The climate class level is determined by the lowest class level among those found for particular climatic characteristics. Then, similarly, the land class is determined by the lowest class level among those found for particular soil characteristics.

Limitation method regarding number and intensity of limitations (LMNI): This method defines land classes according to number and intensity of limitations. The crop requirement tables are requested which define limitation levels for each characteristic. The methodology evaluates, in the first place, the climatic characteristics, regrouped according to radiation, temperature, rainfall and humidity. For each climatic characteristic group, the most severe limitation determines the suitability of climatic class, which is subsequently used as the corresponding limitation level for the total land evaluation. The evaluation is carried out by comparing the actual land characteristics with the limitation levels defined by the crop requirement tables. This method is more difficult than SLM, but the results accuracy is more honest, because it considers the land with several limitations of the same level as belonging to a lower-class land than that with only a single limitation of the same level.

Parametric Methods (PM): The parametric land evaluation consists in numerical rating of different limitation levels of land characteristics according to a numerical scale between the maximum (normalized as 100%) and the minimum value. Lastly, the climatic index, as well

Table 2: Selected soil characterization and classification of profiles

Profile	Horizon	Depth (cm)	C (%)	Si (%)	S (%)	Text. class	Gravel (%)	ECe	pH	OC (%)	$CaCO_3$ (%)	CEC Cmol/ Kg
Loamy, mixed, active, mesic, Sodic Haplocambids												
1	A	0-35	31	31	38	SiCL	0	20	8.8	0.27	24.8	15.84
	B_{w1}	35-80	25	42	33	L	0.11	4.4	10	0.2	25	13.15
	B_{w2}	80-120	23	34	43	L	0.15	1.4	9.9	0.26	24.9	12.42
	C	>120	24	35	41	L	0.13	2.9	9.9	0.23	24	12.46
Coarse Loamy, mixed, active, mesic, Sodic Haplocambids												
2	A	0-9	33	29	38	CL	0	3.9	7.6	0.2	24	16.7
	B_{w1}	9-22	31	32	37	CL	3.16	0.05	7.1	0.2	24.8	16
	B_{w2}	22-64	18	2	80	LS	0.75	11	7.4	0.13	25.5	9.9
	C_1	64-120	34	27	39	CL	60	0.9	7.5	0.27	25.8	17.5
	C_2	>120	32	25	43	CL	60	0.5	7	0.26	23.9	16.6
Loamy, mixed, active, mesic, Sodic Haplocambids												
3	A	0-30	28.3	24.4	47.3	SCL	3.33	6.6	7	0.35	24.8	14.8
	B_{w1}	30-60	21.5	26.1	52.4	SCL	0.9	6.7	7.8	0.22	24.9	11.1
	B_{w2}	60-105	25.5	40.3	34.2	L	0	3.9	7.1	0.19	24.8	13.1
	C	>105	5.1	52.2	42.7	SiL	0.8	1.9	7.3	0.2	25	2.3
Loamy, mixed, active, mesic, Sodic Haplocambids												
4	A	0-35	35	29	43	CL	0.6	14.5	7.4	0.26	24.9	18
	B_{w1}	35-75	22.7	34.1	43.2	L	0.14	4.9	7.1	0.21	24.8	11.7
	B_{w2}	75-120	30.5	40.5	29	CL	0.16	4.5	7.5	0.22	24	15.6
	C	>120	17.9	29.9	52.2	L	0.15	3.4	8	0.24	25	9.4
Loamy, mixed, active, mesic, Typic Haplocalcids												
5	A	0-28	22.6	33	4.44	L	0.02	6.2	7.2	0.25	23	11.7
	Bk	28-55	32	27	41	CL	0.05	5.72	7.4	0.23	23.9	16.7
	C_1	55-120	2.8	54.4	43.8	SL	0.15	5.4	7.8	0.24	21	1.8
	C_2	>120	4	55	41	SL	0.2	5.6	7	0.25	22	2
Fine, mixed, superactive, mesic, Sodic Haplocambids												
6	A	0-25	3.36	33.7	41	SL	0.11	7.9	7.7	0.2	23	18.5
	C1	25-55	9.33	27.8	39.3	SL	0	3	7.2	0.23	23.8	16.9
	C2	55-120	43.1	15.9	41	C	0	0.6	7.5	0.25	20	23
	C3	>120	45	16	39	C	0	0.3	7.4	0.2	22	22.9
Coarse Loamy, mixed, active, mesic, Sodic Haplocambids												
7	A	0-25	13	34	53	SL	2.9	8.4	8.1	0.23	224.9	7
	Bw	25-50	9.6	31.2	59.2	SL	0.1	5.5	8	0.21	24.5	4.6
	C	50-80	4.2	40	55.8	SL	33	5.7	9	0.16	24	3.2
	2AB	80-120	17.1	30.5	52.4	SL	0.2	5.5	10	0.24	23.3	9.3
	3C	>120	14	28	58	SL	0.28	5.2	9.7	0.2	24.4	7.5

Table 3: Qualitative land suitability classes of study area for tomato, potato, onion and bean

Profile	Tomato				Potato			
	SLM	LMNI	Storie	Square root	SLM	LMNI	Storie	Square root
1	N_2fn	N_2fn	N_2	N_2	N_2fn	N_2fn	N_2	N_2
2	S_3sf	S_3sf	N_2	N_1	S_3f	S_3f	N_2	S3
3	S_3cf	S_3cf	N_1	S_3	S_3fn	S_3fn	N_1	S3
4	N_2n	N_2n	N_2	N_2	N_2n	N_2n	N_2	N_2
5	S_3s	S_3s	S_3	S_3	S_3n	S_3n	S3	S3
6	S_3s	S_3s	S_3	S_3	S_2sn	S_2sn	S3	S_2
7	S_3sn	S_3Sn	N_1	S_3	S_3fn	S_3fn	N_1	S3
Profile	**Onion**				**Bean**			
	SLM	**LMNI**	**Storie**	**Square root**	**SLM**	**LMNI**	**Storie**	**Square root**
1	N_2fn	N_2fn	N_2	N_2	N_2fn	N_2n	N_2	N_2
2	S_3s	S_3s	N_2	N_2	N_2n	N_2n	N_2	N_2
3	S_3sn	S_3sn	N_2	N_2	N_2n	N_2n	N_2	N_2
4	N_2sn	N_2sn	N_2	N_2	N_2n	N_2n	N_2	N_2
5	N_2sn	N_2sn	N_2	N_1	N_2n	N_2n	N_2	N_1
6	N_2sn	N_2sn	N_2	N_1	N_2n	N_2n	N_2	N_1
7	N_2sn	N_2sn	N_2	N_2	N_2n	N_2n	N_2	N_2

SLM, Simple Limitation Method, LMNI, Limitation Method regarding Number and Intensity, f, fertility, n, salinity and alkalinity limitations, s, soil limitation.

as the land index, is calculated from these individual ratings. In our case, the indices were calculated following two alternative procedures:

1) The Storie method (Storie, 1978):

$I=A\times B/100\times C/100\times ...$

where I is the specified index and A, B, C etc. are different rating given for each property.

2) Square Root Method (Khiddir, 1986):

$I= A\times B/100\times C/100\times ...$, in which R min is the minimum rank.

RESULTS AND DISCUSSION

The results of soil profile analyzing have shown in Table 2, and the results for land suitability of the study area for selected crops have shown in Table 3. In this study after analyzing soil samples, the requirements for tomato, potato, onion and bean summarized by Sys *et al.* (1993) were used, then the Simple Limitation Method (SLM), Limitation Method regarding Number and Intensity (LMNI) and the Parametric Methods (Storie and Square Root methods) were employed. The classes of land suitability were determined S_1 or highly suitable with production of 80-100% of optimum, S_2 or moderately suitable with production of 60-80% of optimum, S_3 or marginally suitable or 40-60% of optimum and N or non-suitable (N_1&N_2). Economic factors were excluded and moderate level of management was assumed. According to Table 3, most important limiting factors in studied area are organic matter, pH and soil salinity which their effect can appear alone or together. Soil attribute dada such as OM, $CaCO_3$ and salinity for tomato and potato, $CaCO_3$ and salinity for onion and salinity for bean caused marginally suitable to non-suitable. In simple limitation method and limitation method regarding number and intensity 69% of area are marginally suitable (S_3) for tomato and 21.2% of area is S_2, 47.8% are S_3 for potato, but for onion and bean 100% of area are actually unsuitable and non-potentially suitable (N_2). In parametric method (Storie) 35.3% of area is S_3, 19.4% is N1 and 45.3% is N_2 for tomato and potato and 100% is N_2 for onion and bean. In parametric method (Square root) 54.7% is S_3, 14.3% is N_1 and 31% is N_2 for tomato, 21.2% is S_2, 47.8% is S_3 and 31% is N_2 for potato, 35.3% is N_1 and 67.4% is N_2 for onion and bean. The results obtained by the parametric square root method are probably more realistic, as suggested by comparison with other reports (Jafarzadeh *et al.*, 2005; Jafarzadeh *et al.*, 2008; Sarvari and Mahmoudi, 2001) in which different methods were applied in different parts of country. The parametric square root method suggests that the study are possesses optimal climatic conditions for irrigated potato, onion and bean (S_1), while the climatic conditions during the growing cycle make the region only moderately suitable (S_2) for irrigated tomato.

CONCLUSION

In general, the area is highly suitable (S_1) from the climatic point of view for potato, onion, bean, moderately suitable (S_2) for tomato. However the soil fertility characteristics (f), the salinity and alkalinity conditions (n) and in some cases, the soil physical characteristics (s) make the lands marginally suitable (S_3) or even non-suitable (N). Based on obtained results especially results from square root method, the cultivation of tomato and potato can be recommended, but the majority of region is non-suitable for onion and bean. Limitations are posed mainly by the high salinity and low organic matter and high calcium carbonates.

REFERENCES

1- Banaei, M.H. (1998). Soil Moisture and Temperature Regime Map of Iran . Soil and Water Research Institute, Ministry of Agriculture, Tehran, Iran.

2- FAO (1976). A framework for land evaluation system. *Soil Bulletin,* 32, FAO, Rome, 72pp.

3- Jafarzadeh, A.A., Alamdari, P., Neishabouri, M.H., & Saedi, S. (2008). Land suitability evaluation of Bilverdi Research Station for wheat, barely, alfalfa, maize and sanflower. *Soil and Water Research,* 3, 81-88.

4- Jafarzadeh, A.A., Khoshazaman, T., Neyshabouri, M.R., & Shahbazi, F. (2005). *Qualitative evaluation of land suitability in Karkaj Research Station of Tabriz University for wheat, potato, maize, tomato, bean and alfalfa,* pp. 95–102. In: Int. Conference of ICEM, Hyderabad, India. October 28-30.

5- Kamkar, B., Dorri, M. A., & Teixeira da Silva, J. A. (2014).Assessment of land suitability and the possibility and performance of a canola (*Brassica napus* L.) Soybean (*Glycine max* L.) rotation in four basins of Golestan Province, Iran. *The Egyptian Journal of Remote Sensing and Space Science,* 17, 95-104.

6- Khiddir, S.M. (1986). A statistical approach in the use of parametric systems applied to the FAO framework for land evaluation. Unpublished thesis. State University Ghent.

7- Safari, Y., Esfandiarpour-Boroujeni, I., Kamali, A., Salehi, M. H., & Bagheri-Bodaghabadi, M. (2013). Qualitative land suitability evaluation for main irrigated crops in the shahrekord plain, Iran: A geostatistical approach compared with conventional method. *Pedosphere,* 23(6), 767-778.

8- Sarvari, S.A., & Mahmoudi, S.H. (2001).Qualitative land suitability evaluation for irrigated sugar beet in Ghazvin region. *Iranian Journal of Soil and Water,* 15, 66-74.

9- Soil Survey Staff. (1993). *Soil survey manual. agricultural handbook.* No. 18, US Department of Agriculture, Soil Conservation Service, Washington, DC.

10- Soil Survey Staff. (2014). *Keys to soil taxonomy.* 12th Ed., US Department of Agriculture, Natural Resources Conservation Service, Washington, DC.

11- Storie, R. E. (1978). The Storie Index Soil Rating Revised. Davis,CA, University of California, Division of Agricultural Science, Special Publication No 3203.

12- Sys, C., & Verheye, W. (1974). Land evaluation for irrigation of arid regions by the use of the parameter method. Trans. 10th International Soil Congress, Moscow, 149-155.

13- Sys, C., Van Ranst, E., & Debaveye, J. (1991a). Land evaluation. Part I: Principles in land evaluation and crop production calculations. *Agricultural Publication.* No. 7. General Administration for Development Cooperation, Brussels, Belgium.

14- Sys, C., Van Ranst, E., & Debaveye, J. (1991b). Land evaluation. Part II: Methods in land evaluation. *Agricultural Publication.* No.7. General Administration for Development Cooperation, Brussels, Belgium.

15- Sys, C., Van Ranst, E., & Debaveye, J. (1993). Land evaluation. Part III: Crop Requirements. *Agricultural Publications.* No. 7. General Administration for Development Cooperation, Brussels, Belgium.

11

Seasonality and Forecasting of Monthly Broiler Price in Iran

Azadeh Falsafian

Abstract

The objective of this study was to model seasonal behavior of broiler price in Iran that can be used to forecast the monthly broiler prices. In this context, the periodic autoregressive (PAR), the seasonal integrated models, and the Box-Jenkins (SARIMA) models were used as the primary nominates for the forecasting model. It was shown that the PAR (q) model could not be considered as an appropriate method for modeling seasonal behavior of the broiler price. Results of seasonal unit root test indicated that the monthly prices of broiler follow a non-stationary stochastic seasonal process. Accordingly, the regression-based model is an appropriate modeling framework. While SARIMA is an alternative modeling approach, the RMSE of forecasting error suggested the superiority of the regression-based model over the SARIMA model. Therefore, the estimated parameters of the regression-based model can be used to predict the monthly prices of broiler in Iran.

Keywords:
Seasonality, Broiler price, Periodic autoregressive (PAR), Seasonal unit root, Seasonal integrated

Assistant Professor, Department of Agricultural Management, Extension and Education, Tabriz Branch, Islamic Azad University, Tabriz, Iran.
** Corresponding author's email: falsafian@iaut.ac.ir*

INTRODUCTION

Broiler industry is one of the main sub-sectors of agricultural sector and is a growing industry in Iran. The production of broilers has increased from 632000 tons in 1998 to about 1967 000 tons in 2013 with an annual growth rate of 13.2 percent on average (Ministry of Agriculture Jihad, 2014). This industry plays a major role in providing food security in Iran by producing more than one million tons of chicken.

In Iran, the production and distribution of the broiler has been under the direct control of government since the 1979 Revolution until 1998. Government sets the prices of the product and the subsidized prices of inputs. In 1998, following the economic adjustment in Iran, the government lifted its control over the broiler industry and allowed market to determine the prices of inputs and output. Consequently, price fluctuation was added to the production variability in the broiler industry. This liberalization in the boiler industry has led to higher broiler price and its seasonal fluctuations whereas it was relatively stable over the past years. This situation amazes the producers and consumers of broiler and they are not stratified. To overcome this problem and stabilize the produces incomes and to support the consumers, Iranian government implemented the policy of chicken market regulation in 2002 [1]. However, Ghahremanzadeh (2008) and Hosseini *et al.* (2008) pointed out that this policy could not give satisfying consequents whilst if the prediction of the broiler prices is available, it can provide a guide for government to develop and design efficient plan for the storage and redistribution of chicken. Since the Iranian boiler markets exhibit characteristics of significant seasonality in both the prices and quantities, the study of the seasonal behavior in the broiler price series is important for model evaluation and forecasting. As Franses (1991) points out, it is extremely important to determine the nature of the seasonality since this bears heavily on forecasting accuracy. The seasonal movements in the broiler price are direct reflection of seasonality in the boiler's marketing and demand. Seasonality in demand is related to religion ceremonies, the New Year's celebration, consumer's seasonal product preferences and change in calendar. On the other hand, seasonality in the boiler supply is related to seasonal demand of consumers, change of weather and calendar, some biological factors and management practices.

Many economic time series contain important seasonal components and it is a common belief that modeler need to pay specific attention to the nature of seasonality. Seasonality has been a major research area in economics. For example, Gustavsson and Nordstrom (2001), Kim and Moosa (2001 and 2005), Koc and Altinay (2007), Kulendran and King (1997), Kulendran and Wong (2005) and Lim and McAleer (2000 and 2002) analyzed the seasonality in international tourism flow. Seasonality was evaluated by Silvapulle (2004) for the financial market, Franses and Van Dijk (2005) for the industrial productions and Arnade and Pick (1998) and Tiffin and Dawson (2000) for the agricultural market. In spite of the importance of seasonality, there are a few studies on analyzing seasonal behavior in economic variables in Iran and this study is a pioneering work in this field. The main objective of present study was to model seasonal behavior of the broiler price in Iran. The seasonality characteristic of the broiler price was evaluated to develop a forecasting model which can be used to predict the broiler price in Iran. This paper was organized as following: section two presents methodology of models and forecasting of seasonal economic time series and section three describes the data and empirical analysis. The final section contains conclusion and implications.

MATERIALS AND METHODS

One of the most popular techniques for forecasting future outcomes of economic variables is time series models. The selection of the proper time series technique to model the behavior of the series depends on the characteristics of the time series. One of the major characteristics of many economic time series is the presence of

[1] *In this policy, a price rang of broiler is determined. If the broiler price is below this range, the government buys the excess broiler and if the actual price is over the price range, the stocked chicken is resupply.*

seasonal movement (Darne and Diebolt, 2002). The main types of movements are the trend, cycle and irregularity. Hence, the proper time series model must be selected based on type and nature of these components.

Since the Iranian broiler price exhibits characteristics of seasonality, the seasonal time series models should be applied. We resorted to time series technique and utilized the periodic autoregressive (PAR), the seasonal integrating and regression-based time series, and the Seasonal Box-Jenkins (SARIMA) models as the primary nominates for forecasting model.

Seasonal integration

If the seasonality is deterministic, the time series should be modeled with seasonal dummy variables (Arnade and Pick, 1998; Brendstrup *et al.* 2004 and Kim and Moosa, 2001). The application of seasonal dummies may be justified in some cases, whereas many economic time series seem to be characterized by seasonal patterns that evolve over time (Franses and Van Dijk, 2005). A very popular approach is to model the seasonality as a non-stationary stochastic process, i.e. seasonality evolves over time by allowing for seasonal unit roots. In this case, the time series should be differentiate distinguished by appropriate filter to account for the presence of seasonal unit roots and the differenced data are modeled by proper approach.

To determine the presence of unit roots, a seasonal unit roots test should be used. The most widely used seasonal unit roots test is the Hylleberg *et al.* (1990) test for quarterly data that extended to monthly data by Franses (1991) and Beaulieu and Miron (1993). The most popular test in practice is the Beaulieu and Miron (1993) [BM], which is based on the following auxiliary regression:

$$(1-L^{12})P_t = \alpha + \sum_{s=1}^{11}\delta_s D_{s,t} + \beta t + \sum_{i=1}^{12}\pi_i y_{i,t-1} + \sum_{j=1}^{p}\phi_j (1-L^{12})P_{t-j} + \xi_t \quad (1)$$

where, P_t represents the broiler price series; $D_{s,r}$ are monthly seasonal dummy variables equal to 1 if time t corresponds to season s and $D_{s,r}= 0$ otherwise; $_t$ is trend; $y_{i,t-1}$'s are non-singular linear transformations of lagged values of P_t whose details are given by Beaulieu and Miron (1993), and ξ_t is white noise process. In equation (1), the value of p should be determined so that the residuals from the above regression mimic a white noise process.

In order to test for the presence of unit roots at zero and π frequencies, the null hypotheses of $H_{k0}: \pi_k=0$ for k=1, 2 against the alternative hypotheses $H_{k1}: \pi_k<0$ are tested using the conventional *t-statistics*, denoted by t_k. To test the complex unit roots, the joint null hypotheses $H_{k0}: \pi_k= \pi_{k+1}=0$ for k = 3, 5, 7, 9, 11 against the alternative hypotheses H_{k1}: at least one of π_k and π_{k+1} is not equal to zero are tested using the conventional *F-statistic*, denoted by $F_{k,k+1}$. Alternatively, the null hypotheses of $H_{k0}: \pi_k=0$ for k=3, 4… 12 are tested against the alternative hypotheses $H_{k1}: \pi_k<0$ using the t-statistic. The asymptotic distributions of the above statistics are non-standard and the critical values are tabulated in BM (1993). To prove that no unit root exists at any seasonal frequency, π_k must be different from zero for k=2 and for at least one member of each of the sets {3, 4}, {5, 6}, {7, 8}, {9, 10}, {11, 12}.

When the BM test is carried out, the proper differencing filter is determined to obtain stationary data and they are molded by a proper approach. A popular approach to allow for time varying seasonality is the autoregressive model, so-called regression-based model, for seasonally differenced data. For the broiler prices, it can be written as:

$$\phi_p(L)\,\Delta_s P_t=\mu+\varepsilon_t \quad (2)$$

where ϕ_p is a polynomial of order p in the lag operator L, Δ_s is the appropriate differencing filter operator, P_t represents the broiler monthly prices, μ is intercept and ε_t is white noise process.

Seasonal ARIMA model

The traditional multiplicative seasonal ARIMA model [ARIMA (p, d, q)(P, D, Q)$_s$] proposed by Box and Jenkins (1976) was used as alternative forecasting model to forecast the seasonal price variables. Fitting a seasonal ARIMA model for monthly price of the broiler requires the data to be stationary. The number of seasonal differences *(1-L*^{12}*)*, *D*, and the number of regular differences

(1-L), d, are used to reduce the series to stationary so that a seasonal ARIMA model can be fitted. For monthly price of broiler, a multiplicative seasonal ARIMA model can be written as:

$$\phi(L)\phi_{12}(L^{12})(1-L^{12})^D(1-L)^d P_t=\theta(L)\theta_{12}(L^{12})\varepsilon_t \quad (3)$$

where, $\phi(L)$ and $\theta(L)$ are invertible lag polynomials in L of nonseasonal orders p,q, while $\phi_{12}(L)$ and $\theta_{12}(L)$ are invertible lag polynomials in L of seasonal orders P,Q, and ε_t is the white noise term. Notice that the SARIMA model assumes seasonal unit roots at all frequencies.

Periodic autoregressive model

An entirely different approach to allow for flexible seasonal patterns is offered by periodic autoregressive (PAR) models, which suppose that not only is the intercept (and trend if present) seasonal, but also the autoregressive parameters vary across seasons (Franses and van Dijk, 2005). The PAR model assumes that the broiler prices in each month can be described using different autoregressive models, and the same goes for the periodic extensions to the MA and ARMA models. Fallowing Brendstrup *et al.* (2004), the PAR (p) model for the broiler monthly price that is observed for N years (N=10) can be represented as:

$$P_t=\sum_{s=1}^{12}\mu_s D_{s,t}+\sum_{s=1}^{12}\varphi_s D_{s,t}t+\sum_{s=1}^{12}\phi_{1,s}D_{s,t}P_{t-1}+\ldots+\sum_{s=1}^{12}\phi_{p,s}D_{s,t}P_{t-p}+\xi_t \quad (4)$$

where, $t=1,2,\ldots,n$ and $n=12N(12\times 10)$, μ_s, φ_s and $\phi_{i,s}$ are periodically varying parameters, $D_{s,t}$ is seasonal dummy equal to 1 when t falling in s and zero otherwise and $\xi_t \sim NID(0, \sigma^2)$.

However, to apply the PAR (p) model, the data generating process of broiler price must have periodic variation. In practice, we must test whether the periodic variation in some or all of the parameters is significant. Boswijk and Franses (1996) showed that testing for periodicity in (4) amounts to testing the hypothesis H_0: $\phi_{is}=\phi$ *for* $s = 1,2,\ldots 12$ and $i=1, 2,\ldots,p$. This hypothesis can be tested by likelihood ratio test that is asymptotically $X^2(11p)$ under the null, irrespective of unit root in P_t. An important implication of this result is that equation (4) can be estimated directly by P_t series and no need to consider priori differenced P_t series. We determine the order of PAR (q) using a general-to-specific approach based on diagnostic check tests following Franses and Paap (2004). Also, *F-test* version for this hypothesis, denoted here by F_{per}, has the standard $F(11p, n-(12+12p))$ distribution in the case of a PAR (q) series with S seasonal intercept (Boswijk and Franses, 1996). If the null hypothesis is not rejected, we may proceed with BM analysis of seasonal unit roots; if significant periodicity is found, we may test for periodic unit root tests (Boswijk and Franses, 1996).

Brendstrup *et al.* (2004) and Franses and Paap (2004) pointed out that fitting a PAR (q) model does not prevent the finding of a non-periodic AR process if the latter is, in fact, the DGP. Thus, we will start by selecting a PAR (q) model, and then it will be tested whether the autoregressive parameters are periodically varying using the method described above.

Forecast evolution criteria

In this paper, the regression-based model, SARIMA and PAR (q) and their comparative performances are considered. The accuracy of these forecasting models needs to be evaluated so that the model that generally works the best and which produces the smallest error could be identified. The selection of the most accurate forecasting model should be based on the out-of-sample forecasting performance. In this context, we use the root mean square forecast error (RMSE), mean absolute forecast error (MAE) and mean absolute percentage forecast error (MAPE) criterion and final model selected based on minimizing these criterion.

RESULTS AND DISCUSSION

Our application is to Iran's broiler monthly prices from 1998:2 to 2013:2. The monthly prices of broiler were collected from Ministry of Agriculture Jihad (MAJ). Statistical properties of the broiler monthly prices are reported in Table 1.

Figure 1 illustrates the variations of broiler price over 1998-2013. By visual inspection of Figure 1, it is clear that the monthly prices of

Table 1: Statistical properties of the monthly prices of broiler

Mean	SD	CV[a]	Min	Max	Skewness	Kurtosis
13068	8082	0.618	3874	42673	1.619	5.516

a: Coefficient of variation= SD/Mean.

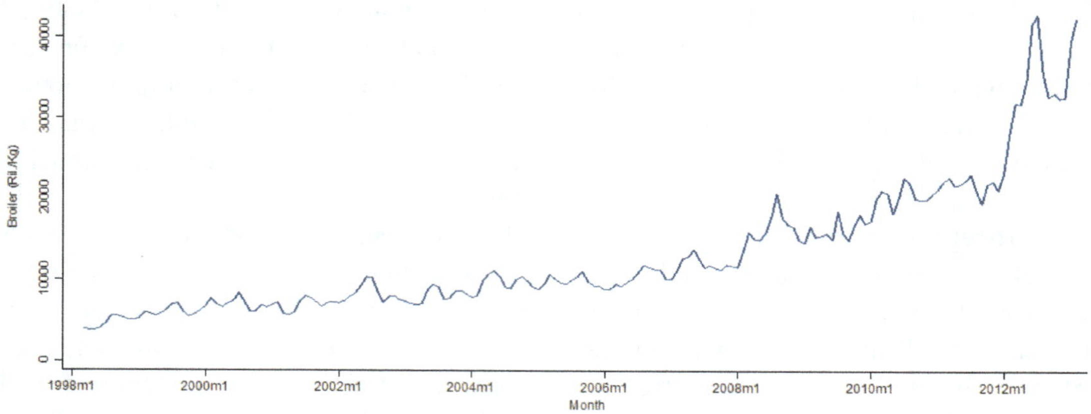

Figure 1: Broiler monthly prices in Iran (1998-2013)

broiler potentially have seasonality and exhibit an upward trend during whole period.

The periodicity test

In this section, the tests for periodic integration were applied to the logarithmically transformed of broiler price series. The pursued model selection strategy amounts to estimating periodic autoregressive models as in (4) of order *p*, where p is the initially set equal to 12, and *p* is decreased when diagnostic tests indicate no obvious signs of misspecification. The diagnostic checks that were used included LM tests for first-order (F_{ser}) and first to twelfth order (F_{12}) residual autocorrelation, LM tests for first- and twelfth -order ARCH effects (ARCH(1) and ARCH(12)) and a LM test for first-order periodic autocorrelation, F_{pser}. The specification search yields that the lags in the PAR model are 1 and 12 (PAR (1, 12). Results of the diagnostic check test for PAR (1, 12) are presented in Table 2.

As can be observed, the diagnostic test values of F_{pser}, F_{ser}, F_{12}, ARCH (1) and ARCH (12) are statistically significant at the 5 percent level. It is important to note that the F_{per} test resulted in the value of 1.46 with the *p-value* of 0.112. Consequently, we cannot reject the null hypothesis of non-periodicity. That is, there is not significantly periodic variation in the parameters suggesting that the PAR model is not suitable to model and generate future outcomes of the broiler price. Therefore, we must consider the regression-based and SARIAM models.

The seasonal unit root test

The BM test was applied to the broiler monthly pieces. The appropriate lag length in the auxiliary regression was selected in the same way as for the PAR (q) models. Results from the BM test are shown in Table 3. As can be observed, the

Table 2: Results of the periodicity test

F_{PER}	Diagnostic check tests				
	ARCH(12)	ARCH(1)	F12	F_{PSER}	F_{SER}
1.46 (0.112)	14.43 (0.298)	0.043 (0.871)	1.87 (0.185)	1.65 (0.175)	3.45 (0.098)

Notes: values in parentheses are p-values

Table 3: Results from seasonal unit root test

Series	Lags	0	π	π/2	2π/3	π/3	5π/6	π/6
		t1	t2	F3.4	F5.6	F7.8	F9.10	F11.12
Data in level	1,12	-2.98	-2.45	3.23	2.65	3.65	4.32	4.87
Differenced data	1,12	-4.32	-3.02	11.32	7.98	9.34	8.54	11.45

Note: the critical value at 1% level are: t_1=-3.37, t_2=-3.21 and $F_{k,k+1}$=7.86. ; at 5% level are: t_1=-3.19, t_2=-2.65 and $F_{k,k+1}$=5.77; and at 10% this level are: t_1=-2.91, t_2=-2.36 and $F_{k,k+1}$=4.86 (Franses and Hobijn, 1997)

results indicate that the null hypothesis of unit root is not rejected at the 5 percent significance level at all frequencies. In conclusion, the test shows strong evidence for unit root at long run and seasonal frequencies; hence, seasonal differencing, , is an appropriate filter for removing the unit roots from the monthly prices of broiler. However, to see whether the price series become stationary after imposing the identified filter, the differenced series were tested for unit roots again. The results revealed that this filters leads to a stationary series.

Estimation of the forecasting model

The regression-based model, defined in Equation 2, and seasonal ARIMA models in Equation 3 were estimated as competitive models to select the best forecasting model for the broiler price. Appropriate lag length of regression-based model, p, is determined based on the same rule for the BM auxiliary regression model. As the appropriate lags in the regression-based model are 1 and 12, therefore the AR (1, 12) is the most preferred model. The seasonal ARIMA models were estimated based on the Box – Jenkins technique (Enders, 2004). The residuals were subjected to diagnostic tests, including visual inspection of residual sample autocorrelation function and the use of the Liung-Box statistics (1978). The SARIMA models were estimated and evaluated at the multiplicative and additive functional forms. According the minimization of the AIC and SBC criteria and the procedure explained above, the ARIMA (2, 1, 1) (0, 1, 1)12 was selected as a most preferred model. Table 4 presents outcomes of the estimated regression-based model and ARIMA (2, 1, 1) (0, 1, 1) 12 models. As can be observed, the estimated parameters are statistically significant and there is no serial correlation at the 5% level.

Evaluation of forecasting performance

In order to select the best forecasting model for the broiler price, the root mean square forecast error (RMSE), mean absolute forecast error (MAE) and mean absolute percentage forecast error (MAPE) were calculated. We considered short forecast horizons of up to four

Table 4. Estimates of regression-based and ARIMA(2,1,1)(0,1,1)12 models for the broiler price

Regression-Based model

$$(1 - L^{12})LnP_t = \underset{(4.13^a)}{0.5510} + \underset{(10.32)}{0.726}(1 - L^{12})LnP_{t-1} - \underset{(-6.21)}{0.312}(1 - L^{12})LnP_{t-12}$$

F_{ser}=0.1662 (0.69) ARCH(1)[b]=0.512 (0.43)

F_{12}=1.54 (0.146) ARCH(12)=13.42 (0.54)

ARIMA(2,1,1)(0,1,1)12 model:

$$(1-\underset{(4.65)}{0.675}L+\underset{(-6.54)}{0.432}L^2)(1-L)(1-L^{12})LnP_t=\underset{(-1.66)}{-0.032}+(1+\underset{(6.34)}{0.674}\varepsilon_{t-1})(1+\underset{(4.65)}{0.745}\varepsilon_{t-12})$$

Q(6)[c] =7.54(0.28) Q(12)=9.61(0.38) Q(24)=22.82(0.12) AIC=-6.61 SBC=-5.92

(a): values in parentheses are t-test statistic. (b): ARCH(1) and ARCH(12) denote to LM tests for first- and twelfth-order ARCH effects, respectively, and the associated p-values are in the parenthesis. (c): Q(p) indicates the Ljung-Box statistic for the joint significance of the residual autocorrelations of up to lag 6, 12 and 24 and the associated p-values 4

Table 5: Out of sample forecasting performance of broiler price

Model	Measure	1 month	2 month	Horizon 4 month	6 month	8 month
Regression-Based	RMSE	53	77	85	81	92
	MAE	47	69	71	68	81
	MAPE	6.7	7.1	7.4	6.1	4.3
	RMSE	78	113	131	81	121
ARIMA(2,1,1)(0,1,1)12	MAE	64	96	108	67	102
	MAPE	7.4	11.1	12.3	7.3	11.2

years and the identified models were estimated on data up to 2009:12. Then, 2009:12 was taken as the forecast origin for forecasting 1,2,4,6, and 8 steps ahead. The model was, then, re-estimated on data up to 2010:1, with the unchanged form of the model. 2010:1 was taken as the forecast origin, and, so on, subject to constrain that we have data on the period being forecasted (the sample ends in 2013:12). The forecasts were, then, transformed from the differentiated form to forecasts in levels. By comparing the forecasts in level, we do not need to calculate the Generalized Forecast Error Second Moment (GFESM) measure developed by Clements and Hendry (2004).

Table 5 illustrates the RMSE, MAE and MAPE measures for the regression-based and ARIMA $(2, 1, 1)(0, 1, 1)_{12}$ models according to the forecasting horizon for the broiler prices. The results indicate that the RMSE, MAE and MAPE measures for the regression-based model are smaller than the ARIMA $(2, 1, 1)(0, 1, 1)_{12}$ model in the forecasting horizon. It can be concluded that the regression-based model is selected as the best model to model and forecast future value of the boiler prices.

CONCLUSION

This study was aimed to analyze the seasonal behavior of broiler prices in Iran using monthly data during the 1998-2013. The BM test shows strong evidence for unit root at long run and seasonal frequencies and forecasting model were established based on this result. Although forecasting is vital to the planning of all activities, it is particularly crucial in the broiler industry due the perishable nature of the broiler product. Choosing the right model to forecast price series requires not only sufficient knowledge about the nature and characteristic of the time series but also a good understanding of the theory behind the models.

There are a number of factors used to evaluate the effectiveness of a forecasting method, such as forecasting accuracy, costs associated with the application of a forecasting procedure such as installation and operating costs, and ease of application and interpretation of the output from a forecasting method. In this context, we found that the regression-based model is most proper method to forecast and model the broiler price. This model is relatively easy to apply and install in the public and private sectors. The government can use this model to predict broiler price and consequently, make better plans to regulate the broiler market. Furthermore, the broilers producers can simply apply this model to generate future outcomes of the broiler prices and therefore, increase own income through organizing the production plans.

ACKNOWLEDGEMENT

The author would like to thank the Ministry of Agriculture Jihad (MAJ) for making the needed available data.

REFERENCES

1- Arnade, C., & Pich, D. (1998). Seasonality and unit roots: the demand for fruits. *Agricultural Economics*, 18 (1), 53-62.

2- Beaulieu, J. J., & Miron, J. A. (1993). Seasonal unit roots in aggregate U.S. data. *Journal of Econometrics,* 55 (1-2), 305-328.

3- Boswijk, H. P., & Franses, P. H. (1996). Unit roots in periodic auto regressions. *Journal of Time Series,* 17 (3), 221-245.

4- Brendstrup, B., Hylleberg, S., Nielsen, M., Skipper, L., & Stentoft, L. (2004). Seasonality in economic models. *Macroeconomic Dynamics,* 8 (03), 362–394.

5- Celements, M. P., & Hendry, D. F. (2004). *A Companion to Economic Forecasting,* USA, Blackwell Publishing.

6- Darne, O., & Diebolt, D. (2002). A note on seasonal unit root tests. *Quality and Quantity*, 36 (3), 305-310.

7- Enders, W. (2004). *Applied econometrics time series*, Second edition. USA, John Wiley & Sons, Inc.

8- Franses, P.H. (1991). Seasonality, non-stationary and the forecasting of monthly time Series. *International Journal of Forecasting,* 7 (2), 199-208.

9- Franses, P. H., & Hobijn, B. (1997). Critical values for unit root tests in seasonal time series. *Journal of Applied Statistics,* 24 (1), 25- 47.

10- Franses, P. H., & Paap, R. (2004), *Periodic time series models, advanced texts in Econometrics*, USA, Oxford University Press.

11- Franses, P. H., & Van Dijk, D. (2005). The forecasting performance of various models for seasonality and nonlinearity for quarterly industrial production. *International Journal of Forecasting*, 21 (1), 87– 102.

12- Ghahremanzade, M. (2008), *Developing a revenue insurance model for Broiler industry of Iran*. Unpublished dissertation, Department of Agricultural Economics, University of Tehran.

13- Gusstavsson, P., & Nordstrom, J. (2001). The impact of seasonal unit roots and vector ARMA modeling on forecasting monthly tourism flows. *Tourism Economics*, 7(2), 117-133.

14- Hossaini, S. S., Salami, H., & Nikoukar, A. (2008). Price transmission model for Iranian chicken industry. *Iranian Journal of Agricultural Economics*, 2(1), 1-22.

15- Hylleberg, S., Engle, R.F., Granger, C.W.J., & Yoo, B.S. (1990). Seasonal integration and cointegration. *Journal of Econometrics*, 44(1), 215-238.

16- Kim, H., & Moosa, I. (2001), Seasonal behavior of monthly international tourist flows: specification and implications for forecasting models. *Tourism Economics,* 7 (4), 381–396.

17- Kim, J. H., & Moosa, I. (2005). Forecasting international tourist flows to Australia: a comparison between the direct and indirect methods. *Tourism Management,* 26 (1), 69–78.

18- Koc, E., & Altinay, G. (2007). An analysis of seasonality in monthly per person tourist spending in Turkish inbound tourism from a market segmentation perspective. *Tourism Management*, 28 (1), 227–237.

19- Kulendran, N., & King, M. L. (1997). Forecasting international quarterly tourist flows using error correction and time-series models. *International Journal of Forecasting*, 13 (3), 319-327.

20- Kulendran, N., & Wong, K. F. (2005). Modeling Seasonality in Tourism Forecasting. *Journal of Travel Research*, 44 (2),163-170.

21- Lim, C., & McAleer, M. (2000). A Seasonal analysis of Asian tourist arrivals to Australia. *Applied Economics*, 32 (4), 499-509.

22- Lim, C., & McAleer, M. (2002). Time series forecasts of international travel demand for Australia. *Tourism Management*, 23(4), 389-396.

23- Silvapulle, P. (2004). Testing for seasonal behavior of monthly stock returns: evidence from international markets. *Journal of Business and Economics,* 43(1 and 2), 93-106.

24- Tiffin, R., & Dawson, P. J. (2000). Structural breaks, cointegration and the farm-retail price spread for lamb. *Applied Economics,* 32 (10), 1281-1286.

12

Financial Comparison of Management Methods of Rice Blast Disease in Langerood County, Iran

Mohammad Reza Safari Motlagh [1*] and Farzaneh Fallahpoor Salkooyeh [2]

Abstract

The objective of the present study was to financially compare chemical and agricultural management methods of rice blast disease, as a prevalent disease in paddy fields of Langerood in Guilan Province, Iran. Thirty-three rice experts and 373 paddy farmers were selected by random sampling procedures. Data were collected by a research made questionnaire and were analyzed by SPSS. Benefit Cost Ratio (BCR) was utilized for analysis to recognize the most efficient disease management method. Given the costs in 2012 and the evaluation of economic variables of chemical disease management procedures in paddy fields per acre, it became evident that agricultural disease management methods can replace the chemical ones financially. Furthermore, the comparison of farmers' demographic, economic and agricultural characteristics by Chi-squared test among agronomic method and chemical method users showed no significant relationship between paddy farmers' gender, age and farming experience, but there was a significant relationship between cultivation area and management method at the 0.05 significant level. However, there was a significant correlation between disease management methods and the size of the area under cultivation. Furthermore, there were significant differences between experts and farmers' professional viewpoints in their preferences of disease management methods. While experts insist on an integrated application of chemical and agricultural procedures, most farmers apply chemical disease management methods.

Keywords:
Rice blast, Financial comparison, Agriculture management, Chemical management

[1] *Department of Plant Protection, Faculty of Agriculture, Rasht Branch, Islamic Azad University, Rasht, Iran*
[2] *Graduate student of Agricultural Management, Rasht Branch, Islamic Azad University, Rasht, Iran*
* *Corresponding author's email: ssafarimotlagh@yahoo.com*

INTRODUCTION

Rice blast disease is one of the most prevalent fungal diseases in rice fields. This fungus has been reported in more than 85 countries which grow rice in a commercial scale. The disease can attack the rice fields in transplantation, tillering and clustering phases. Various studies have focused on the issue. Nevertheless disease is still considered as the most unpredictable and detrimental pestilence of rice fields (Padasht-Dehkayee, 1999).

In Iran, the disease was first recognized in 1966 in the paddy fields. Due to this disease, irregular spots appear on the leaves of rice seedlings and shrubs. Then, rice leaves start shrinking at tillering phase of cultivation resulting in the annihilation of rice shrubs. Cluster contamination causes consequential damages to the rice fields which results in sudden deterioration of the number and the quality of clusters and the weight of cluster kernels (Okhovat *et al.,* 1990).

The economic significance of rice can be seen in the fact that it is a main regional source of food. Approximately 80 percent of worldwide rice is produced in small domestic fields, whose objective is to maintain economic subsistence of rural families. Eighty percent of paddy farmers live in poor economic conditions. In Iran, agriculture plays a key role in utilization of social justice, economic variables, food supply, job opportunities and environmental preservations and therefore, should be paid attention to efficiently (Najafi Khotbeh Sarraii, 2011).

Today, sustainable agriculture is emphasized in most countries. Such agriculture emphasizes on farmers' benefits and facilitates preservative procedures for soil, water and other natural resources. Soil fertility, promotion of a healthy society and utilization of opportunities for future generations are some of the components of sustainable agriculture (Allahyari, 2012). In order to facilitate the realization of sustainable agriculture, three key factors should be taken into consideration: environment conservation, proper social conditions and economic profitability, should be taken into consideration. Economic success should be achieved without harming soil, water resources, natural landscapes and native flora. In most cases, water and soil resources are over-exploited due to the lack of professional knowledge and discernment. Therefore, intensive farming is considered one of the major threats to soil stability (Karimi, 2002).

Today, food supply has become one of the serious challenges for governments and according to a statistical report of the United Nations, millions of people annually die of hunger and malnutrition since only one out of three is financially capable of providing food for him/her qualitatively and quantitatively. One reason behind such devastating observation can be the fact that the level of food production in countries with considerable population growth decreases significantly. World population is projected to reach eight billion by 2025. Today, hunger has become a social challenge whose implications can be politically observed in countries suffering from starvation. Such a threat negatively affects the economics of countries whose agricultural methods have remained traditional for years and therefore, cannot cultivate their lands efficiently. Such technical backwardness in agriculture will make these countries incapable of providing sufficient food for their growing populations (Pajuhandeh, 2013).

One of the most effective disease management procedures of rice blast is agricultural management and rotation. Water preservation is another procedure in paddy fields. The possibility of the emergence of rice blast is considerably high in fields with low water levels. Finally, it should be noted that the utilization of high-quality healthy seeds is of utmost importance so that soil contamination can be prevented (Tebeest *et al.,* 2007).

A number of researchers have proposed a number of chemical and biological management procedures (Shaymala and Sivakumaar, 2012). Chemical management procedure is another method for the management of rice blast fungus. For instance, Tricyclazole fungicide is proposed by a number of agriculture researchers (Peterson, 1990).

The side effects of pesticides compounds were economically evaluated in cotton fields of Pakistan's Punjab Province. The results revealed the detrimental effects of pesticides compounds. Therefore, preventive procedures should be implemented (Azeem Khan *et al.,* 2002).

In order to evaluate the economics of various methods of implementation of methyl bromide in disinfection of soil resources of greenhouse strawberries, partial budgeting of Benefit Cost Ratio (BCR) of various methods was analyzed. The results showed that the utilization of methyl bromide was economically justifiable. Furthermore, it should be noted that agricultural efficiencies of chemical compounds were not similar in all regions and therefore, in some regions, the utilization of Chloropicrin was more efficient than methyl bromide and acquired more than $1670 efficiency per acre (Sydorovych *et al*., 2006).

The application of fungicides on diseases of root vegetables in preservative fields was economically evaluated in Egypt. The results indicated that new fungicide compounds mitigated the level of rhizomatous decay of greenhouse cucumber, tomato and pepper as compared to conventional pesticides and facilitated the level of BCR. Such boost in BCR level facilitated agricultural revenues of cucumber, tomato and pepper fields up to 68.8, 98.5 and 45.8 %, respectively and reduced expenses up to 42.5, 54.5 and 35.3, respectively (El-Mougy *et al.*, 2013).

In peanut fields of Nigeria, economic evaluation of management procedures against weed growth was administered during the cultivation seasons of 2010 and 2011 to estimate their efficiency levels. It was found that the best agricultural treatment of fields was the procedure in which rice straws were implanted at the depth of 10 cm. After cultivating a bundle of weeded rice clusters in six weeks, this method was considered to be economically justifiable. Furthermore, the level of utilization of pesticide was reduced considerably (Etejere *et al*., 2013).

In another study, various irrigation and cultivation methods were economically evaluated in potato fields (Soleimanipoor *et al*., 2010). Three methods of sprinkling irrigation, tape irrigation and furrow irrigation were evaluated technically and economically. At the beginning and ending of the growth periods, the level of water consumption was estimated. The ultimate results of the study depicted that in the sprinkling method, BCR index acquired higher statistical rate than other two methods proving the superiority of sprinkling irrigation in the studied region so that BCR was estimated to be 50-60% in this method.

Economic and technical variables of chemical and biological management of Bemisia tabaci were analyzed in cotton fields of Fars Province. According to the findings of this study, production sectors, the level of consumed pesticides and their spraying frequency per acre significantly influenced the overall efficiency level of field cultivation of the region. Furthermore, considering annual cotton price and consumption of four liters of pesticides per acre, the application of biological management instead of chemical procedures was proposed due to its technical and economic justifications (Zare *et al*., 2007).

Hasani Moghaddam *et al.* (2007) analyzed the efficiency of management variables in disease management of rice diseases in rice fields of Mazandaran Province. Data were collected by interviews and questionnaire. The results revealed that efficient cultivation of agriculture fields of both >2-acre and <2-acre farms were affected by less extensive application of fungicides and more extensive application of pesticides. After surveying social, economic, physical and administrative variables in pesticide waste correlatively, it became apparent that there were correlations between the number of cultivated fields, the size of each cultivated fields, the amount of applied chemical fertilizers and farmer's behavioral characteristics in pesticide wastage. Finally, this research indicated that economic, social and technical variables affected mean management costs of rice diseases. Furthermore, in order to economically measure the level of consumption of pesticides as compared to the BCR rate of the fields, a systematic spraying mechanism with variable parameters was applied in corn fields. The results indicated considerable reduction in the level of consumed pesticides in average, intense and low grass fields. BCR rate was reduced to less than 1 indicating that the mechanism was economically justifiable. Furthermore, this rate depicted that BCR rate would be diminished statistically with the increase in weed coverage (Safi-Yari *et al*., 2011).

In another research, the most economical and efficient method against canker Rhizoctonia disease of potato was evaluated (Hasani Moghaddam *et al.,* 2012). Technical analyses of seedling, flowering, post-flowering and harvest phases were administered. After recognizing BCR rate as an economic index, it was depicted that all the implemented disease management methods could be economically justified.

In another study, in another study, disease management methods against Chilo suppressalis of rice were economically evaluated. The results indicated no correlation between farmers' personal, agricultural, and economic characteristics and the biological-chemical management method. However, there was significant difference between rice plantation experience, agricultural experience, and age (Javadpoor, 2013).

The main objectives of the present study were to analyze the most efficient management method of rice blast disease, to survey farmers' opinions on present management methods of rice blast, to study farmers and agriculture experts' professional opinions, and to analyze farmers' personal and economic characteristics in relation with application of biological and chemical management methods.

MATERIALS AND METHODS

The present study was administered in Langerood County during the growing seasons of 2013 and 2014. Data were collected by a questionnaire and field survey, and were analyzed by SPSS Software Package.

The questionnaire was composed of nine sections: respondents' demographic information, farming system characteristics, paddy farmers' economic attributes, knowledge of rice blast management, paddy farmers' social collaboration, farmers' educational-extension activities, paddy farmers' technical knowledge, and awareness of environmental impacts of rice blast management.

Statistical population was comprised of 33 experts of the Organization of Agriculture Jihad and 13,000 active farmers of this county. Research sample was selected through administration of random sampling methods and Cochran's formula. So, 373 individuals were selected as research sample, and 367 questionnaires were considered as verified sources of research data. Research questions covered issues such as disease management methods of rice blast, production rate per acre, costs of biological and chemical management, the amount of applied pesticide, and harvest expenses, and proposed a number of research hypotheses.

Data analysis was administered by SPSS software application, and variance analysis and descriptive and inferential statistics were carried out by mean rates. At the first step of data analysis, descriptive statistics were used, in which the features of the sample were studied including the tables of frequency distribution, percentage, cumulative frequency, central tendency measures like mean, median and mode, and standard deviation as dispersion measure. Also, the means of two groups – users of agronomic methods and the users of chemical methods – were compared by Chi-square test. In addition, data collected by questionnaire and from plant protection databank of Guilan Province were used for economic assessment of agronomic management method as compared to chemical method. Then, the economically best management method was determined by BCR.

RESULTS AND DISCUSSION

Analysis of demographic data indicated that participants' average age was 55.5 years (SD = 9.98). Respondents were mostly male (97.2%). With respect to participants' academic degree, high school and diploma had the highest frequency of 30.5%, and academic degrees had the lowest frequency of 14.2%. Analyzing participants' dialects, Gilaki dialect had the highest frequency level of 89.3%. Among participants, 60.75% of families had three members. In terms of job experience, 11-20 years of experience had the highest frequency level (34.75%). In the research, mean job experience in rice fields was 27.32 years (SD = 13.33). Study of participants' agricultural characteristics indicated that the highest and lowest frequency levels of area of rice fields were less than one acre (with 68.39%) and more than three acres (with 1.36%),

Table 1: Characteristics of farmers farming systems in Langerood (n= 367)

Variable	Group	Frequency	Percent	Cumulative percent
Total area of land	≤ 1 acre	251	68.39	68.39
	1.1- 3 ha	111	30.25	98.64
	≥ 3 acres	5	736	100
Land leveling	Land logged	267	66.8	66.8
	No leveling Caspian	133	33.3	100
Varieties of rice	Khazar	1	0.30	0.3
	Hashemi	333	83.3	83.6
	Ali Kazemi	29	7.3	90.9
	Jamshid joo	37	9.1	100

respectively. Furthermore, most participants (more than 66.8%) believed that their rice fields had been leveled. Being planted in 83.3% of farms, cultivar Hashemi was ranked the first among all cultivars (Table 1).

Analysis of economic characteristics showed that mean rice yield was less than one tons per acre with the frequency of 63.8% and more than three tons per acre with the frequency of 2.91% . The highest frequency level of participants' income was recognized to be 1,000,001-50,000,000 IRR[1] (with 44.97%). The lowest frequency level of participants' income was recognized to be less than 10,000,000 IRR (with 1.59%). 26.5% of participants utilized bank credits to support their agricultural activities. Furthermore, most participants predicted that they would sell their product between 60,001 to 70,000 IRR in 2014 (Table 2).

Also, the study of disease management knowledge of rice blast indicated that more than 80.3% of participants utilized chemical methods for rice blast management (Figure 1).

Participants' personal, agricultural and economic characteristics (divided into two groups, one utilizing chemical and the other agricultural methods) were comparatively analyzed by Chi-square test. The results showed no correlation between gender, age, job experience, and the method chosen by farmers. However, there was significant correlation between disease management procedures and the size of the area

Table 2: Farmers' economic characteristics

Variable	Group	Frequency	Percent	Cumulative percent
Average yield per hectare (t)	≤ 1 tons	228	60.32	60.32
	1-1.3 tons	139	36.77	97.09
	≥ 3 tons	11	2.91	100
Annual income of rice growing (IRR)	≤ 10000000 IRR	6	1.59	1.59
	10000001-50000000 IRR	167	44.18	45.77
	50000001-100000000 IRR	170	44.97	90.74
	≥ 10000000 IRR	35	9.26	100
The average price of what (IRR)	≤ 60000 IRR	166	41.5	41.5
	60001 -70000 IRR	225	56.25	97.75
	≥ 70000 IRR	9	2.25	100
The need for banking facilities	Needed	106	26.45	26.5
	Not required	294	73.5	100

[1] *$1 = 31,979 IRR*

Table 3: Comparison of the characteristics of farming and agriculture groups using chi-square test

Variable	Group	Number	Chi-square	p-value
Sex	Crop	309	0.981	0.322
	Chemical	58		
Curry rice history	Crop	309	4.069	0.131
	Chemical	58		
Cultivation	Crop	309	10.290**	0.006
	Chemical	58		
Group	Expert	33	144.464**	0.000
	Farmer	367		

**p<0.01

under cultivation. Furthermore, there were significant differences between experts and farmers' professional viewpoints about disease management procedures (Table 3). The results indicated that high costs of chemical management practices in cultivation season make agricultural management methods more economical (Tables 4 and 5). It was found that more than 62% of participants had social influence in their communities as farmers, and when it came to social collaboration, their level of collaboration was more than 76% (n = 280). In terms of educational activities and courses, more than 64% of participants acquired medium level willingness to participate in such activities. Most participants (56%) had high technical efficiency rate. 87% of participants (n = 318) showed high level of environmental knowledge (Table 6).

The comparison of farmers' personal, economic, and agricultural characteristics and their applications of chemical and/or agricultural management practices indicated no correlation between pest management practice and some variables including family size, farmers' performance level and annual income rates. However, there was significant correlation between disease man-

Table 4: Calculation of the costs (in IRR) of chemical and agricultural controls

Method	Plowing	Nitrogen fertilizers	Potash fertilizers	Herbicides	Chemical pesticides	Total
Crop Control	8261000	2669100	1600420	1605390	0	14135910
Chemical Control	8261000	2669100	1600420	1605390	2779000	16914910

Table 5: Economical comparison of chemical and agricultural controls based on the ratio of benefits to costs

Factor	Price	The ratio of benefits to costs
The annual income from rice farming (IRR)	63515600	-
The cost of field control (IRR)	14135910	4.49
The cost of chemical control (IRR)	16914910	3.76

Table 6: Social penetrability of farmers' social accountability, participation, extension, technical knowledge and familiarity with environmental risks of pesticide application

Level	Percent				
	Social penetrability	Community involvement	Extension	Technical knowledge	Understanding environmental risks
Weak	7.40	3.00	12.5	0.00	0.00
Average	22.9	13.1	13.4	6.30	4.40
Good	38.3	76.3	63.8	37.3	9.00
excellent	100	7.60	15.3	56.4	86.6

agement procedures and farm area. In an economic evaluation of disease management methods of Chilo suppressalis in paddy fields, Javadpoor (2013) indicated no correlation between family size, farmers' performance level and annual income rates and their utilization of either chemical or agricultural management methods. In the present study, there was significant correlation between the area of cultivation and the type of disease management method and therefore, it was in direct discordance with Javadpoor (2013)'s study.

The estimation of latent costs of chemical management methods is complicated because they inflict ineluctable damage to ecosystem and mankind in long-term periods. If true costs of the application of chemical management procedures were tallied, one could realize their severe impact on ecosystem and mankind. Yusefnia Pasha *et al.* (2012) studied various management methods of weeds in rice fields. They concluded that double weeding mechanism was economically more justifiable than chemical management methods, for they cause lower ecological contamination and financial expenditures. In bestowing priority to an agricultural management method, that study is in direct agreement with the present study.

In a study on detrimental effects of pesticides on cotton fields, Azeem Khan *et al.* (2002) concluded that the application of agricultural preventives was preferable to the application of chemical pesticides. By such an implementation, expenditures can be cut considerably and farmers can spend more on sanitary maintenance, agricultural recycling and research activities. Etejere *et al.* (2013) economically analyzed weeds management methods in peanut fields. They also confirmed the priority of agricultural method over chemical procedures, in that they proposed manual weeding procedures over chemical and mechanical manipulations. Zare *et al.* (2007) economically analyzed chemical and biological management methods of Bemisia tabaci and believed that if proper biological elements were utilized against this fungus, its population and detrimental impacts could be reduced considerably. Moreover, the substitution of biological measures for chemical pesticides would preserve ecosystem from nefarious contamination. It was worth mentioning that if biological management was administered efficiently, the quality level of agricultural products would be facilitated. With such qualitative outcome, farmers can justify higher expenditure rates of biological management procedures. El-Mougy *et al.* (2013) utilized new fungicides against diseases of root vegetables and concluded that the application of these fungicides facilitated biological control and reduced financial characteristics of agricultural fields.

As mentioned earlier, participants had a good level of social collaboration. Consultation about the problems of the village in village meetings was considered an instance of farmers' active role in the social interactions of their community. Most participants acquire medium level educational activities. Therefore, educational courses should facilitate farmers' level of actuation in learning new management methods against rice blasts. In the time of the research, these courses acquired no significant priority for participants.

Participants' level of technical knowledge in administration of rice blast was high. Technical knowledge of participants in application of preventive measures of weeds for rice blast control, plow protocols in fall and thorough stemming procedures had low efficiency. Farmers' level of ecological awareness had high efficiency rate. According to the research findings, it was indicated that farmers already knew that utilization of chemical management methods would increase prevalence rate of incurable diseases and annihilate possible natural antagonists against rice blasts in ecology. Furthermore, they were aware of this general premise that through facilitation of their educational activities, they should welcome utilization of biological methods instead of chemical procedures; especially in those regions that this substitution acquires no extra expenditure and maintains ecological priorities. Since farmers acquired high level technical efficiency, a local understanding of agricultural mechanisms should be realized so that indigenous experience could be recognized in learning processes of newcomers and inexperienced farmers of rice fields. Furthermore, in order to control rice blast disease, a true understanding of fungal growth and their activation seasons should be recognized. Due to low level of initiation of preventive measures against rice blast, it was incumbent to facilitate farmers' awareness towards such measures through holding proper educational courses.

According to the economical analyses, the development of biological methods instead of chemical procedures should be constantly promoted and monitored by officials of the Organization of Agriculture Jihad.

CONCLUSION

The present study was administered in Langerood County. It is recommended to study rice producing provinces in Iran more comprehensively so that the devastating issue of rice blast can be tackled multifariously. Furthermore, the present study should be compared with those studies which have already been administered in Mazandaran Province. It was recommended that ecological consequences of the application of chemical pesticides against rice blasts in Guilan Province should be studied and that the challenges of the development of the biological management of rice blast should be analyzed.

Since the present study was administered in a specific time frame, it was not possible to use Net Present Value (NPV) rate, Return on Investment rate and Profitability Index for the research samples and population. Therefore, it is suggested that the present study is administered during four seasons of a year so that the level of devastating consequences of utilization of chemical pesticides can be measured accurately. Furthermore, financial economy of both chemical and biological management methods should be analyzed and compared. Considering the importance of the application of biological management procedure, it is suggested that this procedure is thoroughly studied and compared with agricultural and chemical management methods.

ACKNOWLEDGMENTS

This study was supported by Islamic Azad University, Rasht Branch, Iran.

REFERENCES

1- Allahyari, M. P. (2012). *Extension of sustainable agriculture.* Rasht, Islamic Azad University Press, pp. 1 -48.

2- Azeem Khan, M., Iqbal, M., Ahmad, I., & Soomro, M. H. (2002). Economic evaluation of pesticide use externalizations in the cotton zones of Punjab, Pakistan. *The Pakistan Development Review*, 41(2), 683-698.

3- El- Mougy, N. S., Abdel- Kader, M. M., & Lashin, S. M. (2013). Preliminary economic evaluation of fungicide alternatives approaches against root diseases incidence of some vegetables grown under protected cultivation system in Egypt. *International Journal of Engineering and Innovative Technology,* 3(6), 95- 103.

4- Etejere, E. O., Olayinka, B. U., & Wuraola, A. J. (2013). Comparative economic efficacy of different weed control methods in Groundnut. *European Journal of Biological Sciences*, 7(1), 10- 18.

5- Hassani Moghadam, M., Arjmandyan, A., Sohaily, B., & Azimi, H. (2012). Damage assessment and evaluation of different methods of combating Rhizoctonia disease of potatoes. *Journal of Plant Protection Research Institute of Technology Research News*, 2 (12), 14.

6- Hassani Moghadam, M., Esco, T., Asadpoor, H., & Mohajer, A. R. (2007). *Evaluation of the effectiveness of management agents on control of pests of Mazandaran Province.* Paper presented at 6th Conference of Agricultural Economics October, 30-31. Mashhad, Iran.

7- Javadpoor, S. (2013). *Economic comparison with the control and prevention strategies for rice stem borer in the city of Rasht.* Unpublished thesis agricultural management. Islamic Azad University, Rasht Branch. 99 pp.

8- Karimi, S. (2002). The principles of sustainable agricultural production methods. *Jihad Journal,* 22 (254), 40- 46.

9- Najafi Khotbeh Sarraii, N. (2011). *Evaluation of the role of rice in the economic situation Guilan farmer's observers.* Unpublished thesis agricultural economics, Guilan University, pp. 17- 31.

10- Okhovat, M., Sharify Tehrani, A., & Eshtiaghi, H. (1990). Studying the effects of blast disease on the amount of rice production. *Journal of Agricultural Sciences,* 21(1 & 2), 17-25.

11- Padasht-Dehkaiey, F. (1999). Effects of nitrogen and silica on rice blast disease. *Journal of Agricultural Sciences,* 30(4), 736-742.

12- Pajuhandeh, M. (2013). *Economic evaluation of harvesting rice methods in Guilan Province.* Unpublished thesis agricultural management. Islamic Azad University, Rasht Branch, 53pp.

13- Peterson, L. G. (1990). *Tricyclazole for control of Pyricularia oryzae on rice: The relationship of the mode of action and disease occurrence and development.* Pest Management in Rice. Springer Netherlands, pp.122-130.

14- Safi- Yari, H., Shirzadi- Far, A., & Shojaii Arani, J. (2011). *Economic evaluation system with variable rate applied herbicides used in corn fields.* Paper presented at first National Conference on Agricultural Mechanization and New Technologies February, 17-19, Ahvaz, Iran.

15- Shaymala, I., & Sivakumaar, P. K. (2012). Research article Pseudomonas fluorescens mediate disease resistance in rice against rice blast caused by Pyricularia oryzae. *International Journal of Recent Scientific Research,* 3(10), 853-857.

16- Soleimanipoor, A., Bagheri, S., & Safe, H. R. (2010). *Economic evaluation and comparison of different methods of irrigation and planting the potato varieties.* February, 3-4, 2011. Paper presented at 7th Conference of Agricultural Economics. College of Agriculture and Natural Resources, Tehran University, Iran.

17- Sydorovych, O., Safley, CH. D., Ferguson, L. M., Poling, E. B., Fernandez, G. E., Brannen, Ph. M., Monks, D. M., & Louws, F. J. (2006). Economic evaluation of methyl bromide alternatives for production of strawberries in the southeastern United States. *American Society for Horticultural Science,* 16(1), 118- 128.

18- Tebeest, D. O., Guerber, C., & Ditmore, M. (2007). Rice blast. *Journal of Plant Disease*, 10(2), 109-113.

19- Yusefnia Pasha, H., Tabatabai Kolwer, R., & Hashemi, J. (2012). *Comparison of different methods of weed control in rice production costs.* September, 5-6, 2012. Paper presented at 7th National Conference on Agricultural Machinery Engineering and Mechanization. Shiraz University, Iran.

20- Zare, A., Amin, B., & Boustany Far, M. (2007). Comparison of technical and economic control of chemical and biological cotton whitefly on cotton fields in Fars province. *Journal of Agriculture,* 30(3), 15-28.

13

Farming Techniques, Environmental Challenges and Technical Efficiency of Sweet Potato Production in Abia State, Nigeria

Nnamdi Ehirim [1], Stephen Onyeagocha [1], Glory Ben-Chendo[1], Ubon Essien [1], Emeka Osuji [2] and Melissa Okwara [1]

Abstract

Viable sweet potato production is hard to achieve with indiscriminate use of farm inputs, resulting to wastage and environmental damages, as consequences are common problems of arable crop farming in Abia State. Issues arising from farm input use and their corresponding implications for environment called for a study on suitable farming practices and farm-specific technical efficiency for optimal resource use in sweet potato production in Abia State. Data were obtained from 156 sweet potatoe farmers through a multistage sampling technique using a structured questionaire. Data were analyzed using descriptive statistics such as mean, relative frequency distribution, and the stochastic production frontier. Results showed that using inorganic fertilizer (65.378%) under rainfed production system with a frequent bush burning (66.67%) and low liming (5.8%) were very common practices with leaching, fragile soil, erosion, flooding, and soil acidity as consequences. Mixed cropping (63.46%) with improved varieties like TIS 8164 (71.2%) and 0087 (64.1%) were copping measures to some environmental challenges. Maximum Likelihood Estimates (MLE) showed a decreasing return to a scale of 0.236. The implication is that an increase in farm size and fertilizer application can significantly lead to a less than 0.06538 and 0.08142 proportionate increase in output of sweet potato respectively, or reduces it by less than 0.00413, with interest on borrowed capital. The gamma (0.0403) was less than unity and was significant at $p < 0.05$, implying that about 4.30% discripancies in observed and frontier output was due to technical inefficiencies of sweet potato farmers. The wide disparity in farmers' technical efficiencies ranged from 0.298% to 99.4%, and a mean of 47.1% suggested a need to bridge the gap. Hence, a reduction in household size, farming experience, and sourcing of planting materials from NRCRI or IITA is believed to increase farmers' technical inefficiency, which can be reduced with age and formal educational level of sweet Potato farmers in the area. All in all, the results suggest that reducing bush burning but increasing liming as well as including organic soil ammendments and irrigation practicies, when combined with the use of young and educated farmers, can reduce environmental damages and also increase farmers' technical efficiency when it comes to sweet potato production in the area.

Keywords:
Environment, Technology, Efficiency, Sweet potato production, Stochastic Frontier, Abia State

[1] *Department of Agricultural Economics, Federal University of Technology, Owerri, Imo State, Nigeria*
[2] *Department of Agricultural Economics, Michael Okpara University of Agriculture, Umudike, Abia State, Nigeria*
* *Corresponding author's email: essienua@gmail.com*

INTRODUCTION

Sweet potato (*Ipomoea batatas*) belongs to the family convolvulaceae (Woolfe, 1992). It is an important food crop of both tropical and sub-tropical regions with a cultivation spread across 100 countries (Nwaru, 2003). The crop has high photosynthetic efficiency (Kapinga et al., 1997), a high yielding capacity per unit area and ability to sustain human livelihood during periods of food shortage (Ndukwu, 2010). Farmers engage in massive production of this crop because of its short gestation period and its ability to suppress weeds once it has been fully established, thus reducing the overhead cost of production more than that incurred by cassava and yam (Woolfe, 1992).

Inadequate use of disease resistant varieties and high yielding sweet potato clones from National Root Crops Research Institute (NRCRI) and International Institute of Tropical Agriculture (IITA) could be a reason for low output in sweet potato production. The wide use of unimproved varieties such as *Cylas puncticollis* by farmers due to shortage of improved planting materials at the beginning of every cropping season could be a reason for the low output of sweet potatoes. Studies have shown that the Southeast states of Nigeria are yet to attain their full potential food security status using sweet potatoes, because farmers have not taken advantages of improved crop varieties to maximize output and profit (Njoku et al., 2009; Onwueme, 1978 & Tewe et al., 2001). Sweet potato production is characterized by crop failures due to the frequent use of low disease resistant varieties. There is 80% loss in sweet patotoe's output in Nigeria due to *Cylas puncticollis* (Tewe et al., 2001). This implies that the use of bad techniques in food production does not only place the farmers at a disadvantage of crop failure alone and a low output status but also a sever damage to the environment. A fundamental problem confronting mankind is how to meet an increasing demand for food and environmental services without compromising each other. An understanding of the web of relationships and linkages between the use of environment and agricultural productivity is, therefore, critical in meeting the future challenges.

The environment comprises natural resources in its physical, chemical, and biological forms, which humans and other living species depend on for their life support and social welfare (Hussein, 2004) while an increased standard of living and satisfaction of humans' wants are the attributes of agricultural productivity. The two natural variables challenge each other in trying to achieve their common goals. A change in climate affects farm productivity in particular and other socio-economic activities in the country. Aagriculture in Nigeria is mostly rain-fed and some common agricultural practices challenge the climate as more ozone layer is depleted. This in turn poses some major challenges to agricultural productivity. The impact of climate on farming activities could, however, be measured in terms of its effect on crop growth, low availability of soil water, increased soil erosion, incident of pest and diseases, sea level rises, and decrease in soil fertility (Adejuwon, 2004). There is always a serious challenge for farmers who concurrently seek to satisfy the food demand through increased output and retain the quality of natural stock and a balanced ecosystem (Ehirim, et al., 2013; Oldeman et al., 1991). Farmers lack the requisite piece of technology to manage their production and control the environment in such a way that food production will be sustainable (Ehirim, et al., 2013).

Farmers cropping pattern is associated with erosion and leaching, which reduces crop yield. For instance, the bush burning and felling of trees in the Southeast states seem to be the common and suitable choice for farmers but pose a major environmental challenge. Ehirim et al (2013) also noted that the common choice of livelihood strategies in Nigeria is patterned by the poor socio-economic disposition of farmers. Lack of awarness and skepticism of the workability of some improved soil management techniques as well as low income are reasons for most cropping patterns adopted by farmers in the Southeast states of Nigeria (Ogbonna et al., 2007). Low income makes the use of sustainable soil management technologies impossible in sweet patoto production in Nigeria. Again, the general lack of interest in sweet potato production was due to poor motivation from low returns per hectare. Farmers poor socio-economic dispositions are having overwhelming influence on their input mix decisions. Input-output process in arable crop production is important in four major areas like; the distribution of farmers income, allocation of farm input resources, the relation between stocks and flows, as well as the measurement of efficiency or productivity (Olayide & Heady, 1982). Therefore, an increasing rate of investment in agricultural production is corresponding to increasing rate of returns with a high production efficiency (Assa et al., 2012). Yet, farmers input mix decisions on sweet potatoe farming are not only negatively affecting their input-output processes and returns per hectare but also the environment, as productive efficiency is compromised in the long run.

Efficiency is described as the extent to which time, effort, or cost is well managed for an intended task or purpose. Aigner (1977) defined "efficiency" in three related terms: First, he defined "technical efficiency" as the measure of a firm's success in producing maximum output from a given set of input. This implies that undisputed gain can be obtained from adequate input mix process by simply organizing management in a better pattern. Second, "price or allocative efficiency," which measures a firm's success in choosing an optimal set of input based on their relative prices. The gains are possible by varying the input ratios on certain assumptions about their future price structure. Third, is the "overall or economic efficiency," which is simply the product of both technical and price efficiencies? Adequate input combination blended in skills, knowledge of choice of techniques and their interactions with the environment, as well as farming experience among other socio-economic factors influencing a famer's life can set up effectiveness with which a given set of inputs can produce the optimum quantity that gradually brings farmers to the best practiced frontier without denaturing the environment (Ehirim, 2013). The ability to employ the best practices for increased output measures the farmers technical efficiency, but when it ensures a steady natural stock, the production process is potentially viable (Ehirim, 2013). Most empirical studies have demonstrated that farmers are grossly inefficient in the use of farming inputs to increase output and returns (Meeusen & Van-den Broeck 1977; Ndukwu, 2010 & Oni et al., 2009).

This study is centered on achieving an input mix process that can set up an optimal sweet potato production without denaturing the environment. It isolated the factors sweet potato production, as well as the effect of environmental consequencies of the choice of farming techniques on technical efficiency of sweet patoto production in Abia State. The study is set to place farmers on the right footing to reduction in waste of farm inputs use but attain a maximum output and reduction in food security in Imo State. Hence, a stable foundatiion is laid for optimal production, increase in returns and profit in sweet patoto production in Abia State.

In this study, the researchers consider only the stochastic frontier models based on the estimate of the frontier production function. In Greene's (1997) words, "the frontier production function is an extension of the familiar regression model based on the microeconomic premise that a production function represents some sort of ideal, the maximum output attainable given a set of inputs" (Greene, 1997). In practice, the frontier production function is a regression where the estimation of the production function is implemented with the recognition of the theoretical constraint that all observations lie below it, and it is generally a means to another end, that is, the analysis of efficiency. A measure of efficiency emerges naturally after the computation of the frontier production function, since it corresponds to the distance between an observation and the empirical estimate of the theoretical ideal. The estimated models provide a means of comparing individual agents, either with the ideal production frontier or with each other, and also provide point estimates of the effects of environmental variables on efficiency (Francesco, 2009). This study therefore seeks to examine the contributions of suitable farming practices and farm specific technical efficiency to optimal resource use in sweet potato production in Abia State.

MATERIALS AND METHODS

The study was carried out in Abia State which is one of the 36 states in Nigeria. The State lies between longitude 040 45' and 060 07' North, and latitude 070 00' and 080 10' East. It is situated in the Southeast geo-political zone of Nigeria and is bounded by Imo State on the West, Ebonyi and Enugu States on the North, Cross Rivers and Akwa Ibom States on the East, and Rivers State on the South. The State has a total land area of 6320 square kilometres and a population size of 4,222, 476 persons (National Population Commission, 2007). The State is divided into three agricultural zones, namely Umuahia, Ohafia, and Aba Zones.

A multi-stage sampling procedure was employed in this study. In the first stage, of the three agricultural zones of the state, namelyAbia North, Abia South, and Abia Central, a purposive selection of three local government areas, each from an agricultural zone of the state, was made. The areas selected were predominantly sweet patoto producing areas that had also benefited from at least one of the improved sweet potato varieties from National Root Crop Research Institute Umudike. Again, the selection was made across the three agricultural zones, ensuring that all the areas in the state would be represented. The areas selected were: Ikwuano local government area in Abia central, Osisoma local government area in Abia South, and Bende local government area in Abia North agricultural zones. The second stage involved a random selection of two communities from each of the selected local goverment areas, making it six communities for the study. The final stage was a random selection of 35% of the total sweet potato farmers from the list of farmers with each of the community's model farmer. The selection gave 58 farmers from Bende local government area, 76 farmers from Ikwuano local government area, and 76 farmers from Osisoma local government area. This brought the final sample of the farmers to be interviewed to 210 using a well-structured questionnaire. However, only 156 responses were found relevant and were subjected to data analyses in the study.

The questionaire was structured to elicit information on socio-economic features of the farmers, quantity, types, and prices of improved sweet potato adopted by the farmers, the quantity and unit prices of farming inputs and output of sweet patotoes, as well as perceived effect of some farming techniques in sweet patoto production on the environment. Data were analyzed using both descriptive statistics and econometric tools. The level of use of improved planting materials and some technologies of Sweet Patoto Production in the area was analyzed using descriptive statistics such as frequency and relative frequency.

The level of use and the corresponding percieved effect of some management practices (in soil ammendment practices, cropping systems, varieties planted, and sources of farming inputs) on the environment by the farmers were articulated and described using descriptive statistics.

Farmers level of technical efficiency and factors affecting technical efficiency were analyzed using the stochastic production frontier. A stochastic frontier production function is defined by:

$$Yi = f(X_i; \beta) \exp (V_i - U_i), i = 1,2 \ldots n \qquad (1)$$

where;

Y_i is output of the i-th farm

X_i is the vector of input quantities used by the i-th farm

β is a vector of unknown parameters to be estimated

The term V_i is a symmetric error, which accounts for random variations in output due to factors beyond the control of the farmer such as weather, disease outbreaks, measurements errors, and so on. The random error V_i is assumed to be independently and identically distributed as $N(0, \dot{\rho}^2)$. The term U_i is a non-negative random variables representing inefficiency in sweet potato production relative to the stochastic frontier.

The technical efficiency of an individual farmer is defined in terms of the ratio of the observed output to the corresponding frontier output, given the available technology.

$$TE = \frac{Y_i}{Y_i^*} = \frac{f(X_i;\beta)\exp(V_i - U_i)}{f(X_i;\beta)\exp(V_i)} = \exp(-U_i) \quad (2)$$

where Y_i is the observed output of sweet patotoes and Y_i* is the frontier output which the farmer is expected to attain given his input level. The parameters of the stochastic frontier production function are estimated using the maximum likelihood method. Empirically, the stochastic production frontier function is expressed as:

$$\text{In}Y_i = \beta_0 + \beta_1 \text{In}X_1 + \beta_2 \text{In}X_2 + \beta_3 \text{In}X_3 + \beta_4 \text{In}X_4 + \beta_5 \text{In}X_5 + V_i - U_i \quad (3)$$

where Y_i is output of sweet potatoes in kg.,

X_1 = Farm size in measured hectares

X_2 = Labour used in sweet patoto production measured in mandays,

X_3 = Sweet potatoes setts and or vines planted in kilo grammes

X_4 = Quantity of fertilizer used in kilogrammes

X_5 = Rate of interest on borrowed capital in percentage

It is expected that all the included explanatory variables, apart from rate of interest, will have a positive sign. Therefore, $\beta_0 >$; $\beta_1 > 0$; $\beta_2 > 0$; $\beta_3 > 0$; $\beta_4 > 0$ but $\beta_5 < 0$

V_i and U_i are as defined earlier. In addition, U_i is assumed, in this study, to follow a half normal distribution. The farm specific efficiency is given as 1 – TE values (Assa et al., 2012). Again, the determinants of technical inefficiency in sweet potato production follow the model formulated and estimated jointly with the stochastic frontier model in a single stage maximum likelihood estimation procedure using a specialized computer software called Frontier Version 4.1 (Coelli, 1996). This is expressed as:

$$TIE_i = b_o + b_1 Z_1 + b_2 Z_2 + b_3 Z_3 + b_4 Z_o + b_5 Z_5 + b_6 Z_6 + b_7 Z_7 + b_8 Z_8 + b_9 Z_9 \quad (4)$$

where TIE_i is the technical inefficiency of the i-th farmer

Z_1 = Farmers age in years

Z_2 = Household size farmers level of formal education in years

Z_3 = Farmer's farming experience in years

Z_4 = Level of Formal Education

Z_5 = Number of technical advice and packages to the farmer by the agents in a planting season

Z_6 = Source of planting material (Dummy; '1' if it is from NRCRI or IITA and '0' otherwise)

Z_7 = Social capital, number of associations and co-operatives the farmer belongs

Z_8 = Access to Credit (Dummy variable; '1' represents access to credit and '0' otherwise

Z_9 = Depreciation of fixed inputs item measured in naira.

We expect that b_1, b_2, b_3,...b_9 are negative.

RESULTS AND DISCUSSIONS

Technologies of Sweet Potatoe Production and Their Environmental Implications

Table 1 shows the result of different technologies used in sweet patoto production in Abia State. Though inorganic fertilizers such as NPK and urea have some environmental consequences, as they have some residual effects in underground water, they appear to be the most common technologies for soil ammendment practice in sweet potato production in the area, with 65.4% of the farmers applying them. This is because it is easy to apply and farmers-friendly with nutrients present in their simplest forms and adequate proportion and easily given out to the plants roots. This corroborate Akande's (2008) finding that inorganic fertilizer seems to be common, because it is easy to apply and the nutrient is readily available to the plants, however, the high cost of acquiring it and the its relative scarcity have always made it inaccessible to farmers during planting seasons. Few (12.8%) farmers used organic fertilizer. This type of soil ammendment, though cheap and affordable within local means, may not have its nutrients easily available to the plant root. It requires some time to mineralize and release its nutrients to plants, hence the technical knowledge of its preparation and application could be the reason for its low use in the area. The result further re-

Table 1
Distribution of the Respondents Based on the Technologies Involved in Sweet Patoto Production in the State.

Technologies	Management Practices	Frequency	Percentages
Soil ammendment	Use of drainages	0	0.0
	Irrigation	11	7.74
	Organic fertilizer	20	14.08
	Inorganic fertilizer (Urea and NPK)	102	71.83
	Liming	9	6.33
Cropping systems	Bush burning	104	44.82
	Shifting Ccultivation	2	0.86
	Mixed farming	99	42.67
	Continous cropping	25	10.77
	Organic farming	2	0.86
Varieties planted	TIS 8164	111	38.94
	TIS 0087	110	38.59
	TIS 2271	39	13.68
	Ogonegurunwa	11	3.85
	Dukuduku	14	4.91
Source of planting material	ADP/MA&NR	109	65.26
	Root Crop Research Institute	16	9.58
	Market	31	18.56
	Co-operative Society	11	6.58

vealed that no farmer is practicing any form of drainage systems in the area, hence there is a possibility of excessive flooding due to excessive rainfall that comes with climate change challangies in recent times. This finding is consistent with Ebuzoeme (2015) observation that poor drainage system, where there is excessive previous rainfall, could lead to overflooding and erosion with predominanat leaching of soil nutrients. Leaching becomes excessive with the low (5.8%) use of lime. Sweet patotoes do well in soil with a higher PH; accordingly, liming is very important if output is to be increased in the area.

Again, few (7.1%) famers practice irragation system in the area, thus all year round sweet patoto production may be difficult and food insecurity status may increase in the area. Instead of irrigation systems, rainfall predominantly provides water for agricultural activities in the area, and Igwe and Stahr (2004) have emphasised excessive rainfall as the major cause of soil degradation in Eastern Nigeria. Others include the fragile nature of the soil with farming activities and deforestation, but they expose soil to erosion. As a consequence of soil erosion by the agency of water, soil nutrients are depleted, leading to a decline in crop production (Age et al., 2012). Ebuzoeme (2015) noted that a rain-fed production system that is characterised by erosion and flooding is usually prone to excessive leaching, leaving the soil more acidic. This findings seem to have characterized the soil as a sort of acidic soil and unsustainable with the type of soil management practices common in the area. Ehirim et al. (2013) noted that the relative suitability of land for root crop production is dependent on the sustainable soil management practices farmers adopt to gradually return improvised farm lands to a suitable land for crop production. Proper soil management practices such as liming, and timely use of organic amendments such as animal manures and sewage sludge compost on lands can restore increased land productivity (Al-Kaisi, 2012; Hornick & Parr, 2010).

The result further showed that bush burning is a very common practice with more than 66.7% of sweet patoto farmers engaging in it.

This practice seems to be cost-effective especially in the short run and leaves some ash deposites in the soil, which becomes lime and may increase soil PH level. However, the practice may not be a sustainable soil management practice; the heat from the burning can quickly denatures structure and other soil physical properties, or release the soil nutrients to the atmostphere thus exposing it to erosion and excessive leaching. Bush burning leaves a permanent damage to the soil, as plant nutrients that are volatile are easily converted to gases, while the practice itself increases the level of green house gases in the area. It kills soil micro and macro organisms that help in organic mater decomposition and soil erosion The gases disolve in rain water to form acid rain, which later increases soil acididty. The soil in such areas may become predominantly acidic and infertile. Another form of practice that leaves the soil in the area infetile and may lead to crop failure in the long run is continous cropping with only 28.2% of the farmers engaging in it. Continous cropping is mediated by population growth that forces farmers to cultivate on smaller (fragmented) plots where the soil eventually becomes depleted, or expands into fragile hillsides and could lead low productivity (Scherr & Yadav, 1996). Mixed cropping is another croping system that is predominant in the area with 63.5% of the farmers engaging in it. This is a copping strategy for farmers, as it is used to expand revenue and mitigate crop failure due to pest and diseases. In addition, it intensifies the use of land. Organic farming is very low with only 1.3% of the farmers engaing in it.

The common varities planted by the farmers in the area include TIS 8164, TIS 0087, and TIS 2271. Other local varieties include Ogonegburunwa and dukuduku. More than 71.2% of the farmers palnted TIS 8164, while 64.1% palnted TIS 0087. Few (25.0%) farmers planted TIS 2271, while local vareties like *Ogonegburunwa* and *Dukuduku* were planted by only 7.05% and 9.0% of the farmers. A vast majority of (69.9%) sweet patoto vines and setts were obtained from ADP, while the market supplied only 19.9% of the vines. Few (7.1% and 10.3%) of them sourced from co-operative and root crop reasearch institutes respectively. The most currently introduced improved variety of sweet patoto (TIS 2271) has not been widly used, as only 25.0% of the farmers are planting it. The study suggests that extension agencies should persuade farmers to use these varieties because of its short production time and productivity per unit area of land.

Factors of sweet patoto production in Abia State

Table 2 shows the parameters and related statistical test results obtained from the stochastic frontier production function analysis using Maximum Likelihood Estimates (MLE). Apart from labour and planting materials, all other variables like farm size, fertilizer, and rate of capital bor-

Table 2
Maximum Likelihood Estimates for the Stochastic Frontier Production Function of Sweet Potatoes Production.

Production variables	Model parameters	Estimates	SE	t-value
Constant	β_0	4.129***	0.25220	16.37
InFarm size	β_1	0.06538**	0.03020	2.165
InLabour	β_2	0.09322	0.05526	1.687
InPlanting material	β_3	0.00038	0.00670	0.057
InFertilizer	β_4	0.08142**	0.04083	1.994
Rate of Interest on Loan	β_5	-0.00413***	0.00205	2.011
Returns to Scale		0.23627		
	Diagnostic statistics			
Loglikelihood Function	δ^2	-105.601	7.621	
Sigma squared	μ	0.01469***		
Lamda	γ	0.20807	9.218	
Gamma		0.04303***		

*p<0.1, **p<0.05, ***p<0.01

Table 3
Technical Efficiency Levels of Sweet Patoto Farmers in Abia State

Technical Efficiency Limits	Frequency	Percentages
Technical Efficiency ≤ 10.0	4	3.2
10.1 – 20.0	7	5.6
20.1 – 30.0	8	6.4
30.1 – 40.0	28	22.4
40.1 – 50.0	35	28.0
50.1 – 60.0	15	12.0
60.1 – 70.0	10	8.0
70.1 – 80.0	10	8.0
80.1 – 90.0	4	3.2
Technical Efficiency > 90	4	3.2
Total	125	
Mean efficiency	47.1	
Minimum efficiency	0.298	
Maximum efficiency	99.4	

rowing are significant at 0.05 alpha level ($p < 0.05$). Again, apart from the rate of capital borrowing, all other variables are positive and consistent with a piriori expectation in the area. Therefore, an increase in farm size and fertilizer application by 1.0 unit will give a less than 0.06538 and 0.08142 proportionate increase in output of sweet patotoes, respectively. Fertilizers play a supportive role in sweet patoto production; accordingly, their effective distribution is imperative. The study suggests the inclusion of organic fertilizers to reduce cost and environmental problems caused by the use of synthesized fertilizers in the area. Likewise, an increase in the rate of borrowing by 1.0 unit will give a less than 0.00413 decrease in output of sweet patotoes. Labour and planting materials such as sweet potatoe setts and vine, though not significant even at 10.0%, is positive; hence, it will continue to increase the level of sweet patoto production in the area. The model shows a decreasing returns to scale of 0.236 in sweet patoto production in the area. This implies that an increase in the use of aggregate farm inputs used in sweet potatoe production by 1 unit can give a less than 0.763 unit of sweet patoto output in the area. At aggregate level, farmers are advised to respond to production by reducing input use to an equal proprotion of output by 76.3% in the area.

The functional parameters of maximum likelihood estimates has a sigma square (σ^2) value of 0.01469, significant at $p < 0.05$ critical level. The variance parameters (lamda), which showed the ratio of standard error {u (∂_u)} to the standard error {v (∂_v)} is 0.208 or 20.8%. Furthermore, the gamma ratio estimated from the lamda value is 0.04303 and significant at $p < 0.05$ critical level. This value is less than 1.0 which is consistent with the postulation that a true gamma value should be less than 1.0 and significant (Assa et al., 2012). The ratio captures the level of discrepancies in technical efficiency by farmers and about 4.30% discrepancies in observed and frontier output are due to potato farmers technical inefficiencies. Gamma ratio according to Ogundari and Ojo (2006) is the relative magnitude of variance (δ^2) associated with inefficiency effect. Therefore, goodness of fit and correctness of the specified distributional assumptions of dominance of U on V can be acertained, provided the value is significantly different from zero.

Technical efficiency in sweet patoto production in Abia State

The mean technical efficiency distribution of the sweet patoto farmers as shown in Table 3 is 47.1%, implying that, on average, the technical efficiency of the farmers in the area is about 47.1%. This value ranges from a minimum efficiency level of 0.298% to a maximum technical effeicent level of 99.4% in the area. It could be deduced from this result that there is a wide disparity in farmers' technical efficiencies and

Table 4
Determinants of Technical Inefficiency in Sweet Patoto Production in Abia State

Inefficiency Effects	Parameters	Co-efficient	SE	t-value
Constant	b_0	-0.2305***	0.0245	-9.424
Age	b_1	9.7760**	4.4360	2.203
Household size	b_2	-1.8020***	0.0208	-8.676
Experience	b_3	-9.4710**	4.1350	-2.290
Education	b_4	2.6160***	0.1353	19.320
Technical Service/Advice	b_5	-0.0540	0.1544	0.348
Source of Planting Material	b_6	-0.9188**	0.3614	2.542
Social Capital	b_7	9.3901	0.2772	0.034
Access to credit	b_8	-0.1353	0.4059	-0.333
Depreciation	b_9	-3.4060	6.7890	-0.502

*p<0.1, **p<0.05, ***p<0.01

there is room for bridging their technical efficiencies. Many farmers are found below the overall mean technical efficiency. For instance around 22.4% are less than 40% but greater than 30% technically effcient. Again, about 15.2% of farmers are below 30.0% technical efficiency. However, only 28.0% of the sweet patoto farmers are between 40.1 to 50% technically efficient, whereas 34.4% of them are more than 50% technically efficient in sweet patoto production in the area. This shows that with efforts made by farmers toward efficient technology use in sweet patoto production, high technical efficiency can be achieved in the long run, and this can enhance output.

Determinants of technical inefficiency sweet patoto production

The technical inefficiency model as shown in Table 4. has age and formal education level positive and significant at $p < 0.05$ critical level. Others like household size and farming experience, though significant at $p < 0.05$ critical level, are negative to technical inefficiency of the farmers in the area. It can, then, be contended that apart from age and education, an increase in household size and farming experience can reduce farmers technical inefficiency. This is consistent with a priori expectation and akin to the findings of (Kapinga et al.,1997). Farming experience increases farming skills that can increase technical efficiency. In a similar vein, household size, though increases the level of family labour, does not increase the quality of labour required for increased efficiency.

In antithesis to a piriori expectation, age increases with technical inefficiency. This implies that farmers with old age are more technically inefficient than the young farmers. Younger farmers are more competent, risk loving and friendly, accept inovations and technology that can enhance farmers efficiency in potato production in the area. The study therefore suggests that young farmers should be encouraged in potatoe farming system to help reduce technical inefficiency. Similarly, the farmers level of formal education, tends to increase with increase in technical inefficiency. This is contrary to a piriori expectation, as education, which is expected to increase farming skills and educate farmers on the use of improved farming skill, cannot increase technical efficiency in the area. Access to credit and depreciation of fixed input, though not significat at 0.05 critical level, are still negative. This implies that increasing access to credit will reduce technical inefficiency just like depreciation. It could be deduced from the findings that while access to credit provides enough capital for acquisition of farm inputs that can in turn produce more output at a reduced cost, increased depreciation increases the overall cost of production, and hence inefficiency.

CONCLUSION AND RECOMMENDATIONS

The study x-rayed the technologies of sweet potato production such as soil amendment practices, cropping systems, improved crop varieties planted, and sources of planting materials used in the area. The use of inorganic fertilizer is very common with about 65.378% of the farmers' engaging in the use of inorganic fertilizers like NPK in the planting of sweet patotes. About (66.67%) of the farmers practice bush burning which seems to reduce the overhead-cost incurred in clearing the land at the beginning of every planting season. The common varieties of potatoes planted by the farmers in the area include TIS 8164, TIS 0087, and TIS 2271. Other local varieties include *Ogonegburunwa* and *dukuduku*. The results showed that all significantly estimated coefficients were negative except age and education. It can, then, be deduced that apart from age and education, increase in household size and farming experience reduces farmers technical inefficiency. The wide disparity in farmers' technical efficiencies ranging from 0.298% to 99.4% and a mean of 47.1% suggested a need to bridge the gap. Hence, a reduction in household size, farming experience, and sourcing of planting materials from NRCRI or IITA is believed to increase farmers' technical inefficiency, which reduces with age and formal educational level of sweet patoto farmers in the area. Reducing bush burning but increasing liming, inclusion of organic soil ammendments, and irrigation practicies when coupled with the use of young and educated farmers reduce environmental damages and increase technical efficiency in sweet patoto production in the area. The study suggests young farmers to dominate the production so that adoption of improved technologies for improved output will become easy. Further, education should be given utmost priority for easy adoption of innovation in the sector.

ACKNOWLEDGEMENT

The authors appreciate the National Root Crops Research Institute, Umudike, Nigeria, for their providing essential information and data required for the study.

REFERENCES

1- Adejuwon, S. A. (2004, September). Impacts of climate variability and climate change on crop yield in Nigeria. In stakeholders' workshop on assessment of impacts and adaptation to climate change Obafemi Awolowo University, Ile-Ife, Nigeria.

2- Age, A. I., Demenongu, T.S., & Obinne, C.P.O. (2012). Overview of the effects of climate change on agricultural productivity in Nigeria: Adaptation and Mitigation Strategies. Proceedings of the Seventeenth Annual Conference of the Agricultural Extension Society of Nigeria.

3- Aigner, D. J., Lovell, C.A.K., & Schmidt, P. (1977). Formulation and estimation of stochastic frontier production function model. Journal of Econometrics, 1 (1), 21-37.

4- Akande, S.O. (2008). Prospect for agricultural sustainability and growth. A paper presented at the National Productivity Workshop. A Newsletter of the Nasarawa Agricultural Development Programme, 2(1), 16-28.

5- Al-Kaisi, M. (2012). Conservation practices save soil and sustain agriculture. Integrated Crop Management News,j Paper 198, http://lib. dr.iastate. edu/cropnews/198

6- Assa, Maganga M., Edriss, Abdi-Khalil, & Matchaya Greenwell C. (2012). Unexploited Profit among Smallholder Farmers in Central Malawi: What are the Sources? International Journal of Applied Economics, 9(2), 83-95.

7- Coelli, V.J. (1996). Guide to Frontier Version 4.1: A computer program for stochastic frontier production and cost function estimation. Department of Economics, University of New England, Armidale, Australia.

8- Ebuzoeme, O. D. (2015).Evaluating the Effects of Flooding in Six Communities in Awka Anambra State of Nigeria. Journal of Environment and Earth Science, 5 (4), 26-38.

9- Ehirim, N. C., Okere, R. A., Onyeagocha, S. U. O., Rahji, M.A.Y., Awoyemi, T.T., Oluwatayo B. I., & Salman, K. K. (2013). Econometric analysis of suitability and marginal value productivity of farmlands for cassava production in Imo State, Nigeria. Journal of Development and Agricultural Economics, 5(11), 450–456. DOI 10.5897/JDAE 2013.0505 ISSN 2006 – 977 Academic Journals Retrieved from http://www.academicjournals.org/JDAE

10 Greene, W. (1997). Frontier production functions. In Handbook of Applied Econometrics, Vol. 2, Microeconomics, H. Pesaran and P. Schmidt (Eds.), Oxford University Press, Oxford.

11- Hornick, S.B., & Parr, J.F. (2010). Restoring the productivity of marginal soils with organic amendments Ecological Agriculture Projects, McGill University (Macdonald Campus) Ste-Anne-de-Bellevue, QC, H9X 3V9 Canada. American Journal of Alternative Agriculture, 2(2), 64-68.

12- Hussein A. (2004). Principles of environmental economics (2nd ed.). UK: Routledge, ISBN-13:978-0415275606, Pp. 384.

13- Igwe, C.A., & Stahr, K. (2004). Water–stable aggregates of flooded inceptions from southeastern Nigeria in relation to mineralogy and chemical properties. Australian Journal of Soil Resources, 42,171-179.

14- Kapinga, R.E., Ewell, P.T., & Jeremiah, S.C. (1997). Sweet potato in tanzanian farming and food systems. Implications for Research (p. 47), Ministry of Agriculture and Livestock Development, Tanzania and International Potato Centre (CIP), Nairobi Kenya.

15- Meeusen, N., & Van-den Broeck, J. (1977). Efficiency estimation from Cobb Douglas production functions with composite error. International Economic Review, 18 (2), 123-134.

16- National Population Commission (NPC) (2007). Details of the breakdown of the national and state provincial population totals 2006 census. Federal Republic of Nigeria Official Gazette, 94(24), 1-26.

17- Ndukwu, P.C. (2010). Gender and relative efficiency in sweet potato (ipomoea batas) production in Imo State, Nigera. Unpublished thesis, Department of Agricultural Economics, Michael Okpara University of Agriculture, Umudike, Nigeria.

18- Njoku, J.C., Muoneke, C.O., Okocha, P.I., & Ekeleme,F. (2009). Effect of propagule size and intra-row spacing on the growth and yield of sweet potato in Humid Agro-Ecological Zone. The Nigerian Agricultural Journal, 40(1-2), 115-124.

19- Nwaru, J. C. (2003).Gender and relative production efficiency in food crop farming in Abia State of Nigeria. The Nigerian Agricultural Journal, 34, 1 – 10.

20- Ogbonna, K.I., Idiong, I.C., & Ndifon, H.M.(2007). Adoption of soil management and conservation technologies by small scale crop farmers in south eastern Nigeria: Implications for sustainable crop production.Medwell Online Agricultural Journal 2(2),294-298.

21- Ogundari, K., & Ojo, O.S. (2006). An examination of technical, economic and allocative efficiency of small farms: The case of cassava farmers in Osun State. Journal of Central European Agriculture, 7(3), 423-432.

22- Olayide, S.O, & Heady, E.O (1982). Introduction to agricultural production economics.Department of Agricultural Economics University of Ibadan. University Press Ibadan. Pp. 9-247.

23- Oldeman, L. R., Hakkeling, R. T. A., &. Sombroek, W. G. (1991). World map of the status of human-induced soil degradation: an explanatory note. Global Assessment of Soil Degradation (GLASOD), International Soil Reference Information Centre (ISRIC), and United Nations Environment Program (UNEP).

24- Oni, O., Nkonya, E., Pender, J. , Phillips, D., & Kato, E. (2009). Trends and drivers of agricultural productivity in Nigeria. Nigeria Strategy Support Program (NSSP) Report (001). International Food Policy Research Institute IFPRI Headquaters, 2033 K Street NW Washington, DC 20006-1002 USA Supported by CGIAR.

25- Onwueme, I.C. (1978). The tropical tuber crops. John Wiley and Sons, New York. (Pp. 187).

26- Porcelli, F. (2009). Measurement of technical efficiency: A brief survey on parametric and non-parametric techniques. University of Warwick, 11, 1-27.

27- Scherr, S.J., & Yadav, S. (1996). Land degradation in the developing world: Implications for food, agriculture, and the environment to 2020. Discussion paper 14, International Food policy Resource Institute, Washington DC. Pp. 36.

14

Perceived Effects of Poor Road Transportation Network on Crop Production in Kaiama Local Government Area of Kwara State, North Central Nigeria

Lateef Lawal Adefalu [1], Oluwasogo David Olorunfemi [1], LatifatKehinde Olatinwo [2] and Yusuf Olatunji [1]*

Keywords:
cropproduction, Kwara State, perceived effects, poor road transportation

Abstract

This paper examines the effect of the poor road transportation network on crop production in one of the rural agrarian local government of Kwara State, Nigeria. A well-structured interview schedule was conducted to elicit information from 120 crop farmers using a two-stage sampling technique. Findings from the study revealed that the poor road transportation network in their area had led to a reduction in their income, longer time in transporting produce to more buoyant markets, as well as incurring high transportation costs. Chi-square analysis revealed that there was a significant relationship between age, educational level, years of farming experience, and the respondents' perceived effects of poor road transportation network for their crop production. The study recommends an urgent intervention from government and stakeholders in terms of construction and rehabilitation of roads in these rural agrarian areas in an attempt to help improve the livelihoods of the farmers through enhancing transportation of produce to more buoyant markets in good condition and at less cost.

[1] *Department of Agricultural Extension and Rural Development, University of Ilorin, Ilorin, Nigeria*
[2] *Department of Agricultural Economics and Extension, Federal University, Dutsin-Ma, Kastina State, Nigeria*
* *Corresponding author's email: davidsoa2003@yahoo.com*

INTRODUCTION

Roads are thought to be an important factor in the process of economic development of many nations and are equally assumed to play a crucial role in the growth of rural areas (Kiprono & Matsumoto, 2014). The World Bank and lending donor agencies have identified the improvement of roads as an instrument of poverty alleviation in developing nations. Rural road transportation network and infrastructure development in Nigeria have been topical issues and have been identified as crucial components for economic development of the country (Adedeji et al., 2014). Investment in rural roads has resulted in phenomenal growth in agricultural production and productivity, while rapid growth in agricultural productivity has led to a significant trickle-down benefits for the rural poor (Fan et al., 2000). The agricultural sector is essential in the economic development and poverty alleviation drive of many countries (Olorunfemi et al., 2014), and it has been viewed presently as a pillar of the Nigerian economy, because it contributes to the GDP by 40 percent (Adekunle, 2013) and employs about 70 percent of the labour force (FAO, 2012). As reported by Jacoby (2000), a road with an important rural infrastructure provides cheap access to markets for agricultural output, thus stimulating economic development. Agricultural goods in Nigeria are mostly bulky in nature and highly perishable (Ajiboye & Afolayan, 2009). They ought to be conveyed from their area of production to zones of consumption immediately where they will yield more returns for the producer.

However, as reported by Akangbe et al. (2013), a lot of rural farmers in Nigeria and particularly in Kwara State, despite the efforts being put in place to bring about development in agriculture, still expressed the constraints they face in income maximization from their agro-output as a result of poor road network in the area. Therefore, there is aneed to shed light on the effects of the poor road infrastructure network in the rural agrarian areas under study in order to provide information that will give insights, fast-track to, and re-strategize the efforts of governments and other stakeholders in their agricultural transformation and development drive in the area. The broad objective of the present study is to determine the effects of the poor road transportation network on crop production in Kaiama LGA of Kwara State, Nigeria. The specific objectives were to describe the socioeconomic characteristics of the crop farmers in the study area; determine the perceived effects of the poor road transportation network on their crop production; and identify the coping strategies adopted by the farmers in ameliorating the agro-transportation problems encountered in the study area.

MATERIALS AND METHODS

The study was conducted in the Kaiama Local Government Area of Kwara State, Nigeria. Its headquarters is in the town of Kaiama. It has an area of 6,971 km^2 and a population of 124,164 as indicated by the 2006 census (NPC, 2006). Kaiama is endowed with natural resources such as Gold, Tantalite, and Granite. The majority of the population are farmers who are known mostly for planting tubers and cereal crops. Crops grown include yam, cassava, groundnut, millet, maize, guinea corn, cocoyam, vegetables, and fruits.

The population under study comprised crop farmers in Kaiama LGA of Kwara State. A well-structured interview schedule was conduced to elicit information from 120 respondents, whowere selected through

Figure 1. Map of Kwara State Indicating the Study Area (Kaiama LGA)

a two-stage sampling procedure from 10 wards found in the LGA. A purposive selection of the most prominent community in each ward was carried out which was followed by a random selection of 12 leading crop farmers notable for being at the forefront in crop production in these communities.

Data were collected on the socioeconomic status of the crop farmers, their perceptions of the effect of poor road transportation network on their production and the coping strategies they adopt to tackle this problem at their level. These variables were measured as follows:

Perceived effect of the poor road transportation network: Respondents were presented with a series of items and they were asked to rate how they perceive them as effects on a 5-point Likert scale ranging fromstrongly agree (5), to agree (4), undecided (3), disagree (2), and strongly disagree (1).

Coping strategies: respondents were asked to indicate the coping strategies they adopt in tackling the effect of poor road transportation network in their area.

Data analysis was carried out using descriptive statistics such as frequency counts, percentages, mean scores, and ranks, while Chi-square analysis was used as part ofinferential statistics to test the proposed hypothesis in the study.

RESULTS

Table1 summarizes the socioeconomic status of the respondents. Findings from the study showed that the majority (74.2%) of the respondents fell within the age range of 22-59 years. The mean age of the crop farmers was 49 years.

Table 1
Socioeconomic Status of the Respondents (n=120)

Socioeconomic Characteristic	Frequency	Percentage	Mean
Age			
22-40	42	35.0	
41-59	47	39.2	
60-78	25	20.8	49 years
79 and above	6	5.00	
Gender			
Male			
Female	116	96.7	
Level of education	4	3.30	
No formal education			
Primary			
Secondary	73	60.8	
Tertiary	17	14.2	
Farming experience	12	10.0	
1-10	18	15.0	
11-20			
21-30			
>30	4	3.3	
Household size	47	39.2	29 years
1-4	16	13.3	
5-8	53	44.2	
9-12			
>12			
Secondary occupation	14	11.7	
Trading	100	83.3	7 members
Artesian	4	3.3	
Civil service	2	1.70	
None			
	14	11.7	
	18	15.0	
	6	5.00	
	82	68.3	

The result obtained further indicated that the contribution of the female gender to the active production of crops in the area was very low, as only 3.3% of them were involved in crop production when compared to their male counterparts (96.7%). More than half (60.8%) of the farmers had no formal education.

On average, the respondents had been in crop production for 29 years. The majority (83.3%) of the farmers had a household size that consisted of 5-8 members, while only a few (1.7%) had household size of greater than 12 members. The mean household size of the farmers was 7 persons. More than two-third (68.3%) of the farmers did not have any secondary occupation, while the remaining 31.7% were either engaged in trading (11.7%), were artesian (15.0%), or were involved in civil services (5.0%).

Effects of poor road transportation network on crop production

Table 2 reveals the perceived effects that poor road transportation network have on the production of the farmers in the area studied. Using mean score to rank the severity of the effects as indicated by the respondents, "Reduced farmers' income through sale at farm gate" (M=4.50), "Longer time spent to transport produce" (M=4.39), "Increased transport cost" (M=4.37); "Quality of farm produce affected during transportation" (M=4.35), "Discourages expansion of production" (M=4.35), "Wastage of highly perishable produce due to unavailability of timely markets" (M=4.26), "Poor accessibility of buyers and farmers" (M=4.01) ranked 1st, 2nd, 3rd, 4th, 4th, 6th, and 7th respectively. The table further showed that the farmers perceived the other effects whose mean scores were below 4.0 as less being severe.

Coping strategies used by the crop farmers

The coping strategies adopted by the farmers were ranked with a view to ascertain those strategies that were prominently used. As shown by Table 3 below, the majority (87.5%) of the farmers indicated that they had used "Sale of their farm produce to any available buyer at reduced cost", as it ranked first among the strategies listed. The table further reveals that 81.7 percent of the farmers indicated that they had sold their produce to intermediaries who came around to buy from them in large quantities during harvest periods. Above two-third (75.0%) of the farmers opted for transporting their produce together to nearby markets in groups. Yet, it is worthy of mention that less than one-third (31.7%) of the farmers made use of "reduction in the bulky nature of their produce through processing initiatives" for easier transportation and income.

Hypothesis Testing Using Chi-Square Analysis

The following hypothesis was tested in the study:

H_o: There is no significant relationship between the socioeconomic status of the respondents and their perceptions of the effects of poor

Table 2

Respondents' Perceived Effects of Poor Road Transportation Network on Crop Production (n=120)

Perceived effects	Mean	Rank
Reduces farmers' income through sale at farm-gate	4.50	1st
Discourages agro-investors	3.20	11th
Longer time spent to transport produce	4.39	2nd
Increased transport cost	4.37	3rd
Reduce productivity	2.19	15th
Poor accessibility of buyers and farmers	4.01	7th
Discourages expansion of production	4.35	4th
Exposure to robbery and theft on roads	2.92	13th
Discourages transporters from plying routes	3.08	12th
Cause quick dilapidation of vehicles	2.64	14th
Quality of farm produce is affected during transportation	4.35	4th
Wastage of highly perishable produce due to unavailability of timely markets	4.26	6th
Rural Urban migration	3.66	10th
Poor accessibility to improved farm inputs and other govt. agro credit schemes & intervention	4.01	7th
Inadequate accessibility to other infrastructural amenities	3.82	9th

Mean Score derived from strongly agree=5, agree=4, undecided=3, disagree=2, strongly disagree=1;

Table 3

Coping Strategies Employed by the Respondents (n=120)

Coping strategies	Frequency	Percentage	Rank
Sale of produce to any available buyer at reduced cost	105	87.5	1st
Consumption and distribution of unsold produce by family and neighbours	65	54.2	5th
Sale of produce to middle men	98	81.7	2nd
Reduction in Land area cultivated	55	45.8	6th
Transport produce in groups	90	75.0	3rd
Embarking on community service efforts to rehabilitate access roads to make it motorable	72	60.0	4th
Reduce bulky nature of produce through processing initiatives	38	31.7	7th

road transport network on crop production.

Table 4 shows that there is a significant relationship between the respondents' perceptions of the effect of the poor road transportation network on crop production and their age, educational level, and years of experience; therefore, the null hypothesis was rejected. The table shows further that there is no significant relationship between respondents' perceptions of the effects of poor road transportation network on crop production and their gender, household size, and their secondary occupation; accordingly, the null hypothesis was accepted.

DISCUSSION

The present study carries the implication that most of the farmers were still middle aged, active, and agile. This suggests that the male respondents mostly practiced crop production, which might be a positive boost to the level of productivity in the area. This attribute may likely stifle the initiative of the farmers. A good educational background may help facilitate farmers' use of coping strategies that will give them the ability to maximize their income despite poor road network conditions.

This indicates that farmers are highly experienced in farming, and this long-time experience might have facilitated their acquisition of good skills in the production, transportation, and marketing of their produce. The finding on household size suggests that the farmers could draw some level of family labour from their household considering the mean household size. This implies a likely reduction in the cost of hired labour, which increases the net gain of the farmers.

Some effects that were perceived as severe by the respondents agree with Akangbe et al.'s study (2013) where they reported that sale of produce at the farm gate and increased transportation cost were severe effects of poor transportation facilities on crop production. Similarly, Inoni and Omotor (2009) also indicated that poor road quality reduces the income of farm households. The implication of this is that farmers need an urgent intervention from government and stakeholders in terms of construction and rehabilitation of roads so as to improve the livelihoods of the farmers through the privilege of transportation of produce to more buoyant markets in good condition and at less cost.

Thus, as revealed by the farmers was because most of the products from their farms are perishable and so need to be disposed of on time in order to make the fewest and quick income they can get from it before spoilage. This was be-

Table 4

Relationship between the Socioeconomic Status of Respondents and their Perceived Effects of Poor Road Transportation Network on Crop Production (n=120)

Variable	df	X^2	p-value	Decision
Age	3	2.353	0.011	Reject H_0
Gender	1	3.284	0.305	Accept H_0
Level of Education	3	1.846	0.025	Reject H_0
Farming Experience	3	5.422	0.003	Reject H_0
Household Size	3	6.905	0.338	Accept H_0
Secondary Occupation	3	11.98	0.122	Accept H_0

cause they cannot afford high costs, time, and rigour of transporting their produce over a long distance before they can get them sold. This serves as a strategy for spreading and reducing the cost of transportation for better income maximization. This implies that most of the farmers in the study area lack the capacity to process their farm produce, which makes it less bulky to transport, enhance the shelf life, and increase the value for the higher income generation. Therefore, there is aneed for the government, extension agents, and other stakeholders to focus more of their attention on building the capacity of the farmers on processing initiatives which will not only help them cope with the rigour of transporting their bulky goods but will also ensure product availability for a longer period in markets and better income for the farmers.

The implication of this result is that the respondents' perceived effect of the poor road transportation network on crop production is most likely to be influenced by their educational level, age, and years of experience, while respondents' gender, household size, and secondary occupation may not have any significant influence on their perceived effects.

CONCLUSION

Based on the findings of the study, the farmers were able to indicate that poor road transportation network in their area handled to a reduction in their income, longer time in transporting produce to more buoyant markets, and incurring high transportation costs. The respondents have, however, tried to cope with the constraints posed by the poor road transportation network by selling their produce to any available buyer at a reduced cost, sale to middlemen, and sometimes, transporting their produce in groups to reduce costs.

The study, therefore, suggests that urgent intervention in terms of construction and rehabilitation of roads should be made by governments and other stakeholders in an attempt to improve the livelihoods of the farmers through the privilege of transportation of produce to more buoyant markets in good condition and at less cost. In addition, in order to expand the coping strategy options available to the farmers, extension agents and other stakeholders should focus more of their attention on building the capacity of the farmers on processing initiatives, which will not only help them to cope with the rigour of transporting their bulky goods but will also ensure product availability for a longer period in markets and better income for the farmers. This suggests that many of the respondents are solely dependent on agriculture for their livelihood, hence the need to ensure they maximize the income-generating potential of the sector.

ACKNOWLEDGEMENT

I appreciate the effort of the staff of the Kwara State Agricultural Development Program who assisted in the process of data collection over the course of this research.

REFERENCES

1- Adedeji, O. A., Olafiaji, E. M., Omole, F. K., Olanibi, J. A., & Yusuff, L. (2014). An assessment of the impact of road transport on rural development: A case study of Obokun local government area of Osun State, Nigeria. *British Journal of Environmental Sciences, 2*(1), 34 – 48.

2- Adekunle, O. A. (2013). "Key to Unlock". The One Hundredth and Twentieth Inaugural Lecture of the University of Ilorin delivered on Friday, 25th January, 2013 at the University Auditorium. 5-6.

3- Ajiboye, A.O., & Afolayan, O. (2009). The impact of transportation on agricultural production in a developing country: A case study of Kolanut production in Nigeria. *International Journal of Agricultural Economics and Rural Development, 2*(2), 49 – 57.

4- Akangbe, J. A., Oloruntoba, O. O., Achem, B., & Komolafe, S. E. (2013). An appraisal of transportation facilities effects on agricultural development in Moro Local Government Area, Kwara State, Nigeria. *Ethiopian Journal of Environmental Studies and Management, 6* (2), 191 – 200.

5- Food and Agriculture Organisation (FAO) (2012). State of food and agriculture: Investment in agriculture for food security, Retrived from www.fao.org.

6- Fan, S., Hazallel, P.,& Haque, T. (2000). Targeting public investments by agro-economic zone to achieve growth and poverty alleviation in Rural India. *Food Policy, 25*(4), 411 – 428.

7- Inoni, O. E., & Omotor, D. G. (2009). Effect of road infrastructure on agricultural output and income of Rural Households in Delta State, Nigeria.

Agricultura Tropica et Subtropica, 42(2), 90 - 97.
8- Jacoby, H. G. (2000). Access to markets and the benefits of rural roads. *The Economic Journal, 110*(465), 713 – 737.
9- Kiprono, P., & Matsumoto, T. (2014). Roads and farming: The effect of infrastructure improvement on agricultural input use, farm productivity and market participation in Kenya. A Paper presented at CSAE Conference, Economic Development in Africa, March 23-25, The University of Oxford, England.
10- NPC (2006). 2006 *Population and housing census of the Federal Republic of Nigeria*. Priority table (vol.1). National Population Commission, Abuja, Nigeria.
11- Olorunfemi, O. D., Ogunlade, I., Fakayode, S. B., & Adekunle, O. A. (2014). Ensuring improved livelihood opportunities for resource-poor Maize farmers through the dissemination of striga control methods in Kwara State, Nigeria. *Albanian Journal of Science, 13*(2), 80 – 88.

15

Investigating Factors that Affecting Citrus Waste Production in Mazandaran Province, Iran

Narges Mirzaie Gorji [1*], ***Hossein Shabanali Fami*** [2] ***and Hooshang Iravani*** [3]

Abstract

Agricultural waste management is one of the important issues that agricultural producers should consider to increase their income and environmental protection. Therefore, the investigation of factors that reduce the rate of agricultural production wastage as well as citrus production wastage is of utmost consequence in order to manage wastes, reduce environmental degradation and ecosystem destruction, and increase food security and economic growth. Inspired by this assumption, the aim of this study was to investigate factors affecting citrus waste production in Mazandaran Province, Northern Iran (N= 7867). A stratified random sampling method was employed to collect data from 165 farmers. The main tool to collect data was a self-designed questionnaire whose content validity was approved by a panel of experts and an exploratory factor analysis (EFA), and its reliability was confirmed by calculating a Cronbach's alpha coefficient ($\alpha \geq 0.7$).The results of the exploratory factor analysis showed that socioeconomic and technical factors altogether accounted for 60.257% of variances in citrus production wastage.

Keywords:
citrus waste, factor analysis, socioeconomic factor, waste management

[1] *M.Sc. Graduate of Rural Development, College of Agriculture and Natural Resources, University of Tehran, Iran*

[2, 3] *Professors of Management and Agricultural Development, University of Tehran, Iran*

* *Corresponding author's email: nmirzaie543@gmail.com*

INTRODUCTION

Traditionally, agriculture has formed the mainstay of rural economic activities, but it has been changed in the second half of the twentieth century in the direction of a much more balanced and diverse economy as more and more labor was driven out of the agricultural sector into employment in alternative sectors (Smit et al., 2015). Also, there is an extensive literature on agricultural productivity and economic growth in the field of spatial economics (Abreu et al., 2005; Bartelsman and Wolf, 2014; Gardiner et al., 2004; Van Oort et al., 2012). However, rising demand for food, increasing resource scarcity, high market volatility and growing environmental pressures challenge the agricultural sector not only to increase productivity but also to do so in a more sustainable manner (Godfray et al., 2010; OECD/FAO, 2012; Tilman et al., 2011). To address these challenges, agricultural producers must deal with multiple issues and often conflicting objectives.

Agricultural waste management is one of the important issues that agricultural producer should consider in their attempts to increase their income and environmental protection. Waste is an inevitable byproduct that arises from various anthropogenic activities and is also considered as one of the major sources of environmental degradation since it causes air, land and water pollution and contributes to global warming (Pagiola, 1995). Therefore, the investigation of factors that reduce the rate of agricultural production wastage as well as citrus production wastage are vitally important in order to manage waste, reduce environmental degradation and ecosystem destruction, and increase food security and economic growth. In this regard, this study attempts to investigate factors affecting citrus production wastage in Mazandaran Province.

Iran, like most developing countries, faces an increase in the generation of waste in agricultural sectors (Ministry of Agriculture, 2007). In this regard, the waste generated in agricultural sector is mostly associated with gardening production. According to the Iranian annualreports about gardening production, yearly 170,000 tons of citrus production in Mazandaran Province is wasted. Researchers argue that waste generated in Mazandaran Province occurred after the harvesting. Also, they reported that factors such as gardens' watering management, timely access to the marketplace, the gardeners' knowledge about marketing and government subsidy (Kasozi and Mwegombi, 2008), citrus sorting based on size, and gardeners' experience and their academic degree (Andrea et al., 2003) can reduce the rate of citrus wastage.

According to the literature (Andrea et al., 2003; Zheng et al., 2010; Srivastava & Singh, 2009; Kasozi & Mwegombi, 2008) we can categorize factors affecting citrus waste production in three broad dimensions: (A) socioeconomic factors (e.g. lack of government subsidy, lack of pricing policy, lack of farmers' knowledge about the citrus marketing and export, etc.); (B) infrastructure factors (e.g. of packing, lack of transporting, lack of storehouse, etc.); and (C) technical factors (e.g. of watering manage).

MATERIALS AND METHODS

The present survey is a quantitative study in nature, an applied study in terms of objectives, and a descriptive-correlation study in term of statistics. Statistical population consisted of all citrus growers in Mazandaran Province. A stratified random sampling method was employed to select 162 orchard farmers by applying Cochran's formula (see equation 1). The main tool to collect data was a self-designed questionnaire whose content validity was approved by a panel of experts and its reliability was confirmed by calculating a Cronbach's alpha coefficient (0.88-0.97). Data were analyzed by SPSS (v20) in two parts. In first part, descriptive statistics such as mean and standard division were used to ranking factors that affecting on citrus waste production; and in second, the explanatory factor analysis were used to identify factors that affecting on citrus waste production in Mazandaran Province.

$$n^2 = \frac{(t.s)^2}{d^2} \Rightarrow d = \frac{t.s}{\sqrt[2]{n}} \Rightarrow \frac{1.96 \times 2.69}{\sqrt[2]{30}} = 0.96 \div 2 = 0.48$$

$$\mathrm{n} = \frac{\mathrm{N(t.s)^2}}{\mathrm{Nd^2 + (t.s)^2}} \Rightarrow \frac{7867 \times (1.96 \times 2.69)^2}{7867 \times 0.48^2 + (1.96 \times 2.69)^2} = 120 \tag{1}$$

% confidence (t) = 1.96
s = Variances
n = sample size
N = Population size

RESULTS AND DISCUSSION

Based on descriptive results, most respondents (n= 126) were male, and the most common age was something in between 51-60 years. More than 93% of landowners were married, and most of them (n=43) had a diploma.

Ranking of citrus waste production items (Table 1) showed that prolonged picking time, the lack of weed management, and inappropriate use of fertilizers had the highest influence on citrus waste production. The lack of skills to transplant trees among gardener, failure to comply with spacing, and selection of an unsuitable location for the orchard have the least effect on citrus waste production.

An exploratory factor analysis was used to determine factors affecting citrus waste production. Statistical analyses showed that the internal consistency of data was appropriate for factor analysis (KMO=0.938) and the Bartlett's test was significant (Bartlett=5937.196; $p<0.01$). Two factors were extracted by using the Kaiser's rule. Factor loadings after Varimax rotation are shown in Table 3. Next, according to the nature of variables, factors were named as socioeconomic and technical factors. Given the amount of extract eigenvalues, technical factor had the

Table 1
Ranking of factors that affecting citrus waste production

Items	Mean	S.D.	Rank
Prolonged picking time	4.44	0.65	1
The lack of weed management	4.36	0.74	2
Inappropriate use of fertilizers	4.28	0.72	3
Failure to do soil tests	4.26	0.76	4
Use of inappropriate pesticides	4.21	0.81	5
The high maintenance cost in refrigerators	4.17	0.84	6
Improper stockrooms	4.13	0.85	7
The use of contaminated gardening equipment	4.12	0.90	8
Lack of attention to garden management	4.11	0.89	9
Emission stain from calamity berry fruits	4.05	0.90	10
Citrus mechanical damages during harvest	4.05	0.95	11
Lack of attention to suitable picking time	4.05	0.93	12
Unsanitary of boxes	4.04	0.93	13
Inappropriate transportation	4.01	0.95	14
Pests	3.97	0.93	15
Inappropriate garden irrigation	3.96	0.94	16
No sorting of fruits	3.95	1.00	17
Lack of refrigeration	3.95	0.98	18
Lack of using preservatives	3.94	1.02	19
Lack of market for citrus	3.93	1.00	20
Improper packing	3.92	0.98	21
Loss of alternate industries	3.91	1.02	22
Unstable weather conditions (continuous rains or long dry)	3.90	1.01	23
Selection of an unsuitable location for the orchard	3.85	1.05	24
Failure to comply with the spacing	3.81	1.04	25
Lack of skills to transplant trees among gardener	3.79	1.08	26

Scale: very little = 1; little = 2; medium =3; high= 4; very high= 5

Table 2
Rotation sums of variables affecting citrus waste production

Factor	Eigenvalues	% of variances	Cumulative variance
1	14.692	36.729	36.729
2	9.411	23.528	60.257

Table 3
Factors and variables related to citrus waste production after the Varimax Rotation

Factor name	Variables	Load factors
Socio-economic	Prolonged the picking time	0.889
	The high maintenance cost in refrigerators	0.866
	Improper stockrooms	0.846
	Lack of attention to garden management	0.789
	Lack of attention to suitable picking time	0.787
	Lack of using preservatives	0.775
	Lack of market for citrus	0.770
	Loss of alternate industries	0.761
	The lack of weed managements	0.739
Technical	Inappropriate use of fertilizers	0.788
	Failure to do soil tests	0.774
	Use of inappropriate pesticides	0.760
	The use of contaminated gardening equipment	0.754
	Unsanitary of boxes	0.749
	Inappropriate transportation	0.747
	Inappropriate garden irrigation	0.745
	Lack of skills to transplant trees among gardener	0.735

highest effect on total variance of citrus waste production (Table 2). In total, 60.257% of citrus waste production variances were captured by these factors.

Agricultural waste management is one of the important issues that agricultural producer should consider in their attempts to increase their income and environmental protection. Therefore, the investigation of factors that reduce the rate of agricultural production wastage as well as citrus production wastage are vitally important form an aging wastage, reducing environmental degradation, ecosystem destruction, and increasing food security and economic growth. In this regard, this study attempted to investigate factors affecting citrus production wastage in Mazandaran Province.

Results of exploratory factor analysis showed that the socioeconomic and technical factors had the highest effect on the total variance of citrus waste production, and 60.257% of citrus waste production variances were accounted for by these factors. This finding confirmed the results of Andrea et al. (2003); Zheng et al. (2010); Srivastava and Singh (2009) and Kasozi and Mwegombi (2007). They argued that socioeconomic and technical factors affected citrus waste production. Based on this finding, the researcher suggested that government should (a) facilitate exportation of citrus; (b) establish the alternate industries in this region (c) educate gardeners about their needs such as weed management, appropriate use of fertilizers for citrus, appropriate garden irrigation, and so on; (d) establish stockrooms with high quality for citrus in this region; and (e) construct roads to facilitate citrus transportation.

ACKNOWLDGMENT

The authors wish to extend their thanks to the participants in this project and to Tehran's Agricultural and Natural Resources College for their important role in supporting this research.

REFERENCES

1- Abreu, M., De Groot, H. L., & Florax, R. J. (2005). A Meta-Analysis of β-Convergence: the Legendary 2%. *Journal of Economic Surveys, 19*(3), 389-420.

2- Andrea, V., Nadia, N., Teresa, R. M., & Andrea, A. (2003). Analysis of some Italian lemon liquors (Limon cello). *Journal of Agricultural and Food Chemistry, 51*(17), 4978-4983.

3- Bartelsman, E. J., & Wolf, Z. (2014). Forecasting aggregate productivity using information from firm-level data. *Review of Economics and Statistics, 96*(4), 745-755.

4- Gardiner, B., Martin, R., & Tyler, P. E. T. E. R. (2004). Competitiveness, productivity and economic growth across the European regions. *Regional Competitiveness, 30*, 55.55-74.

5- Godfray, H. C. J., Beddington, J. R., Crute, I. R., Haddad, L., Lawrence, D., Muir, J. F., et al., (2010). Food security: the challenge of feeding 9 billion people. *Science, 327*(5967), 812-818.

6- Kasozi, M., & Mwegombi, W. (2008). Food Drying/Production Plant. In *2007 Second International Conference, August 20-22, 2007, Accra, Ghana* (No. 52162). African Association of Agricultural Economists (AAAE).

7- OECD/FAO, 2012. OECD–FAO Agricultural Outlook 2012–2021 Available at: OECD Publishing and FAO http://dx.doi.org/10.1787/agrout look-2012-en, (Retrieved in June 2013).

8- Pagiola, S. (1995). *Environmental and natural resource degradation in intensive agriculture in Bangladesh*. Washington DC: World Bank.

9- Smit, M. J., van Leeuwen, E. S., Florax, R. J., & de Groot, H. L. (2015). Rural development funding and agricultural labour productivity: A spatial analysis of the European Union at the NUTS2 level. *Ecological Indicators, 59*, 6-18.

10- Srivastava, A. K., & Singh, S. (2009). Citrus decline: soil fertility and plant nutrition. *Journal of Plant Nutrition, 32*(2), 197-245.

11- Tilman, D., Balzer, C., Hill, J., & Befort, B. L. (2011). Global food demand and the sustainable intensification of agriculture. *Proceedings of the National Academy of Sciences, 108*(50), 20260-20264.

12- Van Oort, F. G., Burger, M. J., Knoben, J., & Raspe, O. (2012). Multilevel Approaches and the Firm-Agglomeration Ambiguity in Economic Growth Studies. *Journal of Economic Surveys, 26*(3), 468-491.

13- Zheng, Y., He, S., Yi, S., Zhou, Z., Mao, S., Zhao, X., & Deng, L. (2010). Predicting oleocellosis sensitivity in citrus using VNIR reflectance spectroscopy. *Scientia Horticulturae, 125*(3), 401-405.

The Driving and Inhibiting Factors of Mechanized Tobacco Production in Iran Using the Delphi Technique

Seyyed Milad Samizadeh [1] and Saeed Firouzi [2]*

Abstract

Increased cost of tobacco cultivation is a major concern for its sustainable production in Iran. Therefore, the development of tobacco mechanized cultivation is a necessity. Given this, the driving and inhibiting factors of the development of mechanized tobacco cultivation were investigated by the Delphi technique in three rounds in Guilan Province, Iran. Nineteen experts from the Jahad Agriculture Organization of Guilan Province and the Agriculture Management Services in Rasht County and Western Guilan were selected as the research panel as recommended by the academic professors of the Guilan Province. The results were analyzed by $SPSS_{21}$. In the last round, 10 superior driving and inhibiting factors were ranked based on the final experts' consistency as the final decision. The results revealed that identification of pioneer farmers and their encouragement to mechanized tobacco cultivation and the import of the latest technologies of tobacco cultivation and their modification in accordance with regional conditions were the most important driving factors as 97.37 and 94.74% of panel experts agreed, respectively. In addition, 98.68% of experts rated insufficient research on tobacco cultivation mechanization and 96.05% of experts rated the small size of tobacco farms as the most important inhibiting factors. Accordingly, the development of supportive policies to support pioneer tobacco growers in mechanized cultivation of this industrial crop, the technical assessment for modern tobacco cultivation machinery, the reduction of imported tobacco machinery prices, and the support of research on mechanization of tobacco cultivation are some important strategies that should be considered by those involved in tobacco production in Guilan Province, Iran.

Keywords:
agriculture mechanization, Delphi technique, tobacco, tobacco cultivation

[1] Department of Agricultural Management, Rasht Branch, Islamic Azad University, Rasht, Iran
[2] Department of Agronomy, Rasht Branch, Islamic Azad University, Rasht, Iran
** Corresponding author's email: firoozi@iaurasht.ac.ir*

INTRODUCTION

As an important industrial crop, tobacco (*Nicotiana tabacum* L.) plays a major role in economics of many countries. It is a wealth creation factor, an economical driver, and one major underpinning of national income in many parts of the world (Zamani, 2010). Tobacco cultivation area in Iran has declined from about 20,000 ha in 1998 to 6,000 ha in 2009 and had slight variations until 2014. Higher production costs and skilled labor unavailability in traditional cultivation system are mainly blamed for the loss of its cultivation area in Iran. The production of this industrial crop not only ensures the subsistence of tobacco growers but also is a source of income for thousands of people involved in processing, distribution, and marketing of final products. The use of tobacco transplanters cuts labor costs, reduces seedling losses, and improves the accuracy and speed of transplanting. So, it is necessary for Iranian Tobacco Company to supply tobacco-growers with appropriate tobacco transplanters in accordance with local requirements (Namvar Rezaei, 2008). Also, the manual topping and harvesting of tobacco leaves and shoots are very costly. So, the movement towards mechanized cultivation of tobacco is unavoidable if its production is intended to be sustainable and competitive with foreign products.

For a successful formulation of a comprehensive plan for the development of local farming mechanization, it is necessary to recognize the driving and inhibiting factors at the first step. Clarke et al. (1993) proposed three factors of "funding farmers to supply farm equipment and machinery", "providing technical consulting", and "training farm machinery users" as the approaches to develop agricultural mechanization in Slovakia. Salokhe and Oida (2003) noted that "the small size of farms in the Philippines" and "farmers' financial limitations" were the most important limitations for mechanized farming in this country. He also stated that agriculture mechanization in the Philippines depended on providing the small-scale, inexpensive, and environmentally-friendly farm machinery. Capareda et al. (2002) also found the implementation of different mechanization projects throughout the Philippines useful for the transfer of agriculture technologies to the farmers. In a study on factors affecting the development of peanut mechanization in Guilan Province, Iran, Zehtab Naebi et al. (2015) identified three factors including "allocating provincial and national funds to develop mechanization", "organizing training programs to increase farmers' technical knowledge", and "conducting pilot and model projects" as the main driving factors of peanut mechanization development in Guilan Province. Firouzi (2015) reported the unsatisfactory state of mechanization of rice weeding operation in Guilan Province of Iran; therefore, an essential was proposed to enhance the availability to rice weeding machinery in the region. Julius (2014) suggests that increasing tractor hiring points and reducing tractor hiring rate are how the state can help the development of agriculture mechanization in small farms of Abuja, Nigeria. Rasouli et al. (2010) examined factors affecting sunflower mechanization development in Iranian farms by the Delphi method and mentioned the small size and dispersion of sunflower farms as the main factors inhibiting the development of sunflower mechanization in Iran. Kutte and Tya (2001) believe that the inconsistency of mechanization approaches with local farmers' requirements and interests is the main challenge for the development of mechanization in Nigeria. So, they mentioned the inattention of the Nigerian public organizations to the technological changes of farming mechanization as a major reason for the undeveloped status of agriculture mechanization. According to Ou et al. (2002), agriculture mechanization as a system engineering depends on the cooperation and coordination of many factors in addition to the quantitative and qualitative development of farm machinery. These factors can be dealt with as economic, environmental, social, and agronomical factors. In a study in Vietnam, Chi (2010) showed that attending in educational courses resulting in the acquisition of technical knowledge by Vietnamese farmers about farm machinery, investment capacity in agriculture sector, increasing labors' knowledge and the performance of the Vietnam Mechanization Development Organization were the most important factors among factors affecting the development of agricultural mechanization in rice fields of this country. In a study on reasons for non-acceptance of new farm machinery in the Andica region of Khuzestan, Iran, Baldaji and Aghapour Sabbaghi (2015) concluded that the farm size, income in recent years, attendance in educational classes, and irrigated to rain-fed land ratio had positive influence on the acceptance of new machinery for the cultivation of wheat in the studied region. In contrast, farmers' experience and land fragmentation negatively affected the acceptance of new machinery. So, they proposed attempts to transfer the ownership of farms, to integrate lands, to hold educational classes, and to increase the irrigated lands as strategies to motivate the adoption of modern technologies, especially in the cultivation of wheat. Capehart (2004) noted that the tobacco farms are becoming less in number and U. S. tobacco cultivation area has been decreased since the 1950s. Mechanization of burley tobacco production in Kentucky State has not been adapted to the same extent due to limits on farm sizes resulting from the topography problems in the region and difficulties in mechanized harvesting of burley tobacco.

An investigation showed that except for land preparation and spraying, all agronomic practices of tobacco cultivation in Guilan Province are performed manually; therefore its production is very costly due to high demand for labor requirement, especially for harvesting of tobacco leaves. Consequently, in order to keep the sustainability of tobacco production, its mechanized cultivation is unavoidable. So, identifying and analysis of factors affecting the development of mechanization of tobacco production in Guilan Province is essential. A review of reports revealed that no attempt has been made to recognize factors affecting the development of mechanized tobacco growing in Iran and in Guilan Province yet. So, given the strategic importance of tobacco production as the most important industrial crop in Guilan Province and its decisive role in the subsistence of Guilanian families, it is necessary to examine and analyze all aspects of sustainable production of tobacco. Accordingly, the objective of the present study was to identify and study the driving and inhibiting factors of the development of mechanization of tobacco cultivation in tobacco farms of Guilan Province.

MATERIALS AND METHODS

The case study

The present study was carried out in Guilan Province in the north of Iran in spring and summer of 2015. Total cultivated area of tobacco in Guilan Province is estimated to be over 500 ha by which about 1,000 families make a living (Guilan's Organization of Jihad Agriculture, 2013). Table 1 shows the type and date of common agronomic operations for tobacco cultivation in Guilan Province. The present study is a qualitative

Table 1

Type and Date of Different Operations for Tobacco Production in Guilan Province of Iran

Operations type	Date of operation*
Fall tillage	5October to 5November
Seeding in tobacco nursery	10 March to 30 March
Spring tillage	10 April to 30 April
Transplanting	30 April to 10 May
Re-planting	15 May to 30 May
Weeding and earth up	25 May to 30 May
Spraying	25 May to 30 May
Topping and first harvesting	15 July to 20 July
Second harvesting	5 August to 15 August
Third harvesting	15 August to 25 August
Fourth harvesting	25 August to 5 September
Fifth harvesting	15 September to 25 September

*based on the tobacco cultivation experts' comments in Guilan Province

research study in terms of research paradigm with diagnostic needs assessment approach aiming at exploring and identifying the driving and inhibiting factors of the development of the mechanization of tobacco cultivation in Guilan Province by the Delphi method (Conner & Gardy, 2013).

The Delphi technique

Delphi is a systematic method in management research based on achieving a consensus through a series of questionnaire rounds filled out by anonymous respondents and providing the research panel with their feedback (Keeney et al., 2001). The main components of the Delphi technique are iteration, questionnaire, experts, controlled feedback, and anonymity of technical panel, analysis of the results, consensus, time, and facilitating team (Dunham, 1996). The results of this method are based on the non-informed but organized consensus of the experts in a specific field. As such, unlike other methods, the validity of its results depends not on the number of participants but on the expertise of the research panel. According to previous works, the participants of the Delphi technique can be in the range of 5 to 20 people. Thus, 19 mechanization experts including the experts of Jahad Agriculture Organization of Guilan Province and Agriculture Managements of Rasht and western Guilan cities were selected as the research participants after consulting with academic professors in the region. Next, they were briefed about the details and the objectives of the study, and they were requested to cooperate in all the three rounds. In the first round of the three-round Delphi study, two following open-ended questions were asked and the respondents were asked to provide detail answers to the following probes:

1. What are the most important driving factors of the development of mechanization of tobacco cultivation in Guilan Province?
2. What are the most important inhibiting factors of the development of mechanization of tobacco cultivation in Guilan Province?

The questionnaires were distributed among participants and were collected before the deadline. The responses given in the first round were summarized after the analysis of their contents. The results were presented in the tables of the first round including frequency, response percentage, and case percentage. Next, a questionnaire including 16 driving and 17 inhibiting factors was designed for the second round on the basis of the content analysis of the first round. In the second round, the participants were asked to express their agreement with each driving and inhibiting factor of the development of mechanization of tobacco cultivation in Guilan Province on a five-point Likert scale (very strongly, strongly, fairly, weakly, very weakly). Once the questionnaires of the second round were collected, the results were analyzed by SPSS in terms of the importance and normalized score for which the following rating scheme was used: 1=very low; 2=low; 3=fair; 4=high; and 5=very high. The degree of importance of each statement was calculated as the sum of the experts' opinions (the equivalence coefficient of the responses) divided by the number of statements. The normalized weight score of the statements was calculated by their importance divided by total sum of the importance of all statements. Next, the Kendall's W test confirmed the need for the third round of the Delphi technique (Heiko, 2012). Accordingly, 10 high priority inhibiting and 10 high priority driving factors were chosen for the questionnaire of the next round. Since fewer statements were left, it was possible to compare individual items and determine the percentage of agreement with each statement more accurately. The final opinions were organized and discussed as percentage in a descending order. As a final consensus, all factors with agreement percentage of >90% were selected as the most important factors (Abedi et al., 2017). At the end of each round, the results were tabulated as driving and inhibiting priorities.

RESULTS AND DISCUSSION

Driving factors

According to the frequency of the responses to the open-ended question of the first round (What are the most important driving factors of the development of mechanized tobacco cultivation in Guilan Province?), 16 driving factors

Table 2
Delphi Study First Round: Driving Factors of Mechanized Tobacco Cultivation in Guilan Province

Factor	Freq.	Percent of responses	Percent of cases	Priority
Conducting pilot projects of mechanized tobacco cultivation in tobacco growing regions of Guilan Province	10	15.1	52.6	1
Using the development pattern of mechanization of tobacco cultivation in the leading tobacco producing countries	8	12.1	42.1	2
Governmental support through timely purchase of tobacco from farmers	8	12.1	42.1	2
Enhancing tobacco growing experts' technical knowledge	7	10.6	36.8	3
Devotion of bank credits especially for the development of mechanization of tobacco cultivation in the Guilan Province	5	7.6	26.3	4
Visiting mechanized tobacco farms by tobacco farmers	5	7.6	26.3	4
Identifying pioneer tobacco growers and supporting them to mechanized tobacco cultivation	5	7.6	26.3	4
Implementation of intelligent irrigation system and improving water use efficiency in tobacco farming	4	6.1	21.1	5
Enhancing the quality of tobacco leaves by the management of pests and plant diseases and other improving agronomic operations	4	6.1	21.1	5
Promoting the knowledge of benefits of mechanized tobacco cultivation system through conducting educational classes	3	4.5	15.8	6
Formulation of national and provincial plans to develop mechanization of tobacco cultivation in the Guilan Province	2	3.0	10.5	7
Modifying tobacco cultivation pattern in order to executing the mechanized cultivation of tobacco in the Guilan Province	1	1.5	5.3	8
Importing the latest technologies of tobacco cultivation and their modification in accordance with the local conditions	1	1.5	5.3	8
Establishing or strengthening tobacco growing cooperatives	1	1.5	5.3	8
Establishing and specially supporting tobacco mechanization service companies	1	1.5	5.3	8
Enhancing the technical knowledge of mechanization of tobacco cultivation among Guilan's craftsmen	1	1.5	5.3	8
Total	66	100		

were identified (see Table 2). Results showed that *conducting pilot projects of mechanized tobacco cultivation in tobacco growing regions of Guilan Province* was ranked the first driving factor with response frequency of 10 and case percentage of 52.6%. *Using the development pattern of mechanization of tobacco cultivation in the leading tobacco producing countries and governmental support through timely purchase of tobacco from farmers* were ranked the second most important factors with response frequency of 8 and case percentage of 42.1%.

Results of Table 2 were used in the second round of the Delphi technique in which the panel's opinions were presented to the respondents as five-choice questions and the results were prioritized in terms of the importance and normalized weight score. Table 3 summarizes the agriculture mechanization experts' agreement with different opinions. The shift of the

Ranks in the second round are a feature of this technique which leads to the convergence of panels' opinions over the steps. According to the results of the second round (see Table 3), *importing the latest technologies of tobacco cultivation and their modification in accordance with the local conditions* was the most important driving factor with normalized weight score of 7.40. This factor was ranked the eighth in the first round with the frequency of 1.

Promoting the benefits of mechanized tobacco cultivation system by educational classes" and "enhancing tobacco growing experts' technical knowledge, which had been ranked the sixth and third with the frequencies of 3 and 7 in the first round, were ranked the second and third most important driving factor in the second round with normalized weight scores of 7.22 and 6.57, respectively.

According to the results of the third round (see Figure 1), 97.37% of panel experts believed that *identifying pioneer tobacco growers and supporting them* is the most important promoting factor. Research shows that there have always

Table 3
Delphi Study Second Round: Driving Factors of Mechanized Tobacco Cultivation in Guilan Province

Motivator	Level of importance	Normalized weight score	Priority
Importing the latest technologies of tobacco cultivation and their modification in accordance with the local conditions	4.7	7.40	1
Promoting the knowledge of benefits of mechanized tobacco cultivation system through conducting educational classes	4.58	7.22	2
Enhancing tobacco growing experts' technical knowledge	4.17	6.57	3
National and provincial plans to develop mechanization of tobacco cultivation in the Guilan Province	4.11	6.48	4
Enhancing the technical knowledge the mechanization of tobacco cultivation among Guilan's craftsmen	4.05	6.38	5
Devotion of bank credits especially for the development of mechanization of tobacco cultivation in Guilan Province	4.05	6.38	5
Identifying pioneer tobacco growers and supporting them to mechanized tobacco cultivation	4.05	6.38	5
Implementing intelligent irrigation system and improving water use efficiency in tobacco farms	3.94	6.20	6
Visiting mechanized tobacco farms by tobacco farmers	3.94	6.20	6
Modifying tobacco cultivation pattern in order to execute the mechanized cultivation of tobacco in the Guilan Province	3.88	6.11	7
Conducting pilot projects of mechanized tobacco cultivation in tobacco growing regions of Guilan Province	3.82	6.01	8
Establishing and specially supporting tobacco mechanization service companies	3.76	5.92	9
Governmental support through timely purchase of tobacco from farmers	3.76	5.92	9
Using the development pattern of mechanization of tobacco cultivation in the leading tobacco producing countries	3.76	5.92	9
Establishing or strengthening tobacco growing cooperatives	3.52	5.55	10
Enhancing the quality of tobacco leaves through the management of pests and plant diseases and other improving agronomic operations	3.35	5.27	11
Total	63.44	100	

been some farmers in Guilan Province who have welcomed modern agricultural systems to reduce production costs and time. Obviously, identifying these farmers and providing them with specific supports can be a good model and motive for mechanized tobacco cultivation for all tobacco growers. According to Zehtab Naebi et al. (2015), this factor was ranked the fifth most important factor promoting mechanized peanut cultivation in Guilan Province.

Importing the latest technologies of tobacco cultivation and their modification in accordance with the local conditions was found to be the second most important driving factor with 94.74% consensus of the panel. Given the specific technical specifications of tobacco planting machinery particularly tobacco transplanters, leaf toppers and harvesters, it seems that the most possible way to enjoy the advantages of mechanized tobacco cultivation in Guilan Province is to plan for the import of appropriate machinery and to conduct studies on their optimization in accordance with the local conditions. Abedi et al. (2017) reported that specifically supporting studies on optimization of olive harvesting machinery as well as supporting harvesting machinery were ranked the second most important driving factor for the development of mechanized olive harvesting in Guilan Province.

Devotion of bank credits to the development of mechanized tobacco cultivation was found to be the third most important driving factor of the development of mechanized tobacco cultivation in Guilan Province as 93.42% of panel experts agreed. Mohammadi and Zarifian (2008), also, described the complicated procedure for borrowing bank credits as one significant barrier against agriculture mechanization in Nishapur, Iran. Some conditions to get loans for the purchase of agriculture implements are to have bank account turnover, to pay back delinquent debts, and to give guarantees with specific conditions. It seems necessary for the authorities in charge of the development of agriculture mechanization to suggest practical approaches to solve this problem.

Pishbin et al. (2007) suggested low-interest

Table 4
Delphi Study First Round: Inhibiting Factors of Mechanized Tobacco Cultivation in Guilan Province

Inhibitor	Freq.	Percent of responses	Percent of cases	Priority
Fragmented status of tobacco farms in Guilan Province	11	18.0	57.9	1
Higher income of tobacco's competing crops in Guilan Province	9	14.8	47.4	2
High costs of mechanization of tobacco cultivation due to its special cultivation conditions	7	11.5	36.8	3
Higher age of tobacco growers in Guilan Province	6	9.8	31.6	4
Small size of tobacco farms in Guilan Province	6	9.8	31.6	5
Inappropriate financial position of tobacco growers and their inability for mechanized cultivation	4	6.6	21.1	6
Insufficient research on mechanized tobacco cultivation in Guilan Province	3	4.9	15.8	7
Rental status of some tobacco farms in Guilan Province	3	4.9	15.8	7
Low attention of agricultural managers and experts to the mechanized tobacco cultivation	3	4.9	15.8	8
Old traditions of tobacco growers about mechanized farming	2	3.3	10.5	8
Negative view towards the use of tobacco products in governmental systems	1	1.6	5.3	9
Low price of tobacco leaves	1	1.6	5.3	9
Insufficient attention of Guilan's tobacco organization to mechanized cultivation	1	1.6	5.3	9
Lack of tobacco mechanization experts in Guilan Province	1	1.6	5.3	9
Complexity and monopoly status of tobacco farming-specific machinery	1	1.6	5.3	9
Low tobacco cultivation area in Guilan Province	1	1.6	5.3	9
Slope of some tobacco farms in Guilan Province	1	1.6	5.3	9
Total	61	100		

loans with long repayment period and simple guarantees as a policy for the development of mechanization in the Jahrom Region of Fars Province. Clarke et al. (1993), also, mentioned financial support as an important factor for the development of mechanization in the Republic of Slovakia and Ghosh (2010) suggested the devotion of organizational credits as an approach to promote mechanization in the Burdwan region, India.

Promoting the benefits of mechanization and technical introduction of tobacco cultivation machinery through conducting educational classes was ranked the fourth most important driving factor of the development of mechanized tobacco cultivation

Table 5
Delphi Study Second Round: Inhibiting Factors of Mechanized Tobacco Cultivation in Guilan Province

Factors	Level of importance	Normalized weight score	Priority
Small size of tobacco farms in Guilan Province	4.64	7.60	1
Higher income of tobacco's competing crops in Guilan Province	4.52	7.39	2
Fragmented status of tobacco farms in Guilan Province	4.52	7.39	2
Insufficient research on mechanized tobacco cultivation in Guilan Province	4.41	7.22	3
Higher age of tobacco growers in Guilan Province	4.11	6.73	4
Inappropriate financial position of tobacco growers and their inability for mechanized cultivation	4.05	6.63	5
High costs of mechanization of tobacco cultivation due to its special cultivation conditions	3.82	6.25	6
Low tobacco cultivation area in Guilan Province	3.76	6.15	7
Rental status of some tobacco farms in Guilan Province	3.58	5.86	8
Slope of some tobacco farms in Guilan Province	3.29	5.39	9
Old traditions of tobacco growers about mechanized farming	3.05	4.99	10
Low price of tobacco leaves	3.00	4.91	11
Negative view towards the use of tobacco products in governmental systems	3.00	4.91	11
Insufficient attention of Guilan's tobacco organization to mechanized cultivation	3.00	4.91	11
Lack of tobacco mechanization experts in Guilan Province	2.94	4.81	12
Complexity and monopoly status of tobacco farming machinery	2.82	4.62	13
Low attention of agricultural managers and experts to the mechanized tobacco cultivation	2.58	4.22	14
Sum	61.09	100	

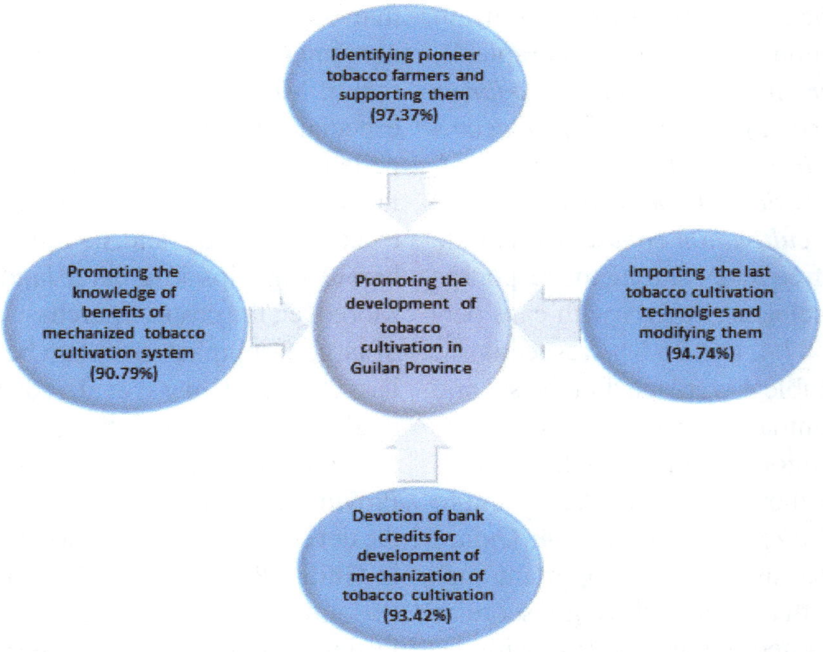

Figure 1. Driving factors of mechanized tobacco cultivation in Guilan Province

in Guilan Province obtaining 90.79% of panel's consensus. In Zehtab Naebi et al. (2015)'s study too, the factor "enhancing peanut growers' technical knowledge by holding specific educational courses" was ranked second most important driving factor of mechanized peanut cultivation in Guilan Province. In Abedi et al. (2017)'s study too, the factor "scientific training and enhancing olive garden owners' technical knowledge about specific machinery for olive harvest" was regarded as the third most important driving factor of the development of mechanized olive harvest in Guilan Province.

Inhibiting factors

According to the frequency of responses to the first open-end question at the first round of the inhibiting factors of the development of

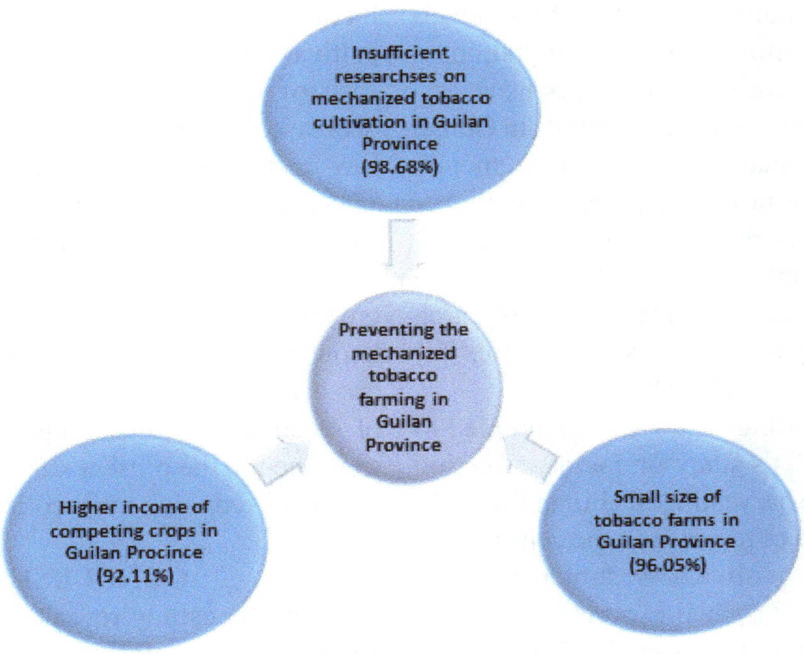

Figure 2. Inhibiting factors of mechanized tobacco cultivation in Guilan Province

mechanized tobacco cultivation in Guilan Province, 17 opinions were identified (Table 4). It shows that *fragmented status of tobacco farms in Guilan Province, higher income of tobacco's competing crops in Guilan Province,* and *high costs of mechanization of tobacco cultivation due to its special cultivation conditions* had the highest response frequency among the responses provided by the selected experts with case percentages of 57.9, 47.4, and 36.8%, respectively.

The results of Table 4 were used in the second round. Table 5 summarizes panel experts' agreement with the opinions provided by the experts of the first round about inhibitors. According to Table 5, *small size of tobacco farms in Guilan Province* was rated as the most important suppressing factor with normalized weight score of 7.60. It had been rated in the fifth rank with the frequency of 6 in the first round. *Higher income of tobacco's competing crops in Guilan Province* and fragmented status of tobacco farms in Guilan Province, which had been ranked second and first with the frequencies of 9 and 11 in the first round, respectively, were commonly rated as the second factor inhibiting the development of mechanized tobacco cultivation in Guilan Province with normalized weight score of 7.39.

According to the results of third round (see Fig. 2), *insufficient research on tobacco cultivation mechanization in Guilan Province* was rated as the first inhibiting factor of the development of mechanized tobacco cultivation in Guilan Province with 98.68% consensus percentage. The review of research reports in Guilan Province shows that they have been mostly conducted on rice production due to its higher importance in this province. So, the Guilan's Tobacco Company and the Guilan's Jahad-e Agriculture Organization must provide research supports dealing with the research topics for mechanized tobacco cultivation.

The same inhibiting factor was mentioned in Abedi et al. (2017)'s study on factors affecting mechanical olive harvesting in Guilan Province as specific support of research on the construction and modification of olive harvesting machinery acquiring 96.23% of panel experts' agreement.

Small size of tobacco farms in Guilan Province that acquired 96.05% of panel experts' final agreement was rated as the second inhibiting factor. Most of tobacco cultivation machinery especially for tobacco harvesters is of self-propelled type having low maneuverability in small farms; hence, the consolidation of small farms to easily turning and handling at the ends of harvesting trips in greater lands is a necessity. Moreover, the costs of purchase and maintenance of large farm machinery can be justified only in large farms (Almassi et al., 2006). Zehtab Naebi et al. (2015) also mentioned this factor as well as fragmented status of peanut farms in Guilan Province as the most important factor suppressing the development of mechanized peanut cultivation in Guilan Province which is in agreement with Rasouli et al. (2010)'s results about factors affecting the development of mechanized sunflower cultivation in Iran and Balachandran (2003)'s study on factors affecting mechanized agriculture in Kerala of India. Lamidi and Akande (2013) found tenure system or small farm holding as one of the main factors for undeveloped agriculture mechanization in Osun State of Nigeria. So, agriculture policy-makers in Guilan Province should consider strategies to prevent the fragmentation of tobacco farms due to inheritance.

There was a 92.11% agreement among panel experts that higher income of competing crops in Guilan Province was the third most important factor inhibiting the mechanization of tobacco cultivation in this province. The increase in the costs of tobacco production in recent years has decreased its cultivation area in exchange for low-cost, high-income crops like kiwifruit. Therefore, the governmental organizations in charge of tobacco cultivation and production should consider strategies to improve tobacco growers' income through enhancing their farming knowledge, crop insurance, and guaranteed purchase. It should be considered that the acceleration of mechanized tobacco cultivation which was examined in the present study can be a decisive factor for reducing tobacco cultivation costs.

CONCLUSION

The present study revealed that various technical, agronomical, skill, economic, social, and policy-making factors are involved in the development of mechanized tobacco cultivation in Guilan Province, among which identifying pioneer tobacco growers and supporting them to mechanized tobacco cultivation, importing the latest tobacco cultivation technologies and their modification, insufficient research on mechanized tobacco cultivation in Guilan Province, and small size of tobacco farms in Guilan Province were recorded as the most important factors. Thus, in order to develop mechanized tobacco cultivation and therefore, sustainable development of its production in Guilan Province, the Ministry of Jahad Agriculture and the Guilan's Tobacco Company should develop operational strategies to encourage the pioneer tobacco growers to use the tobacco cultivation machinery through allocating especial financial and technical assistances; to support the technical researches on modern tobacco cultivation technologies of the leading tobacco producing countries according to the type of tobacco cultivar cultivated in Guilan Province and its local agronomic conditions; to provide all peer reviewed tobacco cultivation machinery from abroad; to support the technical researches on imported machinery to modify them in accordance with the local condition; to codify plenary policies to consolidate the small tobacco farms in the studied region to provide infrastructures of the mechanized tobacco cultivation in Guilan Province.

ACKNOWLEDGEMENT

The authors express their gratitude to Rasht Branch, Islamic Azad University for its financial supports. They also feel deeply thankful to mechanization experts of Rasht and western Guilan for their kindly cooperation.

REFERENCES

1- Abedi, H., Firouzi, S., & Allahyari, M.S. (2017). Analysis of factors affecting the development of mechanical olive harvest in Guilan Province. *Journal of Farm Machinery, 7*(1), DOI: 10. 22067/jam. v7i1.48003 (In Press).

2- Almassi, M., Loveimi, N., & Kiani, S. (2006). Farm machinery management (1sted), Mashhad: Ferdowsi University of Mashhad Press, Iran (In Persian).

3- Guilan's Organization of Jihad Agriculture (2013). Annual statistics of agriculture. Guilan: Department of Planning and Economy, Center of Information and Communication Technology (In Persian).

4- Balachandran, P.G. (2003). Constraints on diffusion and adoption of agro-mechanical technology in rice cultivation in Kerala, Retrieved from www. cds.ac.in/krpcds/publication/downloads/ 59.pdf

5- Baladji, U., & Aghapour Sabbaghi, M. (2015). Investigating the reasons for non-acceptance of new machines by wheat growers in Andica County. *Journal of Scientific Research and Development, 2* (1), 62-66.

6- Capareda, S.C., de Ramos, J.D., del Rosario, A.C., & Castro, S.G. (2002). Mechanization of small to medium scale corn production areas in the Philippines. In: Implementation strategy manual. Agricultural Mechanization Development Program (AMDP), Institute of Agricultural Engineering, College of Engineering and Agro-Industrial Technology, University of the Philippines Los Banos, Philippines.

7- Capehart, T. (2004). Trends in U.S. tobacco farming (TBS-257-02). USDA, United States Department of Agriculture, Electronic Outlook Report from the Economic Research Service, Retrieved from https://www.ers.usda.gov/webdocs/publications/tbs25702/48597_tbs25702.pdf

8- Chi, T.T.N. (2010). Factors affecting mechanization in rice harvesting and drying in the Mekong Delta, South Vietnam. Omonrice, 17, 164-173.

9- Clarke, L.J., Morrison, T.A., Juricek, J., & Studenik, B. (1993). The Slovak Republic: Agricultural mechanization strategy, a review. Retrieved January 21, 2007, from http://www. fao. org /agris/Centre. asp?Content.

10- Conner, N., & Grady, R. (2013) Competencies and experiences needed by pre–service agricultural educators to teach globalized curricula: A modified Delphi study. *Journal of Agricultural Education, 54*(1), 8-17. DOI: 10.5032/ jae. 2013.01008

11- Dunham, R. B. (1996). The Delphi technique. Retrieved from: http://www.medsch. wisc.edu / adminmed/2002/orgbehav/delphi.pdf Firouzi, S. (2015). A survey on the current status of mechanization of paddy cultivation in Iran:

12- Case of Guilan Province. *International Journal of Agricultural Management and Development, 5*(2), 117-124.

13- Ghosh, B.K. (2010). Determinants of farm mechanization in modern agriculture: A case study of Burdwan districts of West Bengal. *International*

Journal of Agricultural Research, 5(12), 1107-1115.
14- Heiko, A. (2012). Consensus measurement in Delphi studies: Review and implications for future quality assurance. *Technological Forecasting and Social Change, 79* (8), 1525-1536.
15- Julius, A. (2014). Factors limiting small-scale farmers' access and use of tractors for agricultural mechanization in Abuja, North Central Zone, Nigeria. *European Journal of Sustainable Development, 3*(1), 115-124.
16- Keeney, S., Hasson, F., & McKenna, H.P. (2001). A critical review of the Delphi Technique as a research methodology for nursing. *International Journal of Nursing Studies, 38*(2), 195-200.
17- Kutte, M.T., & Tya, T.S.K. (2001). Mechanization strategies for sustainable agricultural production in Nigeria. Proceedings of NIAE, 23, 27 - 33.
18- Lamidi, W.A., & Akande, L.O. (2013). A Study of status, challenges and prospects of agricultural mechanization in Osun State, Nigeria. *Journal of Education, Arts and Humanities, 1*(1), 1- 8. Retrieved from https://www.watchpub/jeah/index.htm.
19- Mohammadi, O., & Zarifian, Sh. (2008). Factors affecting the mechanization of agricultural lands (Case city of Nishapur). 5th National Conference on Agricultural Machinery Engineering and Mechanization. Mashhad, Iran, September 2008, pp, 9.
20- Namvar Rezaei, A. (2008). Promotional mechanization tract in tobacco cultivation. Iranian Tobacco Company, Iran, Tehran (In Persian).
21- Ou, Y., Yang, D., Yu, P., Wang, Y., Li, B., & Zhang,Y. (2002). Experience and analysis on sugarcane mechanization at a state farm in China. 2002 ASAE Annual International Meeting/ CIGR 15th World Congress Sponsored by ASAE CIGR, Hyatt Regency Chicago, Illinois, USA July 28- 31.
22- Pishbin, S., Mohamadi, H., & Ejraei A. (2007). Problems of agricultural mechanization in Jahrom Region. *Development and Productivity, 5,* 17-29 (In Persian).
23- Rasouli, F., Sadighi, H., & Minaei, S. (2010). Factors affecting agricultural mechanization: A case study on sunflower seed farms in Iran. *Journal of Agricultural Science Technology, 11*, 39-48.
24- Salokhe, V.M., & Oida, A. (2003). Development of appropriate agricultural machinery for Southeast Asian Countries. Laboratory of Agricultural Systems Engineering. Retrieved 16 April 2005 from www.fsao.kais.kyotou. ac.jp/guest_prof/Salokhe_report.pdf.
25- Zamani, P. (2010). Cultivation and processing of tobacco (1th ed.).Iran: Beh-andishan.
26- Zehtab Naebi, R., Firouzi, S., & Ebrahimzadeh,M.R. (2015). Promoters and deterrents of developing mechanization of peanut cultivation in North of Iran. *International Journal of Agricultural Management and Development, 5*(1), 1-8.

17

Identification of Strategies for Sustainable Development of Rice Production in Guilan Province Using SWOT Analysis

Mohsen Shafieyan [1*], Mahdi Homayounfar [2] and Mehdi Fadaei [2]

Abstract

Sustainable development is one of the most important issues that can be considered from different aspects and in different contexts including sustainable agriculture. This study aimed to investigate the main strategies related to sustainable development of rice production as an important agricultural product in Guilan Province using a SWOT analysis. To this end, different internal (strengths and weaknesses) and external (opportunities and threats) factors affecting rice production development were reviewed in the literature. Next, a questionnaire was designed based on the potential opportunities, threats, strengths and weaknesses in the literature and the initial interview of the experts; then, the experts were asked to determine the importance of factors on a five-point Likert type scale. The results showed that the rice production received the total scores of 2.14 and 2.24 for the External Factor Evaluation (EFE) matrix and the Internal Factor Evaluation (IFE) matrix, respectively, indicating the ineffectiveness of the industry's strategies. These scores mapped the position of rice production in Guilan Province. Finally, a number of strategies for sustainable development of rice crop were outlined in order to reduce the weaknesses, avoid the threats, improve the strengths, and grasp the opportunities for the rice production of Guilan Province.

Keywords:
rice production, strategy formulation, sustainable development, SWOT analysis

[1] *PhD Student, Industrial Management, Department of Management and Accounting, Rasht Branch, Islamic Azad University, Rasht, Iran*

[2] *Assistant Professor, Industrial Management, Department of Management and Accounting, Rasht Branch, Islamic Azad University, Rasht, Iran*

* *Corresponding author's email: mohsen.shns@gmail.com*

INTRODUCTION

Experts and economists believe that in understanding the relationship among economic growth, social, and environmental dimensions of development, unilateral concentration on economic growth makes agricultural development inevitably unsustainable, which means it cannot be continued for a longtime (Soubbotina, 2004). During the last three decades, the concept of sustainability has attracted a lot of attention toward the deterioration of natural resources, consumers' concerns about quality of food, as well as degradation of life quality in rural communities (Shiri et al., 2012).

Sustainable development is a concept that can be considered from many different aspects and in different contexts including exchange values (Allahyari et al., 2008). Jeopardizing environmental quality, providing moral development, changing toward better future, providing social reorganization (Gupta, 1998), producing new capacities, empowering people and increasing knowledge (Bosshaq et al., 2013) that somehow help clarify the concept of sustainable development, which is the development that meets current needs without jeopardizing the ability of future generations in meeting their needs (WCED, 1987). Taylor (2002) believes that the concept of sustainable development is an important stage of environmental theory, because it posits how the society should organize itself. Therefore, sustainable development represents a philosophy with economic, social, and environmental dimensions[1] (Tanguay et al., 2010).

Environmental sustainability is achieved through the protection and effective management of natural resources. Economic sustainability is attained by a mix of occupations that provide long-term and stable incomes. Finally, social sustainability is ensured by means of active community participation and a strong civil society (Goodland, 1995). Furthermore, these three dimensions are fundamentally inseparable aspects of a whole. All three are essential, and thus, no one or two alone would suffice to ensure sustainability (Ikerd et al., 1996).

The production of sufficient food to feed an ever-increasing population is the most serious challenge facing mankind today (Essiet, 2001). The imbalance between the population growth and agricultural production is a real challenge in most developing countries. The growing pressures of population and the limited food supply have aroused global interest in research pertaining to environment, food, and nutrition (Burke et al., 2005). Although agriculture plays a main role in the economy, there have been concerns about the ability of the sector to alleviate rural poverty and to provide food security as well as stable incomes to farmers and other rural people. These concerns have been raised, because the farming environment and rural areas have faced unprecedented environmental problems over, at least, the last two decades (Tatlidil et al., 2009). In Iran, like other developing countries, agriculture is one of the most important economic sectors and contributes a considerably high percentage of production and employment (Ommani et al., 2009).

Sustainable agriculture is one of the important aspects of sustainable development. Basically, sustainable agriculture has no single meaning (De Koeijer et al., 2002). However, as Hansen (1996) argues, there are numerous various scientific conditions for agricultural sustainability that can be interpreted as the ability of an agricultural system to satisfy different requirements over time. It is argued that sustainable agriculture as a philosophy is rooted in a set of values and is based on human targets and long-term understanding of impacts of human activities on the environment and other animal species (MacRae et al., 1990). To ensure environmental, economic, and social sustainability, farmers must adopt different farm-level practices such as judicious use of chemicals, integrated pest management, adequate irrigation, and proper care of plant and animal health (Sadati et al., 2010).

Rice is one of the most important products in the world and a main food for more than 40% of the world's population (FAO, 2013). Rice, as the second most important product of Iran after wheat, has a great importance in nutrition of Iranian community. In Iran, a big share of rice cultivation is contributed by two Northern

[1] *Recently, a fourth dimension has been added:– the institutional dimension of development*

provinces that is, Guilan and Mazandaran. These two provinces are the main hub of rice cultivation and production in Iran. As a case studied in this research, Guilan Province is one of the most important hubs of rice production in Iran and provides, on average, 26 percent of annual rice production in Iran (Iran's Agricultural Statistics, 2013).

In recent years, several problems have occurred in the field of environmental degradation in Guilan Province because of consecutive cultivation of rice paddies without appropriate utilization of productive resources (Daryaee et al., 2011). Inordinate use of fertilizers, especially nitrogen fertilizers (200-300 kg per Hectare); possibility of water pollution, especially underground water and occasional entry of these synthetics into drinkable water resources, has posed many health problems to people and other organisms, in particular, fishes (Khalili et al., 2009). Excessive use of chemical pesticides has caused many harmful effects such as resistant pests, decreasing pesticides impact, extinction of species, and destruction of natural habitats. In addition, burning farm leftovers has caused the loss of organic substances and the increased need for fertilizer. Besides, consecutive plowing has indirectly resulted in the rapid decomposition of organic substances and reduction of soil humus (Daryaee et al., 2014; Nick Nezhad, 2008). Regarding the problems mentioned, regional differences make sustainable development mandatory in order to do proper planning for the existing potentials, and it requires expert assessment and evaluation. Recognition and analysis of the factors affecting each product can be effective for sustainable development of these products in favor of desirable utilization of provided resources, improvement of self-reliance and self-regulation of agricultural ecosystem, lowering of wastes, as well as minimization of environmental, social, and economic impacts. Therefore, the utilization of models related to the identification of agricultural capacities in each region is helpful in preventing the emergence of many problems, because the models are practical tools that help reach widespread perception of facts (Singh & Dhillon, 2004).

There are many models to recognize and analyze factors affecting a product. In the meantime, SWOT analysis is among tools that are used to analyze internal and external strategic factors in the strategy development stage. This matrix evaluates strengths, weaknesses, opportunities, and threats facing organization (Yuksel & Dagdeviren, 2007). SWOT analysis is based on the premise that an effective strategy maximizes strengths and opportunities, while minimizing weaknesses and threats (Dyson, 2004). This simple premise will have a major effect on the selection and formulation of effective strategy in case of proper implementation (Pearce & Robinson, 2003). This method, which is used to a great extent in strategic development, identifies all factors influencing the operating environment of organizations and reflects strategies tailored to current situation (Shrestha et al., 2004).

Falsolayman and Sadeghi (2013), Ghorbani et al. (2015), Javanmard and Mahmoudi (2008) Ommani (2011) and Reihanian et al. (2012) are examples of the studies employing the SWOT analysis for sustainable development. In another study, Vali Poor et al. (2013) formulated strategies for urban agriculture development in Dogonbadan using the SWOT analysis. Sadeghi et al. (2011) formulated advertising strategies for pistachio export in Kerman Province using the SWOT analysis. Sojasi Gheidari et al. (2011) proposed strategies for the development of agricultural entrepreneurship in rural areas using a combination of multivariate analysis (MCDM) and the SWOT. Fakhimi Azar et al. (2011) developed a strategic plan for Jahad-e Agriculture Organization of Eastern Azerbaijan Province using a combination of AHP and the SWOT.

Efficient utilization of resources in a way that is commensurate with their existing capacities seems to be necessary for agricultural sector to benefit from ascending trends of development. This concept lies in sustainable development. Hence, it seems that studying this matter using development models can be effective in providing programs and strengthening economic, social, and environmental aspects for rice farmers in the future. Therefore, the aim of this research is to identify the key internal and external factors affecting rice production in Guilan Province and to review the appropriate strategies for sustainable development of this product in the region using the SWOT analysis.

MATERIALS AND METHODS

The study area

Guilan Province with the area of 14,711 km^2 is located in the North of Iran covering the geographical coordinates of 48°34'25" W to 50°26'42' E and 36°36'3"S to 38°27'7" N. Rasht, the central city of the province, is 355 km west of the capital of Iran (Tehran) (Statistical Yearbook of Guilan Province, 2013).

Guilan Province is covered with coastal and mountainous areas and plains. Lowlands are the vastest area of the province, comprising most of farmland in the region. The soil of Guilan plain is sedimentary because of sediments from rivers in the region. The land in this area in not too steep and has good drainage, which makes them the best lands capable of farming. This type of soil is used to cultivate products such as rice, fruit trees, tobacco plants, and summer products. In Iran, Guilan Province as the case of this study with more than 26 percent of production and 31.7 percent of rice paddy cultivation is ranked the second in terms of production and cultivation of rice crop after Mazandaran Province. In this province, more than 150,000 farmers grow rice in more than 180,000 ha of fertile lands annually, and this is the most important agricultural activity in the province. Economic structure of the province is based on agriculture with a focus on rice production (Iran's Agricultural Statistics, 2013). Figure 1 shows the distribution of rice paddies in Guilan Province.

The sampling method

The research population consisted of the managers and specialists of Jahad-e Agriculture Organization (JAO), Rice Product Research Center (RPRC), as well as rice industry scholars/experts in Guilan Province. According to Eq. (1), the sample size was estimated based on the Cochran's formula for finite population. The formula including the specifications of the initial sample is shown below:

$$n=(N.t^2.s^2)/((N-1).d^2+t^2.s^2) \cong 91 \qquad (1)$$

where:

n is the sample size (?),

N is the population size (147),

t is the test statistic for 0.95 confidence level (1.96),

s^2 is the variance of responses in the pilot study (0.884),

d is the preferred likelihood accuracy (0.12). The expert recommendation for d was 10% which is a common level for a desired precision (Azadi et al., 2013). However, a slightly more precision (12%) was considered.

Data collection

A questionnaire was used to collect data from the experts in a pilot study. It included a list of strengths, weaknesses, opportunities, and threats facing rice industry, which were adjusted according to the indexes of Fifth Development Plan of Agricultural Sector.

A pilot study was conducted to measure the reliability of the questionnaire: Thirty questionnaires in the study area were administered to managers and experts from Jahad-e Agriculture Organization of Guilan Province and professors not mentioned in the investigation. Cronbach's alpha coefficient was estimated at 0.87, which

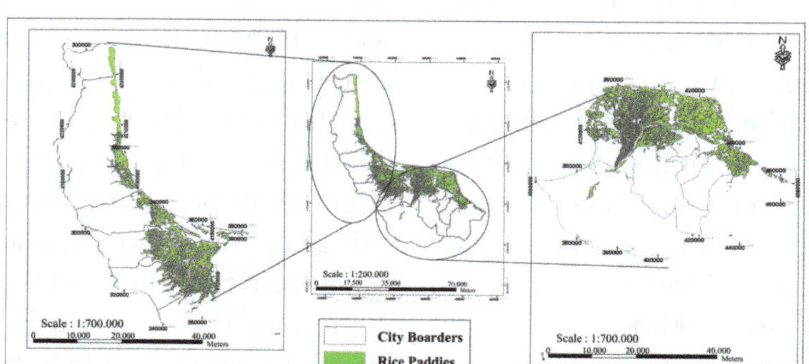

Figure 1: Distribution of rice paddies in Guilan Province

indicates the high internal consistency of the questionnaire.

An empirical study was conducted to answer the research questions, and a questionnaire was developed to collect the data. In the first step, 30 experts participating in the agricultural decisions at the senior level were chosen through a systematic random sampling approach. Next, the experts were asked to fill a questionnaire (developed based on the factors well supported by literature and the experts' opinions from the practical point of view) containing 66 items, 14 for strengths, 17 for weaknesses, 20 for threats and 15 for opportunities. Items were measured on a five-point Likert scale with extreme points of "strongly unimportant (1)" and "strongly important (5)". Next, the final sample size to conduct the research study was estimated at 91 based on the Cochran's formula, and 61 experts out of 91 randomly selected participants attended to fill out the questionnaire. In the second step, the possible list of strengths, weaknesses, opportunities, and threats was analyzed based on the data obtained from the questionnaire. The researchers finalized 43 factors based on the ranking of 66 factors (9 for strengths, 12 for weaknesses, 11 for threats and 11 for opportunities) of the questionnaire. Tables 1 and 2 present the most important factors in sustainable development factors and their normalized importance scores (relative weights) obtained through data collection.

The SWOT analysis

Harvard Business School Faculty developed the SWOT organizing framework in the early 1960s drawing upon Selznick's idea of matching an organization's internal factors (capabilities, resources, and limitations) with its external environment as a first step in the process of strategy development (Bower, 2008).

Identifying its strengths, weaknesses, opportunities and threats, the organization can develop various strategies so as to focus on its strengths, minimize threats, take the greatest possible advantage of opportunities available to organization, and use them to defend against threats. The strengths and weaknesses are revealed by an internal environment appraisal, whereas the opportunities and threats are revealed by an external environment appraisal (Dyson, 2004).

The purpose of a Strengths, Weaknesses, Opportunities and Threats framework is to get managers thinking to anything that can potentially impact the success of the product or industry. The internal environment of the company consists of variables within the company itself, of which the business management of the company does not have an influence in the short term. The external environment consists of variables existing outside the company, which in the short term are not under the control of the company. These variables form the context in which the company exists and functions (Houben et al., 1999).

Strategy formulation is the process of developing long-term plans to effectively respond to the environmental opportunities and threats in the light of the strengths and weaknesses (Houben et al., 1999). The consensus of Strengths, Weaknesses, Opportunities and Threats form a 2*2 matrix which is called SWOT and can be used for developing different strategies to achieve objectives.

The current research study has been done to identify the potential strategies and to prioritize and also to create a common vision for accomplishment of the development strategy. Although SWOT analysis is a research method which has been generally used in business plans, nowadays it is used to evaluate issues and policies related to decision-making and also to evaluate sustainable agriculture in a systematic way (Falsolayman & Sadeghi, 2013). The internal and external environmental factors have important roles in scanning strategic planning and are considered as major components of sustainable development process. This strategic environmental analysis is useful in formulation and selection of a strategy (Reihanian et al., 2012), since it may help researchers to gain insight into the past and to think of possible solutions for existing or potential problems (Ommani, 2011). A SWOT analysis can be carried out for a product, place, industry or person. It involves specifying the objective of the business venture or project as well as identifying the internal and external factors that are either favorable or unfavorable for achieving that objective (Ghorbani et al., 2015).

In this study, according to the strategy formulation models and patterns, the key internal and external factors affecting rice production in Guilan Province were studied and identified. After determining the strengths, weaknesses, opportunities and threats, the study findings were analyzed using the SWOT model that included Internal Factors Evaluation Matrix (IFEM), External Factors Evaluation Matrix (EFEM), and a combination of internal and external factors. Finally, different strategies for the various modes of SWOT (including ST, WO, SO and WT) were adopted to introduce approaches resulting from this analysis as a useful strategy to overcome current undesirable situation and to achieve ascending trend of development which aims at sustainable development. The following section elaborates the scoring process:

To evaluate internal and external factors (to define normalized coefficient), first, weighted scores between zero (unimportant) to one (very important) were assigned to each factor in such a way that the sum of the weighted scores could equal unity (the second column in Tables

Table 1
External Factors Evaluation Matrix (EFEM)

External Strategic Factors	Weight Factor (Normalized	Current Situation Score	Weighted Score
Opportunities			
O_1. Increasing demand for rice product	0.046	2	0.092
O_2. Increasing the production of agricultural products	0.046	2	0.092
O_3. Existence of appropriate capabilities/capacities to develop alternative industries	0.048	3	0.144
O_4. Availability of regional/domestic markets and development of rice export (airports, border markets and etc.)	0.046	1	0.046
O_5. Availability of new technologies to increase resources productivity	0.044	3	0.132
O_6. Possibility of value creation in rice product as well as other agricultural product	0.050	2	0.1
O_7. The distinctive position of Guilan province in rice production (the second largest rice producer in Iran)	0.054	3	0.162
O_8. A considerable number of graduates and experts in the province	0.046	4	0.184
O_9. Availability of considerable workforce in the province	0.041	3	0.123
O_{10}. The ability to manufacture and develop agricultural equipments and to produce high quality chemical, organic and biological agricultural products	0.040	2	0.080
O_{11}. The existence of necessary infrastructure for utilization of rainfalls, surface water and underground streams	0.039	1	0.039
Threats			
T_1. Increasing rate of change in farmland cultivation	0.053	2	0.106
T_2. Fluctuations in price of agricultural products	0.042	2	0.084
T_3. Inappropriate attitude toward agricultural industry and its role in economic development	0.043	2	0.086
T_4. The risk of entering plant pests and diseases to the province	0.042	2	0.084
T_5. Lack of sustainable rural development, workforce retention (especially young people) and increasing the average age of farmers	0.047	3	0.141
T_6. Natural disasters such as drought, flood, frost etc.	0.042	2	0.084
T_7. Inefficiencies of agribusiness	0.044	2	0.088
T_8. Lack of financial resources and investment in agriculture sector	0.042	2	0.084
T_9. Inconsistency of intra-industrial policies with requirements of agriculture industry	0.050	1	0.050
T_{10}. Inadequacy of water management in agricultural industry (poor delivery of water among farmers)	0.047	2	0.094
T_{11}. Decreasing quality and quantity of water resources	0.048	1	0.048
Total	1		2.14

Table 2
Internal Factors Evaluation Matrix (IFEM)

Internal Strategic Factors	Weighted Factor (normalized)	Current Situation Scores	Weighted Scores
Strengths			
S_1. Indigenous experience and knowledge about rice production	0.054	3	0.162
S_2. Fertile rice paddies or lands with susceptible soil	0.056	3	0.168
S_3. Proper climatic conditions and vast rice fields	0.053	3	0.159
S_4. High experienced experts in detection and control of pests and diseases of rice in the province	0.042	3	0.126
S_5. Wide variety of modified rice with high quality in the province	0.042	3	0.126
S_6. Significant institutional structures for agricultural facilities/crops in the province	0.049	3	0.147
S_7. Academic /research and development (R&D)centers in the province	0.055	4	0.220
S_8. The rich databases of various academic and scholarly studies on rice product	0.051	4	0.204
S_9. Wide network of rural cooperative centers and institutes related to the rice product in the province	0.035	3	0.105
Weaknesses			
W_1. Fixed cost of rice in comparison to global price	0.044	1	0.044
W_2. Lack of advanced agricultural machinery and use of traditional ways of rice production	0.041	2	0.082
W_3. Weakness of rice guild and pricing /advertising policies	0.053	1	0.053
W_4. Weak position of consultants in rice marketing	0.06	1	0.06
W_5. Smallholding system and lack of integration and dispersal of rice paddies	0.055	2	0.110
W_6. Insufficient investment in rice production and uneconomical rice cultivation as compared to the other economic activities in the province	0.038	2	0.076
W_7. Low productivity of production facilities	0.055	1	0.055
W_8. Contradiction and inefficiency of some existing procedures, rules and regulations as well as poor supervision over implementation of laws for developing rice product in the province	0.040	1	0.040
W_9. Weak linkage between the findings of applied researches and their implementation in agricultural sector	0.038	2	0.076
W_{10}. Unsuitable insurance services for rice product	0.050	1	0.050
W_{11}. Weakness in integrated management planning to detect harmful factors for rice products in the province	0.037	2	0.074
W_{12}. Nonconformity of farmers with sustainable agriculture principles	0.052	2	0.104
Total	1		2.24

1 and 2). To this end, a five-point Likert scale was used. First, the total scores were calculated for each question based on the judgment of samples. Then, mean total scores were derived in terms of average weight. Next, the average weights were normalized to gain weights between zero and one for each factor and the total weight equaled one, accordingly. Obviously, the higher the total scores of factors, the higher the average weight, and therefore, the higher its normalized ratio. These ratios indicate the importance of each factor based on respondents' opinions and the impact of that factor on current and future success. Next, according to the current condition of each factor, a score ranging from 1 to 4 was allocated to the current situation score column using the Delphi method (the 3rd column in Tables 1 and 2). (It is worth noting that in IFE Matrix, scores 1 and 2 were assigned to weaknesses and scores 3 and 4 were assigned to strengths). Finally, the weighted scores of each factor (multiple product of 2nd and 3rd columns of Tables 1 and 2) and total weighted score of factors were calculated.

RESULTS

Once the required data were collected, a

Table 3
Rice Product Development Strategies Based on SWOT Analysis

SWOT Strategies
SO Strategies
SO_1: Quantitative increase of rice production with regard to the lands with fertile soil, desirable rice paddies and suitable climatic conditions in the region (S_2-S_3-S_5-O_1-O_2)
SO_2: Development of mechanization of agricultural sector with regard to extensive network of cooperatives and unions of agricultural sector in the province (S_9-O_2-O_3-O_5-O_{10})
SO_3: To increase the value of rice product through improving rice quality with regard to study and research facilities as well as utilizing results of studies related to rice product (S_5-S_7-S_8-O_5-O_6)
WO Strategies
WO_1: Development of comprehensive mechanization project and modification of plantation patterns with regard to existing potentials of achieving innovative technologies and capability of manufacturing efficient agricultural equipment in the country (W_1-W_2-W_7-O_3-O_5-O_{10})
WO_2: Increasing irrigation efficiency and productivity of water and soil with regard to related innovative technologies as well as benefitting from graduates of relevant science in the province (W_2-W_7-O_3-O_5-O_8)
WO_3: Using specialists and graduates of province as consultant and study centers in rice product marketing (W_1-W_3-W_4-O_4-O_8)
ST Strategies
ST_1: Strengthening supportive technical and executive laws, regulations and rules to help sustainable development of rural areas, reduction or disposal of rice paddy use changes and fragmentation of lands under rice cultivation (T_1-T_5-T_9-S_6-S_9)
ST_2: Planning to prevent and mitigate the effects of natural disasters, pesticides and diseases through development of early warning systems and control networks with regard to local experience and knowledge as well as research centers (S_1-S_4-S_6-S_7-S_8-T_4-T_6)
ST_3: Developing appropriate mechanisms with scientific and technical criteria for allocating funds and investing in activities related to rice production with regard to extensive areas of activity for this product in the province (S_1-S_2-S_3-T_3-T_8)
ST_4: Cultivation of rice varieties with the approach of improving efficiency of water resources aimed at reducing the quantitative and qualitative loss of water resources (S_1-S_2-S_3-T_{10}-T_{11})
WT Strategies
WT_1: Coordination between cooperation sectors of the province to solve the problems of smallholding system, rice paddy use change and low efficiency of unions related to the rice product (W_3-W_5-W_8-T_1-T_7-T_9)
WT_2: Improving insurance services related to rice product regarding to the risks such as natural disasters and rice pests and diseases (W_{10}-W_{11}-T_4-T_6)
WT_3: Attracting investments to develop facilities and infrastructure required for rice production (W_2-W_6-T_3-T_8)
WT_4: Studying, planning and implementing effective mechanisms to reduce the cost of rice production to reduce fixed price of product and also to stabilize the market (W_1-W_3-W_4-W_9-T_2-T_9)
WT_5: Utilizing methods of extension education in order to promote and teach sustainable agriculture through increasing the level of knowledge and public awareness among farmers on organic agriculture (W_{12}-T_1-T_3-T_5-T_{11})

SWOT analysis was employed to evaluate the feasibility of sustainable development of rice production in Guilan Province through analyzing the results. The results of the analysis are presented in the following sections.

External Factors Evaluation Matrix (EFEM)

As Table 1 shows, the number of 11 external opportunities and 11 external threats were identified as external key factors facing rice production in Guilan Province. "considerable number of graduates and experts" as well as "the distinctive position of Guilan Province in rice production" with the corresponding weighted scores of 0.184 and 0.162 were the most important opportunities facing rice products in the region. In the case of threats facing rice production in the studied region, the factors of "lack of sustainable rural development, workforce retention (especially young people) and increasing the average age of farmers" were the most important factors among the threats, which was weighted as 0.141. In addition, the factor of "increasing rate of change in farmland cultivation" with a score of 0.106 was the second most important rice production threat in Guilan Province.

According to Table 1, total weighted score of

external factors for rice production in Guilan Province is 2.14 that was lower than final mean total scores (2.5) which represents threats overwhelming opportunity (David, 2007). Therefore, rice production in Guilan Province did not have a satisfactory situation in terms of external factors. Taking advantage of opportunities and trying to reduce threats might help leaving the current adverse condition.

Internal Factors Evaluation Matrix (IFEM)

The results of Internal Factors Matrix were analyzed and represented in Table 2. Accordingly, nine internal strengths and 12 internal weaknesses were identified as internal key factors of rice production in Guilan Province. "academic/research and development (R&D) centers in the province" in agriculture sector in Guilan Province was the most significant strength. This factor gained the score of 0.22 among nine introduced strengths. "the rich databases of various academic and scholarly studies on rice product" and "Fertile rice paddies or lands with susceptible soil" were ranked the second and third critical factors with the weighted factors of 0.204 and 0.168, respectively. In the field of weaknesses of rice production in the region, "smallholding system and lack of integration and dispersal of rice paddies" as well as "nonconformity of farmers with sustainable agriculture principles" were ranked the first and second weaknesses with the scores of 0.11 and 0.104, respectively.

According to Table 2, the total weighted score of internal factors matrix was equal to 2.24 for rice production in Guilan Province. This score was lower than the average mean of matrix (2.5), which indicates the dominance of weaknesses of rice over its strengths in the region (David, 2007). It is worth noting that the quantitative results obtained from the internal and external factor matrixes provide the possibility of offering different strategies in effective manner.

SWOT strategies

After selection and evaluation of key internal and external factors and identification of relationships between their attributes, four types of strategies were adopted that can be used effectively for sustainable development of rice in Guilan Province. For example, the relationship between Strengths and Opportunities (SO) could indicate good condition of rice crop of province and provides opportunity to take advantage of an aggressive strategy. These strategies suggest moving toward ideal situation, by the means of which we can use all positive points to maximize the opportunities. Furthermore, interactions between Weaknesses and Opportunities (WO) can be used as a potential for overview strategies. The aim of these strategies is to minimize weaknesses and maximize opportunities. Moreover, the relationship between Strengths and Threats (ST) can provide possibility to take advantage of diverse strategies. In ST strategy which focuses on strengths and threats, the aim is to maximize the utilization of strengths of rice crop in Guilan Province to deal with external threats and to minimize them. Finally, the interactions between Weaknesses and Threats (WT) can be considered as a potential warning and advice to make use of defensive strategies. These strategies seek to minimize weaknesses and threats. In this case, which is the most worrying situation, an urgent need is felt to reevaluate and reform the strategic policies.

As SWOT matrix in Table 3 shows, 15 key strategies for sustainable development of rice production in Guilan Province is set by pairwise matching of SO, WO, ST, WT.

CONCLUSION

This study provides a systematic manner for rice product assessment using SWOT analysis. It did a qualitative assessment considering strengths, opportunities, weaknesses, and threats jeopardizing rice production to obtain strategies for its sustainable development. An important reason for the lack of sustainable development in the region is the data shortage due to the lack of studies on strategies for rice production development. The main issues influencing rice product development in Guilan Province were identified by the SWOT analysis that resulted in criticizing the current and future situation of rice production in Guilan Province. The results from internal and external environment analysis show that the total weighted score of external factors affecting rice production is 2.14, which suggests the predominance of threats over opportunities, which in turn indicates that rice production industry has been reacted somewhat weakly to threats and opportunities. In addition, the total weighted score of internal factors was calculated as 2.24 which shows weakness of rice production industry in terms of internal factors. In the next step, using cross product of strengths with opportunities and threats as well as weaknesses with opportunities and threats in the SWOT matrix, rice production development strategies were categorized for the province in four groups of SO, ST, WO and WT, respectively. Guilan Province, a hub of rice production, has a noticeable potential in this context. Yet, the current situation of its rice production is not in line with the sustainability requirements due to the fact that there are no proper management activities in the context of sustainable development in the region. Notwithstanding, there are still hopes in spite of unsustainable condition of rice production in the region. In this context, policy makers and agricultural managers try to resolve current threats and weaknesses using opportunities and strengths.

Given the objectives of this study, which aimed at identifying key internal and external factors influencing rice production in Guilan Province as well as providing strategies for sustainable development of the rice production at the regional level in accordance with the status quo, effective implementation of the proposed strategies would be useful to achieve development goals and to guarantee the movement toward the desired state. Moreover, for the integration of rice product in the path of sustainability based on previous researches and discussions with senior managers and academics of Jahad-e Agricultural Organization, eight essential actions needed to achieve sustainability of rice production are derived from the studies (Alonge & Martin, 1995; Chaharsughi Amin & Mirdamadi, 2008; Falsolayman & Sadeghi, 2013; Pourzand & Bakhshoodeh, 2014; Saifi & Drake, 2008; Taghdisi & Bosshaq, 2012; Zahedi & Najafi, 2006) that are necessary to be implemented along with the proposed strategies. The following practices can significantly orient rice production of the region toward sustainable development:

1- Creating sustainable (sustainable agriculture) development fund for rice production (Zahedi & Najafi, 2006).

2- Proper management of water and soil resources (Pourzand & Bakhshoodeh, 2014; Saifi & Drake, 2008; Taghdisi & Bosshaq, 2012).

3- Decreasing the use of chemical pesticides (Alonge& Martin, 1995; Taghdisi & Bosshaq, 2012).

4- Using cattle in weed management (Chaharsughi Amin & Mirdamadi, 2008).

5- Decreasing the use of chemical fertilizers (Saifi & Drake, 2008; Taghdisi & Bosshaq, 2012).

6- Using green fertilizers (natural fertilizers and manure) (Alonge & Martin, 1995; Taghdisi & Bosshaq, 2012)

7- Biologically adjusting pest control (Chaharsughi Amin & Mirdamadi, 2008; Taghdisi & Bosshaq, 2012)

8- Organizing smallholding system to provide project development and increase production per unit area (Falsolayman & Sadeghi, 2013).

Finally, the present paper provided a valuable, important, and useful technique for reviewing sustainable development of rice production in the studied region. The present research in this context shows that methods like SWOT analysis can be useful tools for better identification of positive and negative factors affecting sustainable development of rice production in Guilan Province and other rice producing provinces in the country. These results may help managers at Jahad-e Agricultural Organization of the province and other organizations to identify sustainable development activities of rice product in the region.

RECOMMENDATIONS

Some weaknesses are evident in evaluation and measurement steps of SWOT analysis (Yuksel & Dagdeviren, 2007). In conventional SWOT analysis, no constraints are imposed on the magnitude of factors to make it possible to determine the impact of each factor on the strategy or proposed plan (Masozera et al., 2006). In other words, SWOT analysis is not a proper analytical tool to determine the relative importance of factors or capabilities to evaluate appropriateness of available alternatives based on these factors (Kajanus et al., 2004). Also, the results of SWOT analysis is often merely a list or a partial quality examination of the internal and external factors (Kangas et al., 2003). Therefore, it is suggested that in future studies Analytic Hierarchy Process (AHP) be employed to prioritize strategies resulted from SWOT analysis in this study. This combined method was used by Kurttila et al. (2000) for the first time to eliminate the loopholes offered by SWOT analysis in a case study on the certification of planting trees in a forest in Finland. Examples of studies that have followed the method of Kurtilla et al. (2000) include those by Kajanus et al. (2004), Masozera et al. (2006) and Shrestha et al. (2004). These studies, like studies conducted by Kurttila and his colleagues, deal with prioritizing factors and sub-factors of SWOT. AHP is one of the most comprehensive systems designed for decision-makings with multiple criteria and provides the possibility for hierarchal formulation of problems. This process involves different options in decision making and demonstrates the feasibility to analyze the sensitivity of criteria and sub-criteria. In addition, it is binary based which facilitates judgment and calculations and shows the compatibility and incompatibility of decisions, which is the distinctive advantage of this technique in multi-criteria decision making processes. In addition, given the importance of rice production as the second most strategic crop in Iran as well as the need to develop and implement effective mechanisms for sustainable development of this crop in Iran, it is suggested that the present study should be replicated in other provinces of the country that are similarly engaged in rice production and the results be compared.

ACKNOWLEDGMENT

We are thankful to Mr. Adibi, Agriculture Management of "Jahad-e Agricultural Organization" in Guilan Province for his scientific comments on this study.

REFERENCES

1- Allahyari, M.S., Chizari, M., & Homaee, M.(2008). Perceptions of Iranian agricultural extension professionals toward sustainable cultural Sustainability in Terms of Efficiency: The Case of Dutch Sugar Beet Growers. *Journal of Environmental Management, 66*, 9-17.

2- Dyson, R.G. (2004). Strategic Development and SWOT Analysis at the University of Warwick. *European Journal of Operational Research, 152*, 63-640.

3- Essiet, E. U. (2001). Agricultural sustainability under small-holder farming in Kano, northern Nigeria. *Journal of Arid Environments, 48*, 1-7.

4- Fakhimi Azar, S., Malek, A., & Naqinezhad, A. (2011). Strategic Study and Compilation Using SWOT Model and Analytic Hierarchy Process (AHP) (Case Study on Agricultural Organization of East Azerbaijan Province). *Strategic Management Studies Journal, 18* (49), 117-95. (In Persian)

5- Falsolayman, M., & Sadeghi, H. (2013). An Analysis on Capabilities of Agricultural Sector of Southern Khorasan to Have a Sustainable Development Managed by the Application SWOT Model. *Iranian Journal Geography and Development, 11*(30), 139-156. (In Persian)

6- FAO. (2013). *Rice market monitor (monitor (Report),* XVI (4), Trade and Markets Division. Retrieved from http://www.fao.org/economic/est/publications/rice-publications/rice-market monitor rmm/en/

7- Ghorbani, A., Valiollah, R., Rafiaani. P., & Azadi, H. (2015). Ecotourism sustainable development strategies using SWOT and QSPM model: A case study of Kaji Namakzar Wetland, South Khorasan province, Iran. *Tourism Management Perspectives, 16*, 290-297.

8- Goodland, R. (1995). The concepts of environmental sustainability. *Annual Review of Ecology and Systematics, 26*, 1-24.

9- Gupta, A. (1998). *Ecology and Development in the Third World, Second Edition.* London: Routledge.

10- Hansen, J. W. (1996). Is Agricultural Sustainability a Useful Concept? *Agricultural Systems, 50*(1), 117-143.

11- Houben, G., Lenie, K., &Vanhoof, K. (1999). A knowledge-based SWOT-analysis system as an instrument for strategic planning in small and medium sized enterprises. *Decision Support Systems, 26*, 125-135.

12- Ikerd, J., Devino, G., & Traiyongwanich, S. (1996). Evaluating the sustainability of alternative farming systems: A case study. *American Journal of Alternative Agriculture, 11*(1), 25-29.

13- Iran's Agricultural Statistics. (2013). *Comprehensive Agricultural Information. Tehran: Ministry of Agriculture, Planning and Economic Department, Information and Communication Technology Center*, Iran. Retrieved from http:// dpe.agri-jahad.ir/portal/ Home/Default. aspx?CategoryID=117564e0-507c-4565-9659-fbabfb 4acb9b

14- Javanmard, M., & Mahmoudi, M. (2008). A SWOT Analysis of Organic Dried Fig Production in Iran. *Environmental Sciences, 6*(1), 101-110.

15- Kajanus, M., Kangas, J., & Kurttila, M. (2004). The Use of Value Focused Thinking and the A'WOT Hybrid Method in Tourism Management. *Tourism Management, 25*, 499-506.

16- Kangas, J., Kurtila, M., Kajanus, M., & Kangas, A. (2003). Evaluating the management strategies of a forestland estate-the S-O-S approach. *Journal of Environmental Management, 69*, 349-358.

17- Khalili, H., Azizi, M., Cherati, A. & Bahadori, M. (2009). *The Effect of Nitroxin Biological Fertilizer on Growth and Yield of Hybrid Rice in Spring1*. Rice Product Research Institute, Agricultural Promotion, Education and Research Organization of Ministry of Agriculture. Accessible:http://www.berenge.com/ Articles/ ShowArticle.aspx?ArticleID=3292

18- Kurttila, M., Pesonen, M., Kangas, J., & Kajanus, M. (2000). Utilizing the Analytic Hierarchy Process (AHP) in SWOT Analysis-a Hybrid Method and Its Application to a Forestcertification Case. *Forest Policy and Economics, 1*, 41-52.

19- MacRae, R.J., Hill, S.B., Mehuys, G.R., & Henning, J. (1990). Farm-Scale Agronomic and Economic Conversion from Conventional to Sustainable Agriculture. *Advances in Agronomy, 43*, 155-198.

20- Masozera, M.K., Alavalapati, J.R.R., Jacobson, S.K., & Shresta, R.K. (2006). Assessing the suitability of community-based management for the Nyungwe Forest Reserve, Rwanda. *Forest Policy and Economics, 8*, 206-216.

21- Nick Nezhad, D. (2008). *Contamination of Soil and Water Resources with an Emphasis on Sustainable Agriculture*. Paper presented at Second Conference of Environmental Engineering, Tehran,17-21 May, 2008.

22- Ommani, A. R. (2011). Strengths, weaknesses, opportunities and threats (SWOT) analysis for farming system businesses management: Case of wheat farmers of Shadervan District, Shoushtar Township, Iran. *African Journal of Business Management, 5*(22), 9448-9454.

23- Ommani, A. R., Chizari, M., Salmanzadeh, C., & Hossaini, J. F. (2009). Predicting Adoption Behavior of Farmers Regarding On-Farm Sustainable Water Resources Management (SWRM): Comparison of Models. *Journal of Sustainable Agriculture, 33*(5), 595-616.

24- Pearce, J.A., & Robinson, R. B. (2003). *Strategic Management: Formulation, Implementation, and Control*. New York: McGraw Hill.

25- Pourzand, F., & Bakhshoodeh, M. (2014). Technical efficiency and agricultural sustainabilitytechnology gap of maize producers in Fars province of Iran. *Environment, Development and Sustainability, 16*(3), 671-688.

26- Reihanian, A., Binti Mahmooda, N. Z., Kahromb, E., & Hin T. W. (2012).Sustainable tourism development strategy by SWOT analysis: Boujagh National Park, Iran. *Tourism Management Perspectives, 4*, 223-228.

27- Sadati, S. A., Shaabanali Fami, H., Asadi, A., & Sadati, S. A. (2010). Farmer's Attitude on Sustainable Agriculture and its Determinants: A Case Study in Behbahan County of Iran. *Research Journal of Applied Sciences, Engineering and Technology, 2*(5), 422-427.

28- Sadeghi, T., Taleb Beydokhti, A., & Sedaghat, R. (2011). Codification of Advertisement Strategies in Coherence with the Export of Pistachio in Kerman province by SWOT Model. *Agricultural Economics, 5*(1), 123- 142. (In Persian)

29- Saifi, B., & Drake, L. (2008). A co-evolutionary model for promoting agricultural sustainability. *Ecological Economics, 65*(1), 24-34.

30- Shiri, N., Hashemi, S.M.K., Asadi, A., & Motamedinia, Z. (2012). Agricultural Researchers' Attitudes toward Sustainable Agriculture and its Determinants in ILAM province, Iran. *International Journal of Agricultural Science and Research, 3*(1), 17-30.

31- Shrestha, R.K., Alavalapati, J.R.R., & Kalmbacher, R.S. (2004). Exploring the Potential for Silvopasture Adoption in South-central Florida: An Application of SWOT-AHP Method. *Agricultural Systems, 81*, 185-199.

32- Singh, J., & Dhillon, S.S. (2004). *Agricultural Geography (3rd ed.)*. New Delhi, India: Tata McGraw-Hill.

33- Sojasi Gheidari, H., Palouj, M., Roknadin Eftekhari, A., & Sadeghlou, T. (2011). Using multi criteria decision analysis methods (MCDM) and SWOT in presentation the strategies of agricultural entrepreneurship in rural area. *Agricultur-*

al Economic and Development, 19(74), 149-180. (In Persian)

34- Soubbotina, T.P. (2004). *Beyond Economic Growth, Second Edition: An Introduction to sustainable Development.* Washington, D.C., the World Bank.

35- Statistical Yearbook of Guilan Province. (2013). Guilan: Management and Planning Organization in Guilan Province, Iran.

36- Taghdisi, A., & Bosshaq, M. R. (2012). Analysis and Evaluation of Agricultural Sustainability in Rural Areas and the Role of Farmers (Case Study: Rural areas of AZNA city). *Geography (Research and Science Journal of Geographical Society of Iran), 10* (33), 12-147. (In Persian)

37- Tanguay, G. A., Rajaonson, J., Lefebvre, J. F.,& Lanoie, P. (2010). Measuring the ustainability of cities: An analysis of the use of local indicators. *Journal of Ecological Indicators, 10*, 407-418.

38- Tatlidil, F. F., Boz, I., Tatlidil, H. (2009). Farmers' perception of sustainable agriculture and its determinants: A case study in Kahramanmaras Province of Turkey. *Environment, Development and Sustainability, 11*, 1091-1106.

39- Taylor, j. (2002). Sustainable Development a Dubious Solution in Search of a Problem. *Policy Analysis, 449*, 1-49.

40- Vali Poor, S., Akbari, M. R., & Zaker Haghighi, K. (2013). Strategic planning for Urban Agriculture Development by SWOT (Case Study: Dogonbadan city of GACHSARAN). *Urban Management, 5*(15), 45-57. (In Persian)

41- WCED (World Commission on Environment and Development). (1987). *Our Common Future.* Oxford University Press, Oxford.

42- Yuksel, I., & Dagdeviren, M. (2007). Using the analytic network process (ANP) in a SWOT analysis-A case study for a textile firm. *Information Sciences, 177*(16), 3364-3382.

43- Zahedi, S. A., & Najafi, G. (2006). Management of Sustainable Agricultural Development. *Journal of Management Studies, 50*, 3-18. (In Persian)

18

Impact of Energy Subsidies Elimination on Technology Gap Ratio in Cucumber Production

Reza Esfanjari Kenari [1*]***, Zohre Karami*** [2] ***and Seyadeh Sedighe Ahmadzade*** [3]

Abstract

This paper presents an analysis of technical efficiency and technology gap ratio (TGR) in greenhouse cucumber in Fars Province, Iran. Cucumber production was chosen for this study for the reason that greenhouse productions in this province mainly have focused on this product. The data used in this study was obtained from a random sample of 127 greenhouse s in Fars Province for 2010 to 2011. Metafrontier production function model for firms was used within the parametric framework of stochastic frontier analysis (SFA). The frontier models are applied in the analysis of cross-sectional data by assuming a translog functional form. Results indicate that eliminating energy input subsidies has led to significant decrease in greenhouse cucumber production efficiency so that the mean technical efficiency declined from 98% to 67 % during 2010-2011. Furthermore, subsidies elimination has also led to decrease of the mean technology gap ratio in greenhouses from 0.92 to 0.87, in other words, it has caused more distance from efficient production frontier.

Keywords:
metafrontier, stochastic frontier Analysis, translog function, Fars

[1] *Assistant Professor, Agricultural Economics, Faculty of Agricultural Sciences, University of Guilan*
[2] *MSc, Agricultural Economics, Shiraz University, Shiraz, Iran*
[3] *PhD Student, Agricultural Economics, Sari Agricultural Sciences and Natural Resources University*
* *Corresponding author's email: Rezasfk@gmail.com*

INTRODUCTION

Agricultural sector has a significant contribution for economy growth in respect to providing food for population, supplying essential materials required by industry and assisting foreign balance of payments through the exports. In this regard, food security is one of the most controversial issues in developing countries with the high rate of population growth. In order to confront with food shortages and their negative impacts on the economic development, food production capacity in agriculture must be increased. The increasing population growth, reducing water resources and arable lands due to the development of urban areas, warrant more efficient use of existing resources. In this context, greenhouse production technology can be considered as an applicable way to improve the efficiency and productivity of these resources.

Among various approaches to increase production, development of production inputs and making major changes in technologies are faced with several difficulties and limitations particularly in developing countries. Therefore, increasing technical efficiency has been considered as the most appropriate solution. Improving technical efficiency can result in more production outputs from a specified set of inputs and it also can prevent wasting resources. Up to 2010 energy consumption in Iran due to having cheap sources of energy and support from government subsidies in comparison with the international standards has been high. Determining energy prices lower than the world price by the government increases government spending and as a result increases unsustainable budget deficit. Undoubtedly, removing energy subsidies for optimal allocation of limited resources is the most important task that should be done in the Iranian economy. The aim of this study was to investigate the effects of removing energy subsidies on technical efficiency and technology gap ratio.

Monjezi and Sheikhdavoodi, (2011) investigated the energy efficiencies of cucumber producers using a non-parametric method of Data Envelopment Analysis (DEA) for 25 greenhouses in Khuzestan province of Iran. Average technical, allocative and economic efficiency indices were estimated to be 88%, 91% and 96%, respectively. (Taki et al., 2012), applied a parametric (Cobb-Douglas production function) and a non-parametric (DEA) method to analyze the efficiency for 25 cucumber greenhouses in Esfahan province (Iran). The average values of technical, allocative and economic efficiency of greenhouses were found to be 0.90, 0.95 and 0.94, respectively. (Omid et al., 2011) studied the degree of Technical Efficiency (TE) and Scale Efficiency (SE) of 18 selected cucumber greenhouses in Iran during September–December 2008 period. The TE of the inefficient units, on average, was calculated as 91.5%. This implies that, there are possibilities for either increasing total production of cucumber using the same inputs or decreasing input for the current level of cucumber production or a mixture of both by filling the gap between the best producer and other producers.

Mehrabi Boshrabadi et al., (2008) used a translog metafrontier production function to estimate TE and environmental-technological gaps (ETGR) in wheat production in Kerman Province of Iran in five major wheat production regions. Their results demonstrated that in regions with a lack of water resources, there was a wide technological gap compared with metafrontier function.

In another study by (Moreira & Bravo-Ureta, 2010), they estimated the technical inefficiency of pistachio farms in Kerman Province using a translog production function. The estimated TE of different varieties of pistachio trees reported to be from 59.4% to 78.7%, while the farmers experience was found to be the most effective factor. Some studies compared technical efficiency indices for different groups. For instance Battese et al., (1993) measured TE of wheat farmers in some distinct districts of Pakistan. But they did not test whether the frontiers of all districts are the same or not. Villano et al., (2010), discussed about when was meta frontier analysis appropriate; they outline two criteria to justify its use: an inability in farmers to switch between production technologies except in the long term, and satisfaction of statistical tests on metafrontier coefficients. They mentioned farmers may have access to a set of production techniques and their potential for turning inputs into outputs-but for one reason or other are unable to use them, or at least are unable to use them except in the long run. Physical conditions, environmental constraints, capital scarce and are the most important reasons expressed by them that should be taking into account, if farmers want to reach the metafrontier. So far, to the best of our knowledge, no study had tried to measure technical efficiency of the greenhouse cucumber while taking into consideration the varietal difference and their relationship with technology gap.

Cucumber is one of the major greenhouse vegetable products worldwide. In Iran, the planted area for cucumber was 78,000 ha with 1.72 million tones production in 2007. From 2002 to 2007, the total greenhouse area doubled in Iran from 3380 ha to 6630 ha. The shares of greenhouse crops production were as follows: vegetables 59.3%, flowers 39.81%, fruits 0.54% and mushroom 0.35 % (Omid et al., 2010). In Fars Province, the planted area of greenhouse vegetable production was 614523 m^2 (180 units) of which 48 ha was dedicated to cucumber production with 7841.8 tons of production in 2010-2011. The greenhouse cucumber production has been specified the most area and production among the other varieties of vegetables such as tomato and pepper in Fars province (Ministry of Jihad-e-Agriculture, 2010). The purpose of this study was to analyze the technical efficiency and technology gap ratio (TGR) in greenhouse cucumber production in Fars Province Iran, where the cucumber production was the major greenhouse productions. The effect of subsidies elimination was also assessed by comparison of TE and TGR calculated for the two time periods (prior to and after elimination of subsidies).

MATERIALS AND METHODS

In present study the stochastic frontier production function approach is used to estimate the TE In stochastic frontier models, efficiency estimation typically assumes that the fundamental production technology is the same for all farms. Due to differences in technology for greenhouse cucumber production under the survey period (2010 and 2011), efficiency comparison is not valid for categorized groups. The main advantage of stochastic frontier analysis is its ability to detect stochastic effects errors from the resultant errors of inefficiency effects. Given the other inputs, technology and environmental factors, technical efficiency is estimated based on the ratio of realized to expect maximum output.

(Aigner et al., 1977; Meeusen et al., 1977) developed the original model which later has been improved to handle different situations by (Pitt & Lee, 1981; Jandrow et al., 1982; Battese & Coelli, 1992, 1995; Kumbhakar, 2002). In recent studies, excluding the assumption of the uniformity in technology across the firms has been considered as a remarkable progress. In respect to this (Battesse & Rao, 2002; Battesse et al., 2004) have recommended that the stochastic metafrontier framework to compute both the technical inefficiencies of firms and the technology gap.

The Technology Gap Ratio (TGR) measures the ratio of the output for the frontier production function for the k-th group relative to the potential output that is defined by the metafrontier function, given the observed inputs (Battese & Rao, 2002; Battese et al., 2004). According to this method, we assume there are k groups with different technologies in the industry. The standard stochastic frontier model for group-k defined as:

$$Y_{it(k)} = f(X_{it(k)}, \beta_{(k)})e^{V_{it(k)}-U_{it(k)}} \quad (1)$$

Where Y_{it} *(k)* implies the output of the *i*-th farm in the *t*-th period for the *k*-th group; X_{it} *(k)* denotes a vector of functions of the inputs used by the *i*-th farm in the *k*-th group;

$\beta_{(k)}$ is the vector of unknown parameters to be estimated associated with the *k*-th group; $V_{it(k)}$ represents statistical noise assumed to be independently and identically distributed a $N(0,\sigma^2 v_{(k)})$ random variables; and $U_{it(k)}$ are non-negative random variables assumed to account for technical inefficiency in production and assumed to be independently distributed as truncations at zero of the $N(\mu_{it(k)}, \sigma^2 U_{(k)})$ distribution.

The technical efficiency of the i-th farm with respect to the group-k frontier can be obtained using the result:

$$TE_{it}^{k} = \frac{Y_{it}}{e^{Xi\beta + V_{it(k)}}} = e^{-U_{it(k)}} \quad (2)$$

Equation (2) allows us to examine the performance of the i-th farm relative to the individual group frontier. In order to examine the performance of the i-th farm relative to the metafrontier, the stochastic metafrontier production function approach is used. The metafrontier is a function that envelops the stochastic frontiers of the different groups such that it is defined by all observations in the different groups in a way that is consistent with the specifications of a stochastic frontier model (Battese & Rao, 2002). Following a stochastic metafrontier production function model in the industry is defined as:

$$Y_{it}^{*} = f(X_{it}, \beta^{*}) e^{X_{it}\beta^{*}} \quad i = 1, 2, \ldots, N_k \quad t = 1, 2, \ldots T \quad (3)$$

where $i = 1, 2, \ldots, N_k$, $t = 1, 2, \ldots T$; Y_{it}^{*} is the metafrontier output that dominates all group frontiers, and β^{*} denotes the vector of metafrontier parameters satisfying the constraints:

$$X_{it}\beta^{*} \geq X_{it}\beta_{(k)} \quad \text{for all } k$$

The observed output defined by the stochastic frontier for the k-th group in Equation (1) can be alternatively expressed in terms of the metafrontier function in Equation (3), such that

$$Y_{it} = e^{-U_{it(k)}} \times e^{X_{ij}\beta_{(k)}} / e^{X_{it}\beta^{*}} \times e^{X_{it}\beta^{*} + V_{it(k)}} \quad (4)$$

The first term on the right-hand side of Equation (4) is the same as that in Equation (2), which denotes the technical efficiency of the i-th farm in the t-th period relative to the group-k frontier. The second term is what named the Technology Gap Ratio, which is expressed as:

$$\text{TGR} = e^{X_{it}\beta_{(k)}} / e^{X_{it}\beta^{*}} \quad (5)$$

The TGR measures the ratio of the output for the frontier production function for the k-th group relative to the potential output that is defined by the metafrontier function, given the observed inputs (Battese et al., 2004). The TGR has values between zero and one.

The technical efficiency of the i-th farm, given the t-th observation, relative to the metafrontier, is denoted by TE_{it}^{*} and is defined in a similar way to Equation (2). It is the ratio of the observed output relative to the last term on the right-hand side of Equation (5), which is the metafrontier output, adjusted for the corresponding random error. Following Equations (2) and (5), equation 6 can be determined such that:

$$TE_{it}^{*} = TE_{it}^{k} \times TGR_{it} \quad (6)$$

In this study, we estimated the Cobb–Douglas and translog functional forms of equations (1) and (3). In order to choose the appropriate form between two aforementioned functions, the likelihood-ratio test (equation 7) was used (12).

$$LR = -2\left[\ln\left(\frac{L(H_0)}{L(H_1)}\right) = -2[\ln(L(H_0) - \ln(L(H_1))\right] \quad (7)$$

Based upon equation (7), the specification of the translog and Cobb- Douglas functional forms is given by

$$LnY_{(k)} = \beta_{0(k)} + \sum_{j=1}^{4} \beta_{j(k)} LnX_{j(k)} + \frac{1}{2}\sum_{j=1}^{4}\sum_{s=1}^{4} \beta_{js(k)} LnX_{j(k)} X_{s(k)} + V_{(k)} - U_{(k)} \quad (8)$$

where j represents the j-th input ($j = 1, 2, \ldots 4$) of the i-th unit ($1, 2, \ldots, N_k$) in the t-th time period ($t = 1, 2$) in the k-th group ($k = 1, 2$); $Y_{i(k)}$ represents the output for the i-th unit in the k-th group ($k=1,2$) (kilograms); $X_{i1(k)}$ is the pesticides (kilograms); $X_{i2(k)}$ is fuel (liter); $X_{i3(k)}$ represents amount of chemical fertilizer (Kilograms); $X_{i4(k)}$ represents the total other costs (in local currency) and Ln shows natural logarithm. $\beta_{ij}(k) = \beta_{ji}(k)$ for all j and k. Stochastic frontiers were estimated for the individual two groups using FRONTIER 4.1 (Coelli, 1996). While the metafrontier was estimated using SHAZAM following (O'Donnell et al., 2005). A likelihood-ratio test using a mixed chi-squared distribution confirms that the technical inefficiency term is a significant addition to the individual variety and pooled models.

First, considering equation 8, pooled stochastic frontier is estimated using cross-sectional data collected from greenhouses for 2010-2011. Then based on equation 8, estimation of individual stochastic frontier is carried out with respect to the number of groups. SFA models and technical efficiency are estimated with Maximum Likelihood Estimation (MLE).

The generalized likelihood ratio statistic was used for testing the hypothesis $\gamma=0$ as the most important assumption in this method. Having calculated LR statistics using equation 7 and its comparison with critical value chi-square, the hypothesis $\gamma=0$ was examined and is presented in Table 1.

It can be concluded that the null hypothesis (lack of inefficiency effects) is rejected at 10% level and higher in all presented models; hence MLE is preferred to ordinary least square estimation (OLS). This indicates that production difference is to some extent influenced by the management factors. An additional assumption is to choose an appropriate functional form (Cobb-Douglas vs. Translog) for desired data and selected samples. To examine this hypothesis, both Cobb-Douglas and Translog functions were estimated individually utilizing FRONTIER software. The data set of this study consisted of two years (2010 – 2011) which was obtained from 127 greenhouses in Fars province of Iran and further divided into two groups. Group 1 included greenhouses in 2010 (prior to elimination energy input subsidies) and group 2 comprised of greenhouses in 2011(after elimination energy input subsidies).

RESULTS AND DISCUSSION

Results for tests of both the best functional form and the hypothesis that the group frontiers are the same are summarized

in Table 1, it should be noted that the suitable functional form has to be specified by LR test based on the values of log likelihood function. According to the Table 1, the values of likelihood function for null hypothesis (H_0) and alternative hypothesis (H_1) which are related to values of likelihood Cobb-Douglas and Translog functions are more than critical chi-square. Subsequently, the null hypothesis is rejected in both individual and pooled models, which indicates that Translog functions have more consistency and compatibility for existing data. Therefore, all analyses have been performed using Translog function.

As can be seen from Table 2, gamma coefficient is significant at 1% level for both groups. The

Table 1
MLE Test Results

Groups		Null hypothesis	X^2	(χ^2)	decision	Model
Individual stochastic for group1	Cobb-Douglas	$\gamma=0$	9.883	2.7(1)	reject	SFA
	Translog		12.796	2.7(1)	reject	SFA
	Cob-douglas or translog	$\gamma=0$	17.432	15.98(10)	accept	Translog
Individual stochastic for group2	Cobb-Douglas	$\gamma=0$	44.374	2.7(1)	reject	SFA
	Translog		60.085	2.7(1)	reject	SFA
	Cob-douglas or translog	$\gamma=0$	288.066	15.98(10)	accept	Translog
Pooled frontier model	Cobb-Douglas	$\gamma=0$	37.426	2.7(1)	reject	SFA
	Translog		56.445	2.7(1)	reject	SFA
	Cob-douglas or translog	$\gamma=0$	-35.978	15.98(10)	accept	Translog
Individual or pooled model	LR(pooled) with sum LR for two groups		73.81	21.06(14)	accept	Metafrontier

Table 2
MLE Test Results

Groups	Variable	Panel data	Group frontier			Meta frontier
			2010	2011	Total	
β_0	Constant	30.057	22.221	40.452	29.504	51.878
β_1	Pesticide(Rial)	0.135	0.706	0.474	1.302	-0.091
β_2	Fertilizer(Rial)	0.049	8.505	-1.475	-0.674	0.490
β_3	Fuel (Rial)	0.052	-1.582	0.355	1.580	-2.685
β_4	Divi (Rial)	-7.037	-11.769	-8.577	-8.253	-0.420
β_{11}	(Pesticide)^2	0.002	-0.033	-0.273	0.286	0.140
β_{22}	(Fertilizer)^2	-0.412	-0.978	-0.086	-0.024	-1.262
β_{33}	(Fuel)^2	0.014	0.212	-0.863	-0.753	0.181
β_{44}	(Divi^2	0.070	0.192	0.179	0.138	-0.088
β_{12}	Fertilizer * Pesti-	-0.174	-0.468	-0.161	-0.251	-3.089
β_{13}	cide	-0.040	-0.112	0.174	-0.339	0.194
β_{14}	Fuel * Pesticide	0.177	0.440	0.188	0.098	0.720
β_{23}	Divi * Pesticide	-0.100	-0.560	-0.089	-0.089	-6.813
β_{24}	Fertilizer * Fuel	0.672	0.618	0.477	0.386	-0.071
β_{34}	Divi * Fertilizer	0.129	0.653	0.551	0.749	-0.039
Sigma-squard	Divi * Fuel	2.413	0.678***	0.945***	0.849	
Gamma		0.873	0.669***	0.938***	0.812	
Loglikelihood		-2.902	-	-108.528	239.827	
function		-3.610**	118.687	-	-	
Eta			-			

*p<0.1, **p<0.05 and ***p<0.01

estimate of γ parameter using MLE was 0.667 and 0.938 for greenhouses in 2010 and 2011, respectively. These results are consistent with this concept that $\gamma > 0$. The calculated γ value is close to 1, which also denotes a high convergence of stochastic frontier toward metafrontier. This means that the contribution of these errors in production function and interpretation of production variations is lower in respect to included variables.

The statistical significance of η at 5% level indicates that the technical efficiency of cucumber production greenhouses for the study period (for panel data, 2010-2011) did not follow a constant trend and also the mean of the efficiencies for two years are significantly different. Table 3 shows that mean efficiency estimation using panel data has reduced from 98% (2010) to 67% (2011), suggesting that the elimination of subsidies for energy inputs led to significant reduction in greenhouse cucumber production efficiencies.

The results, as shown in Table 1, suggest that the hypothesis that the group frontiers are the same is rejected which means groups are different in applying different technologies. In these cases, stochastic metafrontier function has been considered as appropriate framework for estimating production function as well as comparison of technical efficiencies between groups (Bettesse et al., 2004). Accordingly, coefficients of stochastic metafrontier functions were obtained using the estimated coefficients from individual stochastic frontier functions and also SHAZAM software (calculating linear programming and coefficients), which are presented in the last column in Table 2.

The results of technical efficiency and Technology Gap Ratio are shown in Table 3, wherein TE indicates technical efficiency resulted from pooled stochastic frontier. TEK is technical efficiency of individual stochastic frontier for two groups which are being examined. TE* denotes the technical efficiency of units relative to the metafrontier. As can be seen, the maximum estimated mean TGR for

Table 3
Estimates of TEs and TGRs

Model	Item	Year 2010	Year 2011	Total	Panel data 2010	Panel data 2011
TE	Mean	0.609	0.611	0.610	-	-
	Min	0.028	0.055	0.028	-	-
	Max	0.940	0.844	0.940	-	-
	SD	0.119	0.152	0.136	-	-
TE ^{K}	Mean	0.648	0.609	0.629	0.987	0.674
	Min	0.056	0.031	0.031	0.093	0.078
	Max	0.922	0.915	0.922	0.995	0.829
	SD	0.104	0.174	0.144	0.013	0.148
TE *	Mean	0.594	0.528	0.561	-	-
	Min	0.054	0.027	0.027	-	-
	Max	0.879	0.851	0.879	-	-
	SD	0.109	0.158	0.140	-	-
TGR	Mean	0.917	0.870	0.894	-	-
	Min	0.585	0.344	0.344	-	-
	Max	1.000	1.000	1.000	-	-
	SD	0.088	0.094	0.094	-	-

both groups are equal to unity, which means that stochastic frontier functions of these groups are tangent to the metafrontier.

Our findings show that the mean TE, TE^{K} and TE^{*} are 0.609, 0.648 and 0.594, respectively. This mean of TEK states that firms are only able to produce, on average, 60.9% of the output that could be produced with current input and existing technology. Therefore, through the filling gap between the best producer in the same group, greenhouses could increase their production by 39.1 %.

TGR in different groups are comparable. In fact, estimates of the lower value of TGR demonstrate their greater distance from the superior technology. Prior to the elimination of energy input subsidies (2010), TGR for greenhouses was 0.917, which indicates the proximity of the technology level applied in this group of firms to the estimated technology from metafrontier function. However, after elimination of energy input subsidies (2011), this ratio decreased to 0.870, indicating the more distance to the efficient frontier.

CONCLUSIONS

Over the past years, a government support scheme for energy input price has resulted in an appropriate situation for greenhouse cucumber production and development of planted area in Iran and in particular in Fars Province; however, upon launching new policies for targeted subsidies in 2010, the crucial energy input lost the government support. The results of this study show that eliminating energy input subsidies has led to significant decrease in greenhouse cucumber production efficiency so that the mean technical efficiency declined from 98% to 67% during 2010-2011, when evaluated based on panel data. Furthermore, subsidies elimination has also led to decrease of the mean technology gap ratio in greenhouses from 0.917 to 0.870. In other words, it has caused greenhouses far apart from efficient production frontier. Accordingly, it would be better for energy input subsidies to be decreased gradually. It is also suggested that government provides the greenhouse owners with long term loans in order to equip their greenhouses with more energy efficient equipment.

ACKNOWLEDGEMENT

We would like to thank Professor Mansour Zibaei for his valuable assistance.

REFERENCES

1- Aigner, D. J., Lovell, D., & Schmidt, p. (1977). Formulation and estimation of stochastic frontier production function models. *Journal of Econometrics, 6* (1), 21-37.

2- Battese, G. E., & Coelli, T. (1995). A model for technical inefficiency effects in a stochastic frontier production function for panel data. *Empirical Economics, 20* (2), 325-332.

3- Battese, G. E., & Coelli, T. J. (1992). Frontier production functions, technical efficiency and panel data with application to paddy farmers in India. *Journal of Productivity Analysis, 3* (1), 153-169.

4- Battese, G.E., Rao, D.S.P., & O'Donnell, C. (2004). A metafrontier production function for estimation of technical efficiencies and technology gaps for firms operating under different technologies. *Journal of Business Economics, 21* (1), 91-103.

5- Battese, G. Malik, SJ., & Broca, S. (1993). Production functions for wheat farmers in selected districts of Pakistan: an application of a stochastic frontier production function with time-varying inefficiency effects. *Pakistan Development Review, 32* (2), 233-268.

6- Battese, J., & Rao, D. S. P. (2002). Technology gap, efficiency and stochastic metafrontier function. *Journal of Business Economics, 1* (2), 87-93.

7- Coelli, T.J. (1996). *A guide to FRONTIER version 4.1: A computer program for stochastic frontier production and cost function estimation.* CEPA Working Papers(No. 7/96), Department of econometrics, University of New England, Armadale.

8- Kumbhakar, S.C. (2002). Specification and estimation of production risk, risk preferences and technical efficiency. *American Journal of Agricultural Economics, 84* (1), 8-22.

9- Meeusen, W., & Van Den Broak, J. (1977). Efficiency estimation cobb-douglas production function with composed error. *Journal of International Economics, 18* (2), 435-444.

10- Mehrabi Boshrabadi, H., Villano, R., & Fleming, E. (2008). Technical efficiency and environmentaltechnological gaps in wheat production in Kerman province of Iran: A meta-frontier analysis. *Journal of Agricultural Economics, 38* (1), 67-76.

11- Ministry of Jihad-e-Agriculture (2010). Cultivation and production Database, Agricultural Crops Information. Retrieved from http://www. dbagri. maj.ir/Zrt/year.

12- Moreira, V.H., & Bravo-Ureta, B.E. (2010). Technical efficiency and metatechnology ratios for dairy farms in three southern cone countries: A sto chastic metafronitier model. *Journal of Productivity Analysis, 33* (1), 33-45.

13- Monjezi, N., & M.J. Sheikhdavoodi. (2011). Energy use pattern and optimization of energy consumption for greenhouse cucumber production in Iran using Data Envelopment Analysis (DEA). *Modern Applied Science, 5*,139-150.

14- O,Donnell, C., Battese, G., & Rao, D.S.P. (2005). Metafrontier frameworks for the study of firmlevel efficiencies and technology ratios. Unpublished paper, Centre for Efficiency and Productivity Analysis, University of Queensland, Brisbane.

15- Omid, M., Ghojabeige, F., Delshad, M., & Ahmadi, H. (2010). Energy use pattern and benchmarking of selected greenhouses in Iran using data envelopment analysis. *Energy Conversion and Management, 52*, 153-162.

16- Pitt, M., & Lee, L. F. (1981). The Measurement of Sources of Technical Inefficiency in the Indonesian Weavig Industry. *Journal of Development Economics, 9* (1), 43-64.

17- Taki, M., Ajabshirchi, Y., & Mahmoudi, A. (2012). Prediction of output energy for wheat production using artificial neural networks in Esfahan province of Iran. *Journal of Agricultural Technology 8*(4), 1229-1242.

18- Villano, R., Mehrabi Boshrabadi, H., & Fleming, E. (2010). When is Metafrontier Analysis Appropriate, An Example of Varietal Differences in Pistachio Production in Iran. *Journal of Agricultural Science and Technology, 12*, 379-389.

Impact of Climate Variability on Cool Weather Crop Yield in Ethiopia

Arega Shumetie [1], Belay Kassa [2], Degye Goshu [3] and Majaliwa Mwanjalolo [4]*

Abstract

The research examined effect of climate variability on yield of the two dominant cool weather cereals (wheat and barley) in central highland and Arssi grain plough farming systems of Ethiopia using eight round unbalanced panel data (1994-2014). The stochastic frontier model result revealed that production inputs for producing wheat and barley in the two farming system had significant effect. Crop season rainfall increment had negative and significant effect on technical efficiency of smallholders to produce wheat as to the model result. Technical efficiency of two crops responded differently for cropping season rainfall variability, in which wheat had negative and significant interaction with it while barley had positive. Given this, cropping season temperature had significant and positive effect on technical efficiency of both wheat and barley. Having this into account, yield of the two crops responded similarly for changes in production inputs like working capital, human labor and fertilizer. In general, rainfall inconsistency at the different stages of the production period had strong effect on yield of the two crops. Given this, the study forwarded an assignment to plant scientists in order to have further investigation on how the two crops responded differently to temperature variability.

Keywords:
climate variability, crop yield, technical efficiency, smallholder, rainfall, temperature

[1] PhD Candidate in Haramaya University of Ethiopia and Makerere University of Uganda
[2] Professor Pan African University, Africa Union, Addis Ababa, Ethiopia
[3] PhD, Haramaya University, College of Agriculture and Environmental Science: School of Agricultural Economics and Agri-business: Department of Agricultural Economics, Ethiopia
[4] PhD, Makerere University, College of Agriculture and Environmental Science. School of Forestry, Environmental and Geographical Science: Department of Geography, Geo Informatics and Climatic Sciences, Uganda
* Corresponding author's email: ashueconomist@gmail.com

INTRODUCTION

By the end of the 21st century, climate change would have a substantial impact on agricultural production and aggravate poverty level (Slater et al., 2007). Intergovernmental Panel Climate Change (IPCC), 2007) argued that there have been widespread changes in extreme temperatures and longer droughts and frequency of heavy precipitation over most land areas. Climate change affects developed and developing nations differently. For example, the latter ones suffer from serious food shortage and starvation as of the problem while developed ones face relatively simple problems (FAO, 2008). Land productivity and crop yield have been expected to decline because of climate change more or less continuously in recent times, and have experienced sharp decline in some places (Asha et al., 2012).

Agriculture is the most sensitive sector for climate change and variability (Cruz et al., 2007). Climate variability is currently the dominant cause of short-term fluctuation in rain-fed agricultural production of developing regions. In semi-arid and sub-humid areas, rainfall deficiency reduces crop yields and productivity drastically. Households pursuing crop production as the only source of livelihood are more affected by climate variability (Corinne et al., 2004). The problem is severe in Sub Saharan African (SSA) region specifically in which crop production is the main source of generating food and income for households. A study done by Wolfram and David (2010) argued that yield of many cereal crops will decline by 10% in 2050 nearly in all SSA countries due to rainfall inconsistency. Fractional variation in one of the climate elements would have devastating impact on smallholder production system as well as overall futurity of agriculture in most of African nations. Since Ethiopian agriculture is purely rain-fed type each variation in either rainfall or temperature, or combination of them would have huge effect on crop yield. Variation in national yields due to climate variability is common in most of the rural areas of Ethiopia (Getnet & Mehrab, 2010).

Ethiopia is one of SSA nations that suffer most from climate change and variability. Agriculture contributes 40% to the GDP, 84% of labor force employment and more than 90% to the export of the country. Agriculture survey in 2010 showed that cereal crop production takes the lion's share (85.94%) in the overall grain crop production of the country followed by pulses (10.5%). In the same production year, wheat and barley respectively took 19.8% and 11.27% proportion from the overall cereal production. Nearly 66% of the cereals produced in Ethiopia were used for household consumption (Central Statistics Agency (CSA), 2010), which means variability in yield of those crops may have huge impact on livelihood.

Since agriculture is the crucial sector in employment and livelihoods of Ethiopia, loss of agricultural productivity due to climate change may affect the entire economy strongly. Cereal crop damage sourced from rainfall inconsistency showed increment from 2006 to 2008 production years. Pest infestation and crop diseases outbreak were also high in the previously specified time interval. Major causes of crop damage in Ethiopia as to the response of smallholder farmers were hailstone, too much rain, frost and flood all of which are related to climate variability. CSA (2010) reported that hailstone was the major cause of crop damage followed by too much rain. Rain-fed agriculture of the country encountered different problems sourced from the ongoing climate variability. Each change in one or more of the climate variability elements results in huge production loss and starvation in most parts of the country. Researchers conducted before have not examined crop level responses to climate variability elements by employing time series or panel type of data. Thus, this research identified the impact of climate variability on the cool weather cereal crops (wheat and barley) yield in Central Highlands and Arssi Grain Plough farming systems of Ethiopia.

MATERIALS AND METHODS

Description of the Data Source

ERHS was conducted in the seven dominant farming systems of the country except pastoralism. The survey considered 15 villages and covered different aspects like demographic, consumption, production, asset holding, purchases and sales, landholdings and livestock ownership. This study considered central highland and Arssi grain plough farming systems of Ethiopia in which five villages (out of six for ERHS) were samples due to focusing only on wheat and barley yield.

Sampling Unit and Design

Farming system was an important stratification base in selecting sample villages in ERHS (Dercon & Hoddinott, 2011). The survey collected data for seven rounds from 1994 to 2009 having unbalanced period gap. Each survey followed a similar format of sample units that were already determined in the first survey in 1994. This study directly accepted all sample households from the two sample farming systems, considered by the previous survey in 2009 from each village, to collect the eighth round data in 2014. To have representative sample households from the farming systems, the research considered the whole households (502) in the five villages (Table 1), which were considered by ERHS in 2009, to collect the eighth round data in 2014.

This farming system took 40% of the total population of the country based on CSA (1994) and 56.4% sampling share of ERHS in 1994 (Dercon & Hoddinott, 2011).

Type and Methods of Data Collection

Both primary and secondary data were considered in this study, in which the former one was from sampled villages listed in Table 1. The survey in 1994a[1] considered about 1,477 households and these households have been re-interviewed in 1994b[2] as well as in 1995, 1997, 1999, 2004 and 2009. The eighth round of data in 2014 could help understand the recent circumstances of sample smallholders from selected villages. The primary data collected in 2014 adopted the questionnaire (after having amendment as to the objective of this research) used by the previous agency in collecting the seventh round in 2009. Including the previous seven rounds (1994a-2009) and the recent one in 2014,

Table 1
Sample Distribution in Each Village

Name of village	Sample households in 2009	Current sample households	Percentage
Debre Berhan	168	168	33.47
Korordegaga	106	106	21.12
Yetmen	51	51	10.16
Sirbana	82	82	16.33
Turufe	95	95	18.92
Total	**502**	**502**	100

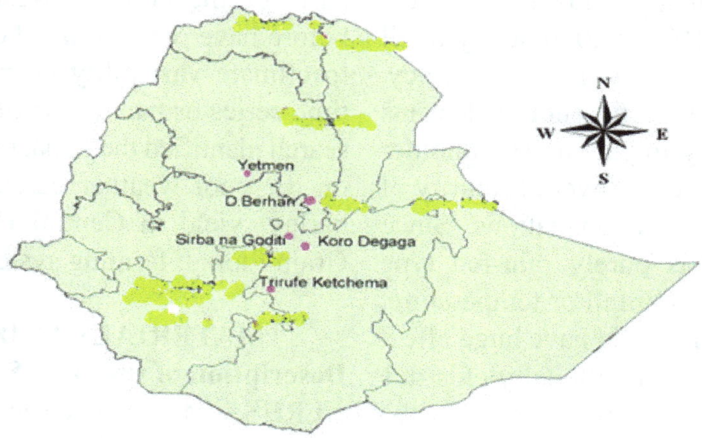

Figure 1. Geographical location of sample villages
Source: Community survey Ethiopian Rural Household Survey 1994

[1] *There were two round data collection in 1994 and this study consider the first one as 1994a*
[2] *The second round data collected in 1994 is named as 1994b*

the research had eight rounded unbalanced panel data for sample households of the five villages. In addition to the panel data collected by ERHS and this research, the study also used other secondary data collected by different national and international organizations.

Methods of Data Analysis

Crops have different optimal temperature and precipitation level for the well performance of them. Movement away from these levels may be damaging for crops, especially in countries where the current temperature and precipitation levels are already close to the tolerance limit. Production per unit of land yield was the dependent variable to analyze the impact of climate variability on agricultural productivity in Greek (Nastis et al., 2012). Gupta et al. (2012) adopted Cobb-Douglas production function for investigating impact of climatic variability on rice, sorghum and millet productivity utilizing panel data. Thus, this research considered wheat and barley yield in analyzing impact of climate variability and adopted Cobb-Douglas production system where the model assumes that agricultural production is a function of many variables like cultivated area, oxen power, fertilizers, labors, working capital, rainfall and temperature.

Functional form of the model is:

$$TP_{iht}=F\{AS_{iht},IFA_{iht},AL_{iht},WC_{iht},DP_{iht},R_{irt},T_{irt}\} \quad (1)$$

where:-i=1, 2,…., 5 (for the five cereals)

h=1, 2, ….,H (the sample household)

t=1, 2, ….,8 (the eight round sample years)

TP_{iht} = Total Production of sample cereal i for smallholder h at year t

AS_{iht} = crop-wise Area Sown by each smallholder h at year t

IFA_{iht} = Inorganic Fertilizer Application for each crop by smallholder h at year t

AL_{iht} = Agricultural Labor of each smallholder h for each crop at year t

DP_{iht} = Draught Power of each smallholder h for each crop at year t

WC_{iht} = the Working Capital budgeted by smallholder h for each crop at year t

R_{irt} = Cropping season rainfall level for each crop i from the nearest meteorological site r at time t

T_{irt} = Average cropping season temperature for crop i from the nearest meteorological site r at time t

Thus, total production (TP) should be divided by area sown (AS) of each sample cereal i to get yield, then the above equation (1) would become:

$$(TP/AS)_{iht}=Q_{iht}=F\{IFA_{iht},AL_{iht},WC_{iht},DP_{iht},R_{irt},T_{irt}\} \quad (2)$$

$(TP/AS)_{iht}=Q_{iht}$ is yield of crop i for each household h at year t. After computing yield of each cereal, then the above equation would take the following form:

$$Q_{iht}=F\{IFA_{iht}, AL_{iht}, WC_{iht},DP_{iht},R_{irt},T_{irt}\} \quad (3)$$

Thus, equation 3 can be rewritten following Battesse & Coelli, (1992):

$$lnQ_{iht} = ln(f(X_{it},\beta)\ exp(V_{it}-U_{it}) \quad (4)$$

And

$$U_{it}=\eta_{it}U_i=\{exp(-\eta[(t-T_i)]\}U_i \quad (5)$$

Q_{iht} represents yield of cereal i for the hth household at the tth period; $f(X_{it},\beta)$ is a suitable function of a vector, X_{it}, of factor inputs associated with production of ith cereal in the tth period, and a vector, β, of unknown parameters. The random effect, U_i is independent and identically distributed (iid) non-negative truncations with $N(0,\sigma_\mu^2)$, V_{it} is also iid $N(0,\sigma_V^2)$ random errors and independent of μ_i. η is an unknown scalar parameter that shows the type of yield variation. Household's technical efficiency (TE) tends to improve, remain constant if not decrease in producing cereals towards the base year, 2014 if $\eta > 0$, $\eta = 0$ or $\eta < 0$, respectively. The time period t = 1, 2……, Ti periods for which observations for the ith household are obtained.

From equation 4 and 5 TE of household h for producing the ith cereal at time t would be:

$$TE_{iht}= Q_{iht}/(Q_{iht}^*)= (f(X_{it},\beta)exp(V_{it}-U_{it}))/ (f(X_{it},\beta)\ exp(V_{it}))=exp(-U_{it}) \quad (6)$$

To deal with factors affecting smallholder's

efficiencies in cereal production, different socio-economic and climate variability elements were considered on the regression using Generalized Least Squares (GLS). This model was employed because the TE scores of all the observations were between 0 and 1, but not 0 or 1, which allowed to have linear type of dependent variable. Since the dependent variable was based on panel data, the appropriate regression model was GLS that consider hetroscedasticity and panel-autocorrelation problems. The functional form of the efficiency equation could be:

$$TE_{iht} = \sum \alpha_i X_{ht} + v_i + \varepsilon_{it} \quad (7)$$

where: TE_{iht} = the TE scores of smallholder h for cereal i at time t

α_i = a vector of unknown parameters ought to be estimated

X_{ht} = a vector of explanatory variables i (i = 1, 2, ..., k) for household h

υ_i= an error term that is iid, N $(0, \sigma_\upsilon^2)$, and it is the time invariant part of error term

ε_{it} = the time variant error term that should be iid, N $(0, \sigma_\varepsilon^2)$ independently of υ_i

RESULTS AND DISCUSSION

Crop yield trend in the two farming systems

Cereal crops that are produced in different agro-ecology of the country contributed around 86% of the national grain production. Major grain crops of the country including *teff*[3], wheat and barley (categorized as primarily cool weather crops), maize, sorghum and millet are warm weather cereals (Dereje & Yilma, 2003).

Table 2 shows that overall average yield of each crop was better in Arssi grain plough as compared to central highland farming system. The test results also revealed that yield of wheat had a significant difference in the two farming systems ($p<0.01$), but there was no significant difference in yield of barley (p-value = 0.317) for the sample production years.

There was no wider gap in the overall yield of sampled crops, and ups and downs of them happened mostly on the same production year. For instance, in 1999, there was yield reduction for both crops and this overall reduction also happened in 2009 exactly after ten years for wheat. Rainfall inconsistency was the main reason for those drastic yield reductions as to the view of respondents.

Yield of the two crops showed drastic increment in central highland for production years of 2009 and 2014. After 1999, yield of the two crops showed better and successive improvement in central highlands than Arssi grain plough farming system. Table 3 reveals that there was a significant difference in yield of wheat and barley for most of the production years.

Average level difference of the two farming systems in yield of the two crops was low for some of the production years. Variability in production of the two crops revealed that wheat and barley respond similarly, though it is not statistically proved, for each production input variation.

Trend of crop season rainfall and temperature in the two farming systems

Meteorological data collected from the nearest sites of each sample village of the two farming systems revealed that starting from 1994 production year rainfall condition of some sites showed similar trend in the ups and downs, es-

Table 2
Average Crop Yield in the Two Farming Systems

Description	Crop type	Central highland	Arssi	p-value	Combined Average
Area in hectare	Barley	0.91	0.30	0.000	0.72
	Wheat	0.43	0.48	0.000	0.44
Production in kg	Barley	736.50	321.58	0.000	617.30
	Wheat	405.78	673.38	0.000	497.57
Yield in kg/ha	Barley	939.56	978.72	0.317	950.81
Yield in kg/ha	Wheat	1068.56	1262.20	0.000	1134.97

[3] *It is an Ethiopian indigenous crop produced only in the country*

Table 3
Crop Yield Trend of the Two Farming Systems

Year	Wheat (kg/ha)		p-value	Barley(kg/ha)		p-value
	Central highland	Arssi		Central highland	Arssi	
1994a	901.91	895.57	0.000	1187.20	1043.74	0.012
1995	803.40	970.39	0.000	1169.93	942.36	0.000
1997	1216.25	1555.67	0.018	995.06	1242.19	0.056
1999	1112.00	921.36	0.029	1028.69	1003.68	0.003
2004	1181.62	1104.09	0.000	987.76	869.02	0.314
2009	1236.58	1308.43	0.020	1373.19	1185.56	0.000
2014	2091.49	1460.48	0.000	2322.04	1178.98	0.000
Average	**1220.46**	**1173.71**		**1294.84**	**1066.50**	

Source: Data from Ethiopian Rural Household Survey (1994a-2009) and researcher in 2014

pecially in those that have geographic proximity. Rainfall trend of sites in Arssi grain plough farming system was lower as compared with others in central highlands.

Crop seasonal rainfall in central higland farming system was higher in all of the production years compared with the other farming system, and it showed greater inconsistency in some of the production years registering the highest amount in 2010 (see Figure 2). Rainfall gap between the two farming systems became wider in recent years, mostly after 2006. There was cyclical type of movement in rainfall of central highlands farming system having reduction after three to five years of interval. Though it was not clear enough like the central highlands', but rainfall trend of Arssi grain plough also had cyclical type of movement to some extent and exhibited extrem reduction in few years like 2002 and 2012 exactly after ten years.

Crop season temperature of Arssi grain plough farming system was consistently higher than central highland, and some ups and downs happened more frequently in the latter farming system. Test of equality of variances showed significant difference in crop season temperature of the two farming systems having p-value of 0.004.

Model Result for Stochastic Frontier

The dependent variable was in physical quantity, yield, then explanatory variables should be also in quantity but not dummy type non-input variables. This research employed different equations to regress yield of wheat and barley that are most common cereals in the two farming systems. The coefficient of μ (mu) was significant for

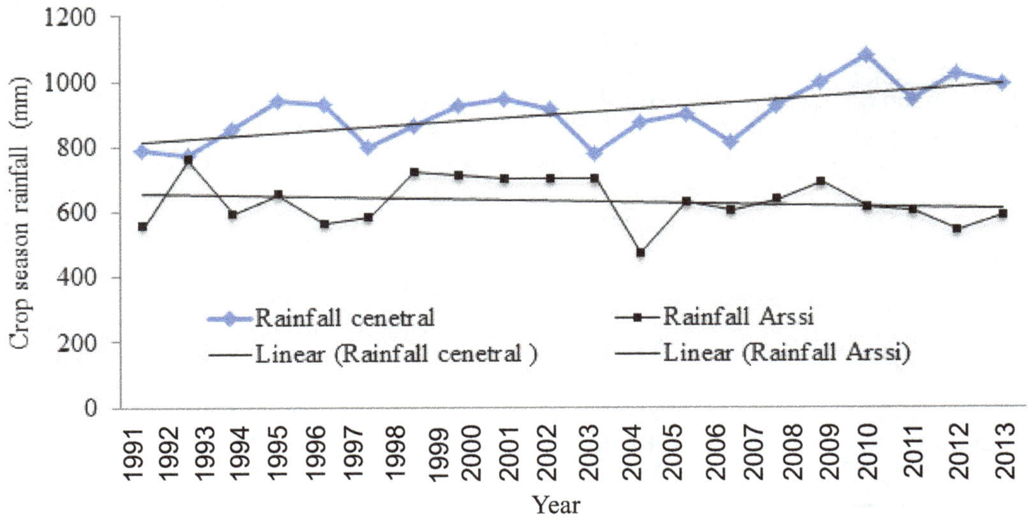

Figure 2. Crop season rainfall of sample farming systems
Source: Ethiopian National Meteorology Agency, 2014.

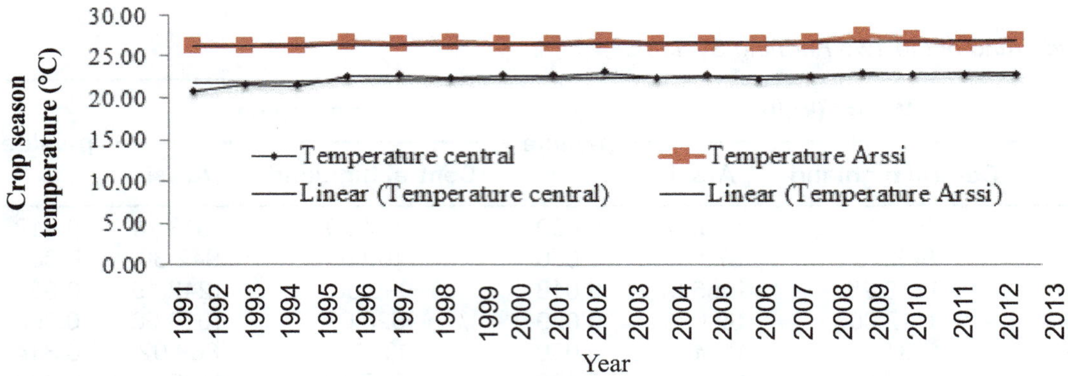

Figure 3. Crop season temperature of the two farming systems
Source: Ethiopian National Meteorology Agency, 2014

the time variant frontier model of the two cereals (Table 4), which showed presences of time variant efficiency variability as well as fitness of the model, Stochastic frontier. The returns to scale of producing those cereals in the two farming systems were not in constant returns since the coefficient of μ was significant.

The relatively large value of μ for wheat testified strong variability of the returns to scale for the across time (Table 4). The value of η (Eta) in the following model result revealed significant production inefficiency reduction towards the base year, 2014. The inefficiency level decays towards the base year, since value of η (Eta) is positive for the two cereals. Technical inefficiency reduction for barley was the higher than wheat (Table 4). The inefficiency for those cereals had significant time variant component that was normally distributed with a constant mean and variance level. Households in the two farming systems were highly inefficient in producing wheat, wherein they had unutilized potential of 61.50%. Households in the two farming systems could increase wheat yield by more than 60% from the given inputs at the existing technology.

Table 4
Time Variant Frontier Model Result

Description	Wheat		Barley	
	Coefficient	Standard error	Coefficient	Standard error
Ln man equivalent	0.129***	0.047	0.105**	0.043
Ln Fertilizer	0.098***	0.012	0.082***	0.011
Ln Oxen	0.050*	0.028	0.027	0.026
Ln Capital	0.041***	0.007	0.028***	0.007
In Cropping season rainfall	-0.124*	0.067	0.332***	0.108
In Cropping season temper	0.984***	0.214	0.564***	0.162
Constant	3.576***	0.846	2.764***	1.043
Mu (μ)	0.901***	0.191	0.466***	0.151
Eta (η)	0.023***	0.004	0.056***	0.011
Ln sigma2	-0.161***	0.044	-0.637***	0.037
Ilgt gamma	-0.919***	0.154	-2.774***	0.317
sigma2	0.852		0.529	
Gamma	0.285		0.059	
sigma_u2	0.243		0.031	
sigma_v2	0.609		0.498	
Log likelihood	-2870.6		-2089.6	
P-value	0.000		0.000	

Note: ***, ** and * implies 1%, 5% and 10% level of significance, respectively
Ln represents natural logarithm Ilgt implies inverse logit
Source: Model result, 2016

Table 5
Determinants of technical inefficiency, GLS regression result

Variables description	Wheat		Barley	
	Coefficient	**Standard error**	**Coefficient**	**Standard error**
Summer rain on time	0.004	0.008	0.159***	0.0119
Harvesting time rain	-0.142***	0.008	-0.178***	0.012
Pests	-0.090***	0.009	-0.123***	0.015
Shocks	-0.050***	0.007	0.0098	0.012
Family size	0.029***	0.008	-0.081***	0.012
Age of head	0.051***	0.011	0.061***	0.016
Ln TLU	-0.009***	0.003	0.033***	0.008
Credit access	0.036***	0.008	0.095***	0.012
Off -farm	-0.034***	0.006	-0.040**	0.014
Constant	-0.878***	0.053	-0.640***	0.080
P-value	0.000		0.000	

Note: The dependent variable was technical efficiency in percent
The coefficients showed two dimensional changes
Source: Model result, 2016

The dependent variable in the random effect GLS regression model was generated based on the stochastic regression result presented below. P-value of each model result, on the last row of Table 5, indicated that the models for regressing technical efficiency were best fit. The independent variables were different from zero and had partial contribution in the technical efficiency variation. The test value enable to reject the null hypothesis with high confidence level of above 99.99% and each of the co-variate had individual effect, either significant or insignificant, on the dependent variable.

Estimation result of technical efficiency [4]

The technical efficiency scores computed from frontier model result were between 0 and 1, and this revealed that none of the sample smallholder was either at the frontier level or producing 0 from the given input combination. The whole sample households were technically inefficient in producing the two crops. Since there were not censored observations, the study rejecting tobit model instead the best alternative model for linear type of observation GLS was employed.

The regression result presented in Table 5 considered three broadly categorized types of explanatory variables that affect smallholders' technical efficiency in producing the two crops. Wheat and barley responded similarly for variation in some of the covariates considered as an explanatory.

DISCUSSION

Interaction of crop yield with crop season rainfall and temperature

The main crop season harvest for some households in Arssi grain plough in 1994a was very low and it was the worst when compared with previous five years (Dercon & Hoddinott, 2011). The study also showed that in 2009 production year there was a poor distribution of rain for many of the households in the two farming systems, which resulted in huge reduction in yield of some crops. Households in Arssi grain plough farming system were suffering from rainfall inconsistency in a great extent and they tried to collect crops they needed by covering larger cropland. Some of the problems related to rainfall condition got extremity and drastic increment in their frequency of happening in each sample village of the two farming systems. For instance, many households in the two farming systems were suffering from untimed rainfall near harvesting period in 2009 production season.

The optimum and maximum crop season temperature for producing the two cool weather crops (wheat and barley) respectively were 25°c and 30.2°c in the sampled farming systems of this research. Given this, maximum crop season rainfall of the two farming systems was 1321mm. For better yields, water requirements should be 350-500mm depending on climate and length of

[4] *To avoid complications for readers the dependent variable was technical efficiency but not inefficiency*

Table 6
Number of Households Reported About Rainfall Inconsistency

Variables description	Number of households				Percentage change	
	In 1999		In 2009		Central highland	Arssi
	Central highland	Arssi	Central highland	Arssi		
Wheat	20	15	22	10	10.00	-33.33
Barley	26	1	12	3	-53.85	200.00
Total	150	117	156	44		

Source: Own data in 1999 and 2009

growing period, which is an amount lower than the average level (676.42mm) in the sampled farming systems of this research. Temperature increment may reduce yield and thereby exacerbate vulnerability in food supply (Joshi et al., 2011).

Rainfall inconsistency in the onset and cessation period showed greater variation from 1999 to 2009 within 10 years as per the data collected. This inconsistency and other elements of climate variability may be the main reasons for the lower crop production. In 1995 households in the two farming systems suffered from problems related to climate and weather variability in which 34.4% of the households who produce wheat were affected moderately or drastically. This resulted in a drastic reduction in yield for some households in central highlands farming system. Spearman correlation test of rainfall with wheat yield showed a significant interaction between them having p-value of 0.094 in central highland farming system. Spring rainfall is becoming highly unreliable for households in the two farming systems because of greater variability and sometimes complete absence, which frequently result in loss of yield (Dercon & Hoddinott, 2011). Sample households replied that spring season production is usually risky because of variability, delay or absence of rain.

Findings of Asha et al. (2012) revealed that reduction in rainfall was the major reason for yield reduction followed by pests and diseases outbreak; and changes in temperature and seasonal patterns of rainfall quoted as other reasons for yield reduction. In the same fashion, many households in Arssi grain plough farming system faced insufficient rainfall for their crop production and this resulted in serious crop damage in some production years. Overall rainfall shortage and poor distribution were the problems that happened in the two farming systems frequently miffed many of the sampled households (Table 6).

Discussion with focal persons in Arssi grain plough farming system indicated that there was production of short seasoned crops in autumn in the last 15 to 20 years, but recently it is not possible due to rainfall inconsistency. Currently, households of the farming system produce only in the main cropping season because of autumn rainfall inconsistency. The number of households miffed by inconsistency in rainfall increased in 2009 production year especially those who produce wheat.

Model Result Discussion

Discussion on determinants of yield

Inorganic fertilizer had significant and positive effect on yield of the two crops as to the model result (Table 4), which means small unit increment in its application could enhance yield significantly. Sibiko et al. (2013) had similar finding about fertilizer usage on bean productivity. Any fertilizer usage increment could enhance wheat, maize, barley and sorghum productivity (Ajay & Pritee, 2013), which was a finding corroborated with the above model result for all cereal crops. The model result in Table 4 shows that inorganic fertilizer and working capital had significant and positive effect on cereals' yield for smallholders in both central highlands and Arsi grain plough farming systems of Ethiopia. Rahman & Umar (2009) and Ehsan et al. (2012) also claimed that there is positive relationship between wheat yield and fertilizer application, which was in support of the model result in Table 4. Increasing the available capital for purchasing inputs like seed, pesticides and insecticides resulted in significant cereal yield increment as to the model result.

Oxen power and human labor had similar effect on cereals yield in the study area. Their complement behaviors may be the reason behind having similar interaction between the two sources of power and cereals yield in the two farming systems. Cereals are laborious crops and smallholders in the two farming systems manage them by human and animal power starting from plowing to harvesting, thus each change in the available labor had significant effect on yield of the two crops. Smallholders' crop yield variation may not purely climate forcing that is determined solely by rainfall, but also by the use of yield-improving external inputs (Amikuzuno & Donkoh, 2012). Similarly, Akinseye et al. (2013) confirmed that optimum yield performance of any crop do not depend only on climatic parameters. Smallholder farmers in less drought-stricken regions may improve land productivity through budgeting more labor and other production inputs (James et al., 2008), which was a finding similar to Table 4 regarding effect of human labor and draught power. Since smallholders of the study area are purely dependent on animal and human power for producing their crop, this positive and significant effect of the inputs on cereal yield was expected.

Cropping season temperature increment had positive and significant effect on yield of wheat and barley as to the model result (Table 4). This indicated that temperature variation significantly changed yield of the two crops. If cropping season temperature of the area increased by 1%, yield of wheat and barley would increase by 0.98% and 0.56%, respectively, wherein the former crop was more sensitive for each temperature variation. This finding was similar to Joshi et al. (2011), wherein cropping season maximum temperature increment has positive contribution on wheat and barley yield variation. Similarly, Sommer et al. (2012) found that wheat yield would increase in the future with the projected climate change in the semiarid zone of Tajikistan.

The result in Table 4 was in contradiction with Geethalakshmi et al. (2011), who identified that mean temperature increment reduced barley yield. Liangzhi et al. (2009) argued that impact of growing season temperature on wheat productivity is region specific, wherein effect of the problem is relatively lower in China as compared to the other regions. Successive temperature increment would not affect crop yield if there were extensive rainfall or water supply to curve down the moisture loss of crops (Kutcher et al., 2010). This indicated that positive interaction between cropping season temperature and yield of those cereals might be due to the excessive cropping season rainfall of the study area. The rampant rainfall of cool and moderate agro-ecology of Ethiopia may reduce negative effect of the temperature rise. In the other direction, successive temperature increment may eradicate waterlogging problem of the heavy summer rainfall and enhance wheat and barley yield. Thus, it may be easy to conclude that yield of the two cool weather crops would increase due to continuous cropping season temperature increment.

Discussion on determinants of technical efficiency (TE) [5]

The regression result presented in Table 5 considered three broadly categorized types of explanatory variables that affect smallholders' TE in producing cereal crops. Wheat and barley responded similarly for variation in some of the covariates considered as an explanatory. Timely onset of summer rainfall in the two farming systems significantly increased smallholders' TE of producing barley, which was in support with Inoussa (2010) who claimed that precipitation intensity affected sorghum yield significantly. Consistent availability of rainfall starting from the beginning may create auspicious environment for producing the crop efficiently. The adequate rainfall at the beginning of summer may enable smallholders to sow cereals on time, which would enhance their TE in producing cool weather crops.

TE of smallholders in producing those cereals responded similarly to harvesting season rainfall. Frequent happening of this problem in harvesting

---

[4] *To avoid complications for readers the dependent variable was technical efficiency but not inefficiency*

season could result in serious yield variability that might decrease smallholders' TE as to the model result (Table 5). The two cereal crops were strongly responsive to this problem. Smallholders' TE in producing barley and wheat respectively, decreased by 17.8% and 14.2% if there was harvesting season rainfall. The coefficients indicated that barley and wheat production were sensitive to one of the dimensions of climate variability, harvesting season rainfall.

Climate change will reduce crop productivity up to 14% by 2020 (Srivastava et al., 2010), and yields are likely to be affected even more in 2050 and 2080, which was a finding highly related to Table 5. Outbreak of pests and diseases significantly decreased TE of households in the study area for producing the two cereals, wherein barley was more sensitive (Table 5). In identifying impacts of current climate variability, Kettlewell et al. (1999) showed that heavy rainfall could result in fungal disease outbreak that would finally reduce yield.

The demographic related factors including age of the household head had similar effects on smallholders' TE of producing the two cereals. Experienced household heads would have significant effect on TE of smallholders in the two farming systems in producing cereals. Having experienced household head would significantly increase TE of households in producing the two crops, which was a finding in support of Hardwick (2009), Zenebe et al. (2012) and Sibiko et al. (2013). One of the critical crop production input, human labor, may be easily available if the household had larger family size, which would improve TE of producing cereals. Family size increment may be a source of labor power for the crop production that would enhance TE of wheat, which was a finding corroborated with Rahman and Umar (2009). Family size increment has both productive and consumption effect, wherein TE of households in producing cereals would be depending on magnitude of the two effects. The latter effect may be the reason for having negative effect of the covariate on TE of producing barley in the study area.

Given those covariates, the study also considered other economic factors as an explanatory variable that have significant effect on smallholders' TE in producing the two cereals. Household's TE of producing barley responded positively to the livestock holding level. Farmers with more livestock units, which are ready to convert into cash, can be able to buy modern inputs than those that own fewer (Fantu et al., 2011). Possession of large livestock could significantly increase TE, which may be due to having additional income and labor for the crop production (Table 5). Participation of households in off/nonfarm activities had negative and significant effect on TE of producing wheat and barley in central highlands and Arsi grain plough farming systems of Ethiopia, which may be due to labor competition between the crop production and those activities.

CONCLUSIONS AND RECOMMENDATIONS

Long run crop season rainfall of Arssi grain plough farming system was becoming lower with the passing of time and there were frequent ups and downs recently in very short period gaps. Similarly, crop season temperature of central highland farming system showed continuous increment with the passing of time. Extreme low and high rainfall as well as severe droughts also became common in the two farming systems. These all notify that the global problem, climate variability, is also happening in Ethiopia and it would have devastating effect on poor smallholders who have nature dependent livelihood.

Data from focus group discussion indicated that wheat and barley production showed continuous improvement in the cool weather areas of central highland farming system especially the former one exhibited successive increment implying that the weather condition is becoming auspicious for its production.

Agricultural inputs like inorganic fertilizer and human labor are critical in improving productivity of the two crops in the two farming systems. However, successive input price increment was a strong threat to apply the inorganic fertilizer as to the standard level. Sample smallholders in the study area affirmed that the current time sky rocketing input price rise is going out of their capacity to purchase the production inputs.

Cropping season temperature increment positively affected yield of producing wheat and barley. Thus, successive increment of this input improved yield of the tow cereals significantly in the two farming systems of Ethiopia. Further increment of cropping season rainfall from the current level resulted in mixed effect on technical efficiency of the two cereals for households in the two farming systems of Ethiopia.

The rainfall inconsistencies especially in harvesting time, and outbreak of pests significantly affected smallholders' technical efficiency in producing wheat and barley. Yield of the two crops significantly affected by those climate variability elements.

Source of additional working capital like credit access have positive and significant effect on yield of the two cereals in the study area.

Smallholders should have easy access to agricultural inputs like inorganic fertilizer and draught power at an affordable price to enhance application of them as to the recommended level for improving yield of the two crops.

The current level of crop season rainfall and temperature may be above optimum of wheat and barley production in the near future even though yield of the two crops currently improved. Thus, there should due attention to keep the temperature on the optimum level.

There should be continuous effort from both the government and private stakeholders to reduce negative and significant effect of pest and disease outbreak in order to have better yield from the two crops in the two farming systems.

To have a better yield of the two crops in the sample farming systems there should be due attention from the concerned body to keep the moisture on its normal level through reducing any form of rainfall inconsistencies at the different stages of the crop production.

Mechanisms like credit access and diversified source of income should be expanded in the study area to have alternative source of capital for producing wheat and barley.

Since rainfall inconsistencies were critical bottlenecks in producing the two crops, there must be brainwashing and awareness creation for smallholders to make them clairvoyant about precautionary measures.

There should be further investigation on the wheat yield both in laboratory and extended field work to examine why it had negative interaction with the crop season rainfall.

ACKNOWLEDGEMENT

The principal author has strong appreciation for the co-authors' unreserved effort to improve the quality of the article. Additionally, Haramaya University of Ethiopia and Makerere University of Uganda have significant contribution, which should be acknowledged.

REFERENCES

1- Ajay, K., & Pritee, S. (2013). Impact of climate variation on agricultural productivity and food security in Rural India. Discussion Paper No. 2013-43: http://www. economicsejournal. org/economics/discussionpapers/

2- Akinseye, F.M., Ajayi, V.O., & Oladitan, T.O. (2013). Assessing the impacts of climate variability on crop yield over Sudano- Sahelian zone in Nigeria. *International Journal of Agricultural Science,1*(7), 91-98. Retrieved from http:// www.accessinterjournals.org/aijas

3- Amikuzuno, J., & Donkoh, S.A. (2012). Climate variability and yields of major staple food crops in Northern Ghana: Tamale, Ghana. *African Crop Science Journal, 20* (2), 349 – 360.

4- Asha, L.K.V., Gopinath, M., & Bha, A.R.S. (2012). Impact of climate change on rain-fed agriculture in India. *International Journal of Environmental Science and Development, 3*(4), 368-371.

5- Baltagi, B. H. (2008). *Econometrics*. 4th Edition. Springer-Verlag Berlin Heidelberg: Syracuse University. USA.

6- Battesse, G.E., & Coelli, T.J. (1992). Frontier Production Functions, Technical Efficiency and Panel Data: With Application to Paddy Farmers in India. *The Journal of Productivity Analysis, 3*, 153-169

7- Corinne,V., Jere, L. G., Christian, J., Robert, Q., & Rigoberto, E. (2004). Coping and adapting to climate variability: The Role of Assets, Networks, Knowledge and Institutions.http://citeseerx.ist.psu.edu/viewdoc/download; jsessionid=1255C45E22E18E1CBE48CDC2 83BE36AF?doi=10.1.1.201.7357&rep=rep1& type=pdf

8- Cruz, R.V., Harasawa, H., Lal, M., Wu, S., Anokhin, Y., Punsalmaa, B., Honda, Y., Jafari, M., Li, C., & Huu, N. (2007). Asia: Climate Change 2007: Working Group II: Impacts, Adaptation and Vulnerability. IPCC (Inter-governmental Panel on Climate Change) Contribution of Working Group II to the Fourth Assessment Report. https://www.ipcc.ch/ publications _ and_data/ar4/wg2/en/ch10.html

9- CSA (Central Statistics Agency) (2010). *Agricultural sample survey 2009/10 (2002 e.c.)* Volume VII Report on Crop and livestock product utilization (private peasant holdings, meher season. Statistical bulletin 468 Addis Ababa, Ethiopia.

10- Dercon, S., & Hoddinott, J. (2011). *The Ethiopian Rural household surveys 1989-2009: Introduction.* International Food Policy Research Institute. Addis Ababa: Ethiopia.

11- Dejene A., & Yilma S. (2003). Causes and variability of Ethiopian agriculture: modeling the relative importance of environment factors, external shocks, and state policies 1980–1997. Regional and Local Development Studies (RLDS), Addis Ababa University, Ethiopia

12- Ehsan, N., Mohammad, B., Mohammad, R. J., Paul, S., & Alireza, F. (2012). Agro-ecological zoning for wheat (*Triticum aestivum*), sugar beet (Beta vulgaris) and corn (Zea mays) on the Mashhad plain, Khorasan Razavi Province. *The Egyptian Journal of Remote Sensing and Space Sciences, 15*, 99–112.

13- Elodie, B. (2012). The impact of climate change on crop yields in Sub-Saharan Africa. *American Journal of Climate Change, 1*, 1-13

14- FAO (Food and Agriculture Organization) (2008). *The state of food insecurity in the world.* Food and Agriculture Organization of the United Nations, Rome, Italy.

15- Geethalakshmi, V., Lakshmanan, A., Rajalakshmi, D., Jagannathan, R., Sridhar, G., Ramara, Bhuvaneswari, A. P., Gurusamy, K. L., & Anbhazhagan, R. (2011). Climate change impact assessment and adaptation strategies to sustain rice production in cauvery basin of tamil nadu. *Current Science, 101*(03), 342-347.

16- Getnet, A. Z., & Mehrab, M. (2010). *Implications of land policies for rural-urban linkages and rural transformation in Ethiopia.* ESSP II Working Paper 15. Development Strategy and Governance Division, IFPRI Ethiopia Strategy Support Program (ESSP), Ethiopia.

17- Govinda, B. (2013). Effect of precipitation and temperature variation on the yield of major cereals in Dadeldhura District of Far Western Development Region, Nepal. *International Journal of Plant, Animal and Environmental Science, 3* (1).

18- Gupta, S., Sen, P., & Srinivasan, S. (2012). *Impact of climate change on Indian economy: Evidence from food grain yields.* Centre for Development Economics Working Paper 218, New Delhi.

19- IPCC (Intergovernmental Panel Climate Change) (2007). *Agriculture in climate change 2007: Mitigation.* Working Group III Contribution to the Fourth Assessment Report of the IPCC. Cambridge: Cambridge University Press

20- Jaiswal, J.P. (2009). Climatic requirement for Wheattemperature. Retrieved from http://agropedia.iitk.ac.in/content/climatic-requirement-wheattemperature

21- Joshi, N., Maharjan, P., Keshav, L., & Piya, L. (2011). Effect of Climate Variables on Yield of Major Food-crops in Nepal: A Time-series Analysis. *Journal of Contemporary India Studies, 1*:19-26

22- Kettlewell, P. S., Sothern, R. B., & Koukkari, W. L. (1999). UK wheat quality and economic value are dependent on the North Atlantic oscillation. *Journal of Cereal Science, 29,* 205–209.

23- Lobell, D.B., Burke, M.B., Tebaldi, C., Mastrandrea, M.D., Falcon, W.P., & Naylor, R. L. (2008). Prioritizing climate change adaptation needs for food security in 2030. *Science, 319*, 607-610.

24- Lott, N., Ross, T., Smith, A., Houston, T., & Shein, K. (2011). Billion dollar U.S. Weather disasters, 1980 - 2010. National Climatic Data Center, Asheville, NC, USA. Retrieved from http:// www.ncdc.noaa.gov/oa/reports/billionz.html

25- Morton, J. (2007). The impact of climate change on smallholder and subsistence agriculture. Proceedings of the National Academy of Sciences, USA 2007;104:19680-19685.

26- Nastis, S.A., Michailidis, A., & Chatzitheodoridis, F. (2012). Climate change and agricultural productivity. *African Journal of Agricultural Research, 7* (35), 4885-4893.

27- Samarah, N.H., Alqudah, A.M., Amayreh, J.A., & McAndrews, G.M. (2009). The effect of late-terminal drought stress on yield components of four barley cultivars. *Journal of Agronomy and Crop Science, 195* (6), 427-441

28- Sara, L. (2012). *Climate change, weather variability and food consumption: A multidisciplinary study of Rural Uganda.* Catholic University of the Sacred Heart. Via Emilia Parmense 84, 29122 Piacenza, Italy.

29- Schlenker, W., & Michael, J. R. (2009). Nonlinear temperature effects indicate severe damages to U.S. crop yields under climate change. *Proceedings of the National Academy of Sciences, 106* (37),15594–15598.

30- Slater, R., Peskett, L., Ludi, E., & Brown, D. (2007). Climate change, agricultural policy and poverty reduction; how much do we know? Natural Resource Perspectives, 109.

31- Wolfram, S., & David, B.L. (2010). Robust negative impacts of climate change on African agriculture. Environmental Research Letters, 5 (1): IOP Publishing: http://iopscience.iop.org/1748-9326/5/1/014010

32- Wollenweber, B., Porter, J. R., & Schellberg, J. (2003). Lack of interaction between extreme high-temperature events at vegetative and reproductive growth stages in wheat. *Journal of Agronomy and Crop Science, 189*, 142– 150. (doi:10.1046/j.1439-037X.2003.00025.x) World Bank, 2004. Memorandum May 24.

33- Zerihun, G. (2012). *An analysis of the impacts of climate change on crop yield and yield variability in Ethiopia*. PhD Dissertation: Graduate School of Social Sciences: Department of Economics and Management. University of Trento: Italy

Seed Potato Management among Potato Growers in Freydunshahr County (Isfahan Province), Iran

Asghar Bagheri* and Mojtaba Sookhtanlou

Abstract

This study investigated the mechanisms of seed potato management by potato growers of Freydunshahr County in Isfahan province. Survey research method was used in this study. A sample of 250 potato growers (N=725) was selected for data collection. Stratified random sampling method was used. The research instrument was a questionnaire that was validated for face validity by a panel of faculty members and agricultural experts of Agriculture Jihad Organization in Freydunshahr County. For determining the reliability, Cronbach's alpha was estimated which turned to be higher than 0.7. Results showed that "a part of tubers produced by farmers" was the first priority of farmers as seed tuber source. Results showed that 89.6 percent of potato growers store and keep seed tubers in inappropriate conditions. Majority of the farmers did not have sufficient information on seed potato management and there was a positive and significant relationship between the farmers' education and potato yield per hectare, but a negative relationship was found for age. Based on logistic regression analysis, the predictive power of the model was 0.639 and the sensitivity of total model to determine the correct percentage was 86.0 percent. Among individual, professional and economic factors, variables of age, number of family labor, area of owned land, area of potato farming, potato yield per hectare, number of owned land parcels and knowledge and information level of respondents significantly impacted on the adoption and use of modern seed potato production systems.

Keywords:
Freydunshahr County, potato seed tuber, potato growers, seed quality

Faculty of Agriculture and Natural Resources, University of Mohaghegh Ardabili, Iran
** Corresponding author's email: a_bagheri@uma.ac.ir*

INTRODUCTION

Potato (*Solanum tuberosum* L.) is the fourth most important food crop after corn, rice and wheat. It is grown all over the world. The current world production of potatoes is about 321 million tons fresh tubers from 19.5 million hectares (Ebrahimi et al., 2015). In many countries in Asia, Africa and Central and South America, there is a need for increased production of stable potato to meet increasing demands for food from human population growth during a period of environmental (including climate) change. Potatoes with improved nutritional and health properties are desirable, but the overriding need is for increased and stable yields to eradicate human hunger and poverty (Vreugdenhil et al., 2007). Potatoes are grown worldwide under a wider range of altitude, latitude, and climatic conditions than any other major food crop-from sea level to over 4000 m elevation. No other crop can match the potato in its production of food energy and food value per unit area (Mohammadi et al., 2008). Potato growers in different parts of the world have the same problems. For example, the lack of yielding amendment and high quality of seed potato tubers and absence of resistance to pests and diseases are known as the main problems for the cultivation of potato in Cameroon (Deffo & Demo, 2003). Hirpa et al. (2010) cited multiple poor distribution systems and low quality of seed potato tubers as major problems among potato growers in Ethiopia. One of the main problems in the potato cultivation is the cost of potato seed production so that it constitutes 30 to 50 percent of the cost of potato cultivation depending on country or region (Bagheri, 2015).

Sthapit et al. (2008) have defined systems of seed production as the methods in which farmers produce, select, store and prepare seed. Globally, systems of local seed production have been categorized in various types as follows: formal and informal (Endale et al., 2008; Struik & Wiersema, 1999), official and local chains (World Bank, 2009), and formal and farmers (Almikinders & Louwaars, 1999). Louwaars (2007) defined the farmers' informal or local seed chain as a system where seed is chosen, produced and distributed locally and defined the formal chain as a system that seed production and distribution in various dimensions from seed system managing by well-defined methods to proliferation control is conducted by private sector specialists, in most cases under national laws and international standards. In Iran, a part of the harvested potato is saved as seed for farmers to prepare tubers for home consumption, and a part is stored as seed for cultivation in the next season (Majidi Harvan, 2003). Farmers who store seed potato without renewing the seed from a reliable source and use it for several consecutive cultivation seasons will face reduction in quality and performance (Gildemacher, 2012).

The main components of a seed production system can be referred to as: 1) seed production and storage, 2) seed tubers quality, 3) seed availability and distribution, and 4) the flow of knowledge (Hirpa et al., 2012). Seed production and storage management refers to all activities conducted in order to produce seeds at any time of planting period. It includes all stages of production and storage (Weltzien & Vombrocke, 2000). Seed tuber quality also includes seed health, physiological age, condition, size, purity and genetic quality (Struik & Wiersema, 1999). Healthy, viable seed of the preferred variety needs to be available at the right time under reasonable conditions so that farmers can use their land and labor resources with the best yield expectations. The wrong variety, sown at the wrong time with infected seed of poor germination potential, will seriously limit a farmer's expectation of production and productivity. Thus, any seed system has multiple functions to be fulfilled for a range of farmers, farming conditions, and crops in a village, region, or country. A seed system can be assessed at any time according to how well it fulfills these functions. Conditions, situations, groups of farmers, or crops can be identified under which the specific system works well. Similarly, factors that constitute stresses to a particular system can be revealed in terms of these four functions (Weltzien & Vombrocke, 2000).

Mathoni and Nyamongo (2008) found that while a formal seed system exists in Kenya, majority of farmers rely on an informal seed system for planting most agricultural commodities, and often continue to recycle seed, resulting in persistent low yields. Major crops as well as commercial crops are dominated by the formal seed system, whereas orphaned crops as well as poor farmers have no place in the formal system. According to Hirpa et al. (2010), three seed potato systems, i.e., informal, alternative and formal seed systems, co-exist in Ethiopia. The informal system, with low quality seed in terms of health features, physiological, physical and genetic attributes is dominant. The formal system in which seed tuber is produced by reliable professionals from the private sector and cooperative institutes is too small to contribute significantly to the improvement of that situation. Major challenges of seed systems of Ethiopia were identified as lack of proper linkage between different actors involved in seed systems, inadequate supply of good quality seed at affordable prices, focus on few crops (maize and wheat) in the formal system and other beneficial crops (such as pulses and oilseeds) remain orphans, low level of private sector involvement in the formal system, inefficient seed promotion, distribution and marketing mechanisms, and weak variety release and seed quality assurance system (Atilaw & Korbu, 2010). Oladele (2010) examined farmers' information seeking and utilization pattern on seeds and planting materials in Lagos and Ogun states. Farmers were exposed to many sources of information on seeds and planting materials. The results suggested the relevance of use of improved seeds and planting materials among farmers. They also suggested the relevance of the characteristics of information sources as drivers of use of improved seeds and planting materials.

In an investigation on seed potato production systems in East Africa, Gildemacher (2012) stated that less than five percent of the seed potatoes used in East Africa sourced from specialized chains and more than 50 percent from farmers' seed storage (local chain) derived from farmers' own farm products. Oyekale et al. (2014) found that compared with other agricultural inputs (fertilizers, pesticides, etc.), high-quality seed management plays the most important role in improving productivity. The availability of suitable new seeds, along with appropriate training to farmers can play the most important role in the success of seed management systems. Development and optimal performance of seed management is influenced by several factors such as farmers' weak technical potential, poor market mechanisms to provide new seeds, lack of supportive legislation for seed management, inappropriate flow of knowledge and information among farmers, insufficient investment and less use of innovations in seed management. Results of a study on the adoption of mini-tuber seed potato in Ardabil province of Iran (Bagheri, 2015) revealed that higher yield was the farmers' main reason for the adoption. On the contrary, higher seed cost and lack of technical knowledge about mini-tuber cultivation were the main reasons for non-adoption. According to the results, adopters had more mechanized farms. Also, the number of parcels held by a farmer, age and years of farming experience were negatively associated with the adoption. On the other hand, farm size, education level and number of literate persons in the farmer' household was positively associated with adoption.

With respect to the importance of seed management especially in potato farming that consumes 3-4 tones seed tubers per hectare, considering 180 thousands hectares of potato acreage and relatively low potato production yield (22 tons per hectare) in Iran (FAO, 2008), seed potato management system needs to be studied. Furthermore, Isfahan province, especially Freydunshahr County, is of the main potato producing regions of the country that serves as a seed source for other potato producing provinces (Bagheri, 2015). However, the problem is that there is no formal seed system in the county and no study has been conducted on seed potato management system among potato growers of the county. This is an exploratory study to find the strengths and weaknesses of the seed potato management system in Freydunshahr County of Isfahan province as one of the main potato farming regions and a good representative of seed potato system in Iran. Conceptual framework of the study is presented in Figure (1).

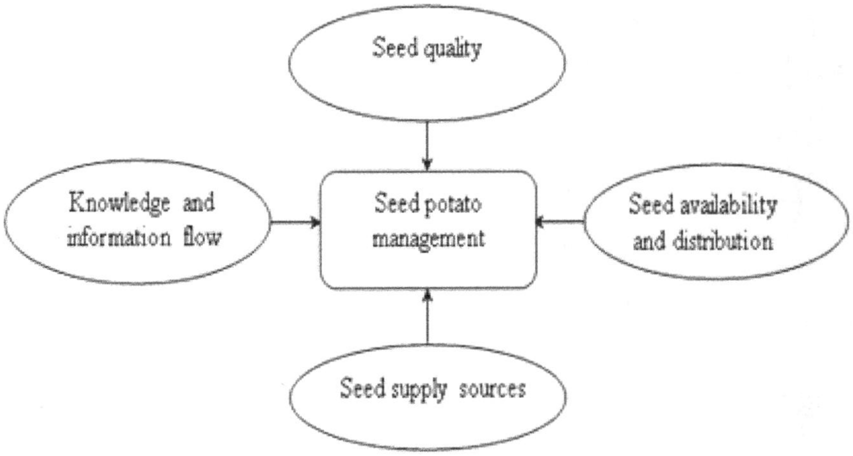

Figure 1: The conceptual model of study for potato seed management

It summarizes the results of previous studies. To examine the seed potato management systems in relation with formal and informal systems (Louwaars, 2007; Sthapit et al., 2008; World Bank, 2009), the following components are studied:

Variables related to seed supply sources, seed quality (including seed tubers purity, size and health), variables related to the availability and distribution of seed tubers (the ease of access seed tubers), and variables related to the knowledge flow (including the knowledge of farmers, information resources and information that farmers seek). So, the main purpose of this study is to investigate seed potato management mechanisms in Freydunshahr County of Isfahan Province. Bearing this reason in mind, demographic and socioeconomic status of the respondents, potato growers' seed supply sources, their agricultural knowledge and information on potato seed management, association of information and knowledge with individual variables and their socio-economic status, and factors affecting adoption of new seed management system were examined.

MATERIALS AND METHODS

The study area

Isfahan province is located in the central part of Iran with an area of 214,503 km^2. It lies within the latitudes of 30.6N and 34.58N and the longitudes of 49.6E and 55.5E. The area has a semi-arid climate with a limited amount of precipitation (130 mm) which generally occurs in winter from December to April. About 3% of total area is cultivated (609,250 ha) from which 95% is assigned to the irrigated lands and 5% is assigned to rain-fed lands (Keshavarz et al., 2014). Freydunshahr County is one of the main potato producing regions which was selected for the study. It is located in the western part of the province (Figure 2). The county has two districts and two cities (Freydunshahr and Barf-Anbar) and 96 villages. It occupies a total land area of 2236.15 km^2, or 2.08 percent of the total land area of the province. It is 175 kilometers far from the province center (Isfahan city). In term of farming population, the county has 4626 farmers or 3.2 percent of total farmers in the province to account for this section and in horticulture sector, 3161 farmers of this county composed of 2.9 percent of all gardens of the province. Potato is the most important crop in the county. In cropping year of 2013-14, potato acreage of the county was equal to 2100 hectares with a production of 44,100 tons meaning 21000 kg per hectare (Weather Bureau of Freydunshahr, 2016).

The dominant approach in this study was survey and field study was used in data gathering for cropping year of 2015. In terms of variables control, it is a cross-sectional descriptive and none-experimental research method. Besides using face to face approach to fill out questionnaire, in-depth interviewing and key informants were used for data collection.

The main instrument of this study was a researcher designed questionnaire in five parts: the first part was related to personal and socio-economic characteristics of potato growers (18

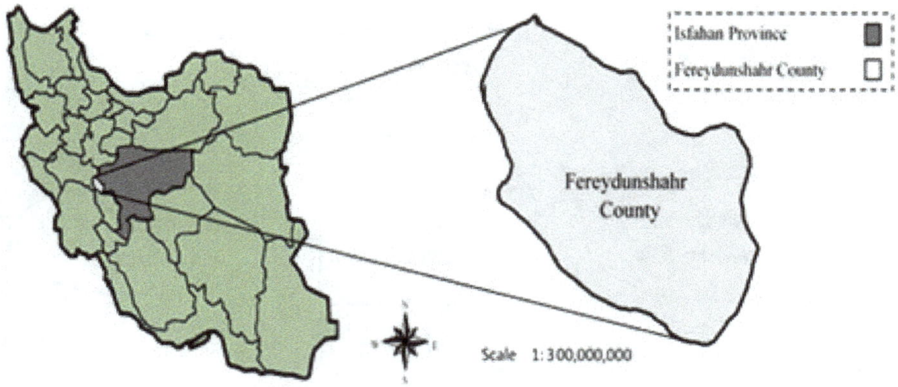

Figure 2. Area of study (Freydunshahr County, Isfahan Province, Iran)

items). Parts 2 to 4 were studied the main components of the seed potato management. These components (according to Atilaw & Korbu, 2010; Bagheri, 2015; Gildemacher, 2012; Hirpa et al., 2010; Louwaars, 2007; Oladele, 2010; Oyekale et al., 2014; Sthapit et al., 2008) included seed supply sources (7 items), availability and distribution of seed (5 items), seed tubers quality (3 items) and flow of knowledge and information (5 items). They were measured by a five-point Likert type scale (1 = very low, 2 = low, 3 = average, 4 = high, 5 = very high). The fifth section of the questionnaire examined the potato growers' barriers to use suitable seeds that had 8 items. Face validity of the instrument was confirmed by a panel of faculty members and agricultural experts of Agriculture Jihad Organization in Freydunshahr County. The statistical population of the research included all potato growers in the crop year 2015 (N=725) in Freydunshahr County of Isfahan province. A sample of 250 potato growers was determined by Cochran's formula. Proportional stratified random sampling was used in this study. For this purpose, based on the geographical location and population distribution in proportion to the distribution of potato growers, population was divided into two separate sectors and samples were taken (Barf Anbar and Freydunshahr regions, rural district of Ashayeri, rural district of Cheshmeh Langan that are among the areas where mini-tuber potatoes are not cultivated). Within each sector, sample farmers were selected using simple randomization method. To ensure reliability, Cronbach's alpha was used and the alpha was obtained to be higher than 0.7 implying the high reliability of the instrument.

$$n = \frac{N * z^2 * \sigma^2}{d^2(N-1)+(z^2 * \sigma^2)}$$

where: n = sample size; N = statistical population; Z = confidence coefficient; = variance; d = percentage of error

To categorize knowledge and awareness of respondents in relation to seed potato management, interval of standard deviation from means (ISDM) was used (Khoshnodifar et al., 2016; Qamar, 2002):

- Low: Min ≤ A < Mean-SD
- Medium: Mean-SD ≤ B < Mean-
- High: Mean ≤ C < Mean+SD
- Very High: Mean+SD ≤ D < Max

Also, to determine factors affecting adoption of modern seed potato production system, logistic regression analysis was used.

RESULTS AND DISCUSSION

Demographic characteristics of respondents

Based on the results, respondents' average age was 50.45 years. The most frequency (38.6%) was related to 50-59 year group. Their average farming experience was 27 years. In terms of education level, about 9 percent of potato growers were illiterate and 32 percent were literate in which the highest frequency (21.6 percent) was related to guidance school (8 years of education). 27.2 percent of them had a farmland area of 11-15 ha. Results also showed that 70 percent of respondents had, on average, five ha of rented land. Respondents'

Table 1
Ranking of Seed Potato Supply Sources among Respondents

Potato seed supply sources	Mean	SD	Rank
A part of potato harvested from their own farms	3.41	0.99	1
Seed potatoes bought from other potato growers	2.99	1.01	2
Local markets	2.50	1.03	3
Tuber planted as seed for next year	1.84	1.16	4
Specialized seed producers	1.79	1.23	5

Table 2
Storage Methods of Potato Seed Tubers

Storage methods	Frequency	Percentage
Stored in gunnysacks	123	49.2
Stored in local warehouses	101	40.4
Delay in harvesting time	15	6
Stored in well ventilated and light places	11	4.4
Total	250	100

Table 3
Availability of Potato Seed Tubers for Potato Growers

Availability	Frequency	Percentage
Not easy access	228	91.2
Easy access	22	8.8
Total	250	100

average potato cultivation area was 9.4 hectares, among which 68.6 percent had less than 5 hectares and only 2.8 percent had more than 16 hectares of potato cultivation land.

Investigation of potato varieties shows that six varieties including Agria, Santa, Marfona, Bamba, Rambos and Spirit are used with the domination of Agria (44.7 percent). The Marfona variety has been cultivated in the lowest acreage (1.2 percent of respondents). Average potato yield per hectare was 34.44 tons for 41.4 percent of respondents. It ranged from 21 to 30 tons while only 4.2 percent of respondents harvested more than 50 tons per hectare. Based on the findings, 55.6 percent of potato growers used both local and mini-tuber seed potatoes. Indeed, 22.8 and 21.6 percent used only local and mini-tubers, respectively.

Potato seed supply sources

As depicted in Table (1), six sources of seed tubers were identified and ranked based on mean and standard deviations. A part of tuber harvested from their own farms was the main source of seed tuber. Seed potatoes bought from other potato growers was their second seed tuber priority. Some potato growers prepared their seed tubers from local market. As shown in Table 1, tuber planted as seed for next year and seed bought from specialized seed producers were two last priorities of potato growers for seed tuber provision, respectively.

Storage methods of potato seed tubers

Results of Table 2 showed that 49.2 percent of potato growers stored seed potato tubers in gunnysacks at home and 40.4 percent stored them in local warehouses. Delayed harvest was another way to store seed potato that six percent of potato growers were used. Other farmers stored seed tubers in well ventilated and light places.

The availability of seed potato tubers

Respondents were asked to identify whether they had easy access to seed potato tubers or not. The results showed that 91.2 percent did not have easy access to seed tubers (Table 3).

Quality of seed tubers

Seed potato tubers were divided into three groups in terms of purity. Based on these categories, 13.2, 66.2 and 20.6 percent of respondents used lowly, moderately and highly pure seed tubers in their farms, respectively. Results also showed that seed tuber size of 5.6, 39.2, 11.6 and 10 percent of potato growers was big, medium, small and very small, respectively. Also, 33.6 percent of them used mixed sizes (accessible) tubers. Regarding seed tubers health, 32.4, 66.4, and 1.2 percent stated that they used desirable, lowly desirable and undesirable seed tubers (Table 4).

Knowledge and information flow

Respondents were asked to indicate their knowledge and information sources and level of knowledge and information regarding seed potato management in 5-point Likert scale statements. Then, they were ranked using mean and standard deviation. Regarding knowledge and information sources as depicted in Table 5, other potato growers were the main information source for potato growers and it was placed in higher priority of their information seeking followed by local markets. Government organization of agriculture and specialized seed producers as the controlled and well managed services of seed tuber production were placed in two last priority of potato growers information and knowledge sources (Table 5). Self-stated knowledge and information level of potato growers

Table 4
Quality of Seed Potato Tubers Used by Potato Growers

Seed tubers quality	Scale	Frequency	percentage
Purity of potato seed tubers	Low	33	13.2
	Medium	165	66.2
	High	52	20.6
Seed tubers size	Big	14	5.6
	Medium	98	39.2
	Small	29	11.6
	Very small	25	10
	Mixed (accessible)	84	33.6
Tuber seed health	Desirable	81	32.4
	Lowly desirable	166	66.4
	Undesirable	3	1.2

Table 5
Knowledge and Information Flow Regarding Seed Potato Management

Cases	Items	Mean	SD	Rank
Knowledge and information Sources	Other potato growers	2.90	0.82	1
	Local markets	2.02	0.85	2
	Government organization of agriculture	0.64	0.90	3
	Specialized seed producers	0.51	0.91	4
Knowledge and information levels	New methods of production	2.47	0.99	1
	waste reduction Methods	2.43	1.12	2
	New potato varieties	2.31	1.01	3
	New methods of storing potato	2.27	1.03	4
	New methods of seed tubers supply	1.99	1.07	5

Mean range: From 1 (very low) to 5 (very much)

Table 6
Classification of the Level of Respondents' Knowledge and Information

Level of knowledge	Frequency	Percentage	Cumulative percentage
Low	43	47.3	47.3
Medium	33	37.2	84.5
High	12	12.7	97.2
Very High	2	2.8	100
Total	90	100	-

regarding seed potato management practices was also investigated (Table 5). As the results show, their highest level of knowledge and information was related to new methods of potato production followed by best methods of potato waste reduction. They were very poorly aware of new methods of seed tubers supply.

In order to classify the level of respondents' knowledge and information on seed potato management, ISDM method was used. Results of Table 6 using ISDM method indicated that near to majority of potato growers (47.3 percent) had low level of knowledge and information about seed potatoes management.

Correlation analysis

To evaluate the relationship between respondents' knowledge of potato seed management, correlation analysis (Pearson and Spearman) was used. Based on the results (Table 7), a significant and positive relationship was found between levels of education, potato production yield per hectare and respondents' knowledge. However, a negative and significant relationship was found between age and respondents' knowledge of potato seed management.

Logistic regression analysis

To determine factors affecting adoption of modern seed potato system, a logistic regression analysis was used. For this reason, respondents were divided into two groups of adopters and non-adopters (Table 8). According to Table 8, the predictive power of the logistic regression was 0.639 (Negelkerke R Square: 0.639). In the other words, the model predicted a high percentage of dependent variable values by variables included in the model and the adoption was accounted for by variables in the equation. -2 Log likelihood values that act as goodness of fit and decreases when model is well matched with data has appeared in one step. It was 174.651, confirming the goodness of fit of the model (Table 8). In another finding (Table 9), the sensitivity of the model for determining the correct percentage about those who did not adopt and use the modern seed potato system production obtained 92.9 percent. Also, for determining the percentage correct about those who adopted and used the modern seed potato system was estimated at 75.0. Overall, sensitivity of the total model to determine correct percentage was 86.0 percent. Based on the results (Table 10), among the individual, professional and economic factors, variables of age, number of family labor, area of owned land, area of potato cultivation, potato yield, number of owned land parcels and knowledge and information levels positively im-

Table 7
Correlation of Knowledge of Potato Seed Management and Socioeconomic Variables

Variables	r
Age	- 0.384**
Experience of potato farming	0.139
Number of household members	0.235-
Number of family labor	0.051
Education	0.223*
Area of owned land	- 0.153
Potato production yield per hectare	0.304*
Area of rented land	0.205
Area of potato under cultivation	0.032
The cost of potato cultivation	0.146
The income of potato cultivation	0.151

* $p < 0.05$; **$p < 0.01$ (2-tailed)

Table 8
Model Summary

Step	-2 Log likelihood	Cox & Snell R Square	Nagelkerke R Square
1	174.651	0.469	0.639

Table 9
Classification Table

Observed			Predicted		
			Adaption		Percentage Correct
			No	Yes	
Step 1	Adaption	No	143	11	92.9
		Yes	24	72	75.0
	Overall Percentage				86.0

Table 10
Logistic Regression for Adoption of New Seed Potato Systems

Variables	B coefficient	SD	Wald test	Exp (B)	p-value
Age (year)	-0.042*	0.292	3.997	0.958	0.046
Experience of potato farming (year)	0.036	0.019	3.556	1.037	0.059
Number of family labor	-0.409*	0.165	6.160	0.664	0.013
Number of household members	-0.072	0.095	0.571	0.931	0.450
Main job (0: non-agriculture, 1: agriculture)	0.264	0.540	0.240	1.303	0.625
Education (year)	0.082	0.128	0.409	0.922	0.523
Area of owned land (hectare)	0.133*	0.059	4.973	1.142	0.026
Area of potato cultivation (hectare)	0.372*	0.100	13.815	0.689	0.011
Potato production yield per hectare (ton)	0.096*	0.028	11.541	0.909	0.021
Number of owned land parcels	0.311*	0.098	10.073	1.365	0.032
Number of rented land parcels	-0.212	0.266	0.633	0.809	0.426
Knowledge and information levels (Pseudo-distance)	0.111*	0.049	5.091	1.117	0.024
Constant value	-3.188	1.361	5.486	24.235	0.019

* $p < 0.05$ (2-tailed)

pacted on the adoption and use of modern seed potato system. Moreover, the Wald statistic indicating the relative contribution of individual variables to the probability of adoption and use of modern seed potato system revealed that area of potato cultivation (13.815), potato yield (11.541) and number of owned land parcels (10.073) were the most important factors predicting the decision to adopt and use of system among potato growers.

CONCLUSION

Systems of seed potato production define the methods in which farmers produce, select, store and prepare seed (Sthapit et al., 2008). High-quality seed management plays the most important role in improving productivity. This study examined mechanisms of seed potato management among potato growers in Freydunshahr County of Isfahan province. Results showed that the main source of potato growers' seed potato supply was "a part of tubers harvested from their own farms". It implies that most potato growers provide seed potato from non-official or local chain. Potato seed tubers supplied by local and informal systems are poor in terms of health, physiological, physical and genetic features (Endale et al., 2008; Gildemacher et al., 2009; Lommen & Struik, 1994; Mulatu et al., 2005). Consistent with this result, Gildemacher et al. (2009) found that 98.7 percent of farmers in Ethiopia had provided seed potato tubers through informal and local seed systems. Based on the results, 91.2 percent of the respondents in the study region had difficulty to obtain seed tubers. Availability of seed means all farmers' access to appropriate seeds in right time.

In terms of potato storage methods, 89.6 percent of potato growers store and keep seed tubers in poor conditions. Only 4.4 percent of them store seed tubers in places with enough light and well ventilated conditions. Potato seed tubers stored in local warehouses with poor light, temperature and ventilation or in gunnysacks at home increase waste and decrease potato growers' income. Consequently, it discourages them for efficient management of seed potatoes. Due to the wide-spread area of potato cultivation in the region, poverty of farmers and lack of ability to build modern warehouses and due to improper maintenance practices of seed potatoes, governmental organizations, cooperatives or farmers' associations should construct standard warehouses.

Regarding quality of seed tubers, the purity of seed tubers used by about 80 percent of respondents was medium to low. 33.6 percent of respondents did not have enough considered the seed tuber size. They used accessible tubers as seed. Except 5.6 percent of potato growers who used adequately large size tubers, seed tuber size used by other potato growers was medium to small. It seems that the target of potato growers who selected small seed tubers as seed is often related to a reduction in seed costs (Endale et al., 2008; Gildemacher et al., 2009; Mulatu et al., 2005). Use of small seed tubers has two main problems: Delay in growing and low germination power because of little food storage (Lommen & Struik, 1994) and increased risk of disease outbreaks from mother seed tubers (Struik & Wiersema, 1999). Results also revealed that 32.4 percent of potato growers used healthy seed tubers and the rest of farmers used unhealthy and undesirable seed tubers for planting. With respect to problems mentioned above and the problem of weak health, physiological, physical and genetic characteristics, it can be concluded that current seed potato system in the region has local or informal nature and it did not develop to a specialized form.

Flow of knowledge includes issues such as farmers' access to knowledge about new varieties and seed sources, and knowledge and information that farmers are seeking for (Weltzien & Vombrocke, 2000). Results showed that majority of potato growers in the region had low level of knowledge of seed potato management. However, most of them were well informed about modern agronomic practices of potato production. Consistent with previous researches (Davoodi & Maghsoudi, 2012), results revealed that potato growers' knowledge and information was significantly correlated with education and potato yield per hectare. Major farmers, who form just a small group of farmers, are more educated and have better access to modern technologies, such as high yielding varieties, machineries and so on, thereby they have higher yield and more income as well as better access to agricultural information and knowledge sources. For this reason, the major group of small farmers who are beneficiaries of small and dispersed units of farm lands (the most arable lands of the country) with low access to modern technologies should be considered by agricultural policy makers and extension services for new knowledge, skills and technologies.

Based on the results of logistic regression model, among personal, professional and economic variables, age (in line sistent with Iqbal et al., 1999), number of family labor (consistent with Bagheri, 2015; Kafle & Shah, 2012; Waller et al., 1998), potato growers' knowledge and information level, number of owned land parcels and area of potato cultivation, potato production yield per hectare and area of owned land (consistent with Abebe et al., 2013; Bagheri et al., 2016; Ekwek & Onunka, 2006; Hirpa et al., 2010; Oladele, 2010) have significant effect on the adoption and use of new seed potato system. However, unlike previous studies, variables such as education, number of rented land parcels, rented land area, number of family members and primary occupation did not significantly determined the adoption of seed potato management practices. Based on the results, the following recommendations are presented:

- To promote production, storage, selection, preparation, certification and distribution of seed tuber, formal seed system should be created with participation of agricultural researchers and extension agents and farmers.
- Village based storehouse should be constructed with the cooperation of potato growers to store seed potatoes of the members in the best conditions for subsequent farming year.
- Extension educations should cover seed

management and train potato growers for seed management.

ACKNOWLEDGEMENT

The authors would like to thank University of Mohaghegh Ardabili (Ardabil, Iran) for financial support of this study. Also, we express appreciation to the respondents who kindly gave their time to make this research possible.

REFERENCES

1- Abebe, G.K., Bijman, J., Pascucci, S., & Omta, O. (2013). Adoption of improved potato varieties in Ethiopia: The role of agricultural knowledge and innovation system and smallholder farmers' quality assessment. *Agricultural Systems, 122,* 22–32.

2- Almekinders, C.J.M., & Louwaars, N.P. (1999). *Farmers' seed production: New approaches and practices.* London: Intermediate Technology Publications.

3- Atilaw, A., & Korbu, K. (2010). Recent Development in Seed Systems of Ethiopia. *Ethiopian Institute of Agricultural Research*: P. O. Box 32, Debre Zeit, Ethiopia. Retrieved from: http:// www.jica.go.jp/

4- Bagheri, A., Allahyari, M. S., & Ashoori, D. (2016). Interpretation on biological control adoption of rice stem borer, Chilo supperssalis (Walker) in North part of Iran: Application of Technology Acceptance Model (TAM). *Egyptian Journal of Biological Pest Control, 26* (1), 27- 33.

5- Bagheri, A. (2015). Determinants of adoption of mini-tuber seed Potato: A case in Ardabil Province of Iran. *International Journal of Agricultural Management and Development, 5*(4), 263-270.

6- Davoodi, H., & Maghsoudi, T. (2012). Analysis of Potato growers' knowledge about sustainable agriculture in Shushtar Township. *Iranian Journal of Agricultural Economics and Development, 42* (2), 265-274.

7- Deffo, V., & Demo, P. (2003). Adoption of two new potato varieties in Cameroon: Progress and constraints. *American Journal of Potato Research, 80*(4), 263-269.

8- Ebrahimi, M., Mokhtari, A., Moradi, K., & Khanahmadi, M. (2015). *Self-sufficiency program for production of virus-free seed tubers of potato varieties*. Agricultural Biotechnology Research Institute. Organization of agricultural research, education and extension. Iran (Tehran), pp.1-66.

9- Ekwek, C., & Onunka, B. N. (2006). Adoption of sweet Potato production technologies in Abia State, Nigeria. *Journal of Agriculture and Social Research, 6*(20), 92-100.

10- Endale, G., Gebremedhin, W., Bekele , K., & Lemaga, B. (2008). *Post-harvest management.* Addis Abeba: Ethiopian Institute of Agricultural Research, Pp. 53-77.

11- FAO. (2008). International year of the potato. Retrieved from: <http://www.fao.org/ potato- 2008/ en/world/asia.html>.

12- Gildemacher, P. (2012). *Innovation in Seed Potato Systems in Eastern Africa.* Unpublished thesis, Wageningen University, Netherlands (Wageningen), Department of Agriculture, Pp. 184-186.

13- Gildemacher, P., Demo, P., Barker, I., Kaguongo, W., Gebremedhin, W., Wagoire, W.W., Wakahiu, M., Leeuwis, C., & Struik, P.C. (2009). A description of seed potato systems in Kenya, Uganda and Ethiopia. *American Journal of Potato Research, 86,* 73–82.

14- Hirpa, A., Meuwissen, M. P. M., Tesfaye, A., Lommen, W. J. M., Lansink, A. O., Tsegaye, A., & Struik, P. C. (2010). Analysis of seed potato system in Ethiopia. *American Journal of Potato Research, 87*(52), 53-75.

15- Hirpa, A., Meuwissen, M. P. M., Van der Lans, W. J. M., Lommen, W. J. M., Lansink, A. O., Tsegaye, A., & Struik, P. C. (2012). Farmers'Opinion on Seed Potato Management Attributes in Ethiopia: A Conjoint Analysis. *Agronomy Journal, 104,* 1413-1424.

16- Iqbal, M., Bashir, A., & Farooq, U. (1999). Factors affecting the adoption of hybrid maze varieties in the irrigated Punjab. *International Journal of Agriculture and Biology, 1* (3), 149-151.

17- Kafle, B., & Shah, P. (2012). Adoption of improved Potato varieties in Nepal: a case of Bara district. *The Journal of Agricultural Sciences, 7,* (1), 14-22.

18- Keshavarz, M. R., Vazifedoust, M., & Alizadeh, A. (2014). Drought monitoring using a Soil Wetness Deficit Index (SWDI) derived from MODIS satellite data. *Agricultural Water Management, 132,* 37-45.

19- Khoshnodifar, Z., Ghonjib, M., Mazloumzadeh S. M., & Abdollahi, V. (2016). Effect of communication channels on success rate of entrepreneurial SMEs in the agricultural sector (a case study). *Journal of the Saudi Society of Agricultural Sciences, 15* (1),83–90.

20- Lommen, W.J.M., & Struik, P.C. (1994). Field performance of potato: mini-tubers with different fresh weights and conventional tubers Crop establishment and yield formation. *Potato Research, 37*, 313-301.

21- Louwaars, N. (2007). *Seeds of confusion: The impact of policies on seed systems*. Unpublished dissertation, Centre for Genetic Resources, WUR, Netherlands.

22- Majidi-Harvan, E. (2003). *Final report of production of maternal Mini-tuber seed potato*. Research project, education and extension organization, ministry of agriculture, Iran.

23- Mathoni, J., & Nyamongo, D. O. (2008). Seed systems in Kenya and their relationship to on- farm conservation of food crops. *Journal New Seeds, 9* (4), 330-342.

24- Mohammadi, A., Tabatabaeefar, A., Shahin, S., Rafiee, S., & Keyhani, A. (2008). Energy use and economic analysis of potato production in Iran a case study: Ardabil province. *Energy Conversion and Management, 49*, 3566–3570.

25- Mulatu, E., Osman, E. I., & Etenesh, B. (2005). Improving potato seed tuber quality and producers' livelihoods in Hararghe, Eastern Ethiopia. *Journal of New Seeds, 7* (3), 31–56.

26- Oladele, O. L. (2010). Determinants of Farmers' Information Seeking and Utilization on Seeds and Planting Materials in Lagos and Ogun States, Nigeria. *Journal New Seeds, 11*(4), 380-389.

27- Oyekale, K. O., Denton O. A., & Adebisi, M. A. (2014). Seed Management Systems in Nigeria: The Gap and the Bridge. Agric Business Trade Show and Conference (Nigeria at 100' Programme), International Conference Centre Abuja, pp:1-18. Retrieved from: http://10 times.com/venues/abuja-international-conferencecenter

28- Qamar, M. (2002). Global trends in agricultural extension: challenges facing Asia and the pacific region. Keynote Paper Presented at FAO Regional Expert Consultation on Agricultural Extension, Bangkok, 16–19 July. Retrieved from: Availablewww.fao.org/sd/cddirect/cdre0055b.htm

29- Sthapit, B., Ram, R., Pashupati, C., Bimal, B., & Pratap, S. (2008). Informal seed systems and on-farm conservation of local varieties. In: Thijssen, M.H. Z. Bishaw, A. Beshir, and W.S. de Boef. (Ed.). Farmers, seeds and varieties: supporting informal seed supply in Ethiopia. Wageningen: Wageningen International, pp. 133-136.

30- Struik, P.C., & Wiersema, S. G. (1999). Seed potato technology. Wageningen: Wageningen Pers. Netherlands, pp. 383 -389.

31- Vreugdenhil, D., Bradshaw, J., Gebhardt, C., Govers, F., MacKerron, D. K. L., Taylor, M. A., & Ross, H. (2007). Potato biology and biotechnology (advances and perspectives). The Boulevard, Langford Lane, Kidlington, Oxford, Amsterdam, the Netherlands, Elsevier, 823 p.

32- Waller, B.E., Hoy, C.W., Henderson, J.L., Stinner, B., & Welty, C. (1998). Matching innovations with potential users, a case study of potato IPM practices. *Agriculture, Ecosystems and Environment, 70*(2–3), 203–215.

33- Weather Bureau of Freydunshahr. (2016). Introducing of Freydunshahr County. About of County. Retrieved from: http://www.freydounshahrmet.ir/ShowPage.aspx?page_=form&order=show&lang=1&sub=17&PageId=334&codeV=1&temp-name=FreydunShahr New Skin

34- Weltzien, E., & Vombrocke. K. (2000). Seed systems and their potential for innovation: Conceptual framework for analysis. In Targeted seed aid and seed system interventions: Strengthening small farmer seed systems in East and Central Africa. *Proceedings of a workshop held in Kampala, Uganda,* 21–24 June 2000, ed. L. Kampala. Retrieved from: https://www.oecd.org/dac/evaluation/CPDE-Launching- Workshop-Report-April-2014.pdf

35- World Bank. (2009). Gender in seed production and distribution. Gender in Agriculture Sourcebook, 764 p. Washington, DC: World Bank.

21

A Survey on the Current Status of Mechanization of Paddy Cultivation in Iran: Case of Guilan Province

Saeed Firouzi

Abstract

This study investigated the status of mechanized power and self-propelled rice cultivation machinery in Guilan Province in northern Iran. The raw data was obtained from 2013 statistics of the Agricultural Jihad Organization of Guilan province. Power per unit of area, mechanization level, area per unit of self-propelled machinery, and mechanization requirement were calculated for eastern, western and central Guilan. The mean power per unit of area for eastern, western and central Guilan was 2.22, 2.07, and 3.09 $hpha^{-1}$, respectively. The total mechanization requirements were determined to be 0%, 73.39%, 99.28%, and 52.47% for land preparation, transplanting, weeding, and harvesting operations, respectively. The area per self-propelled rice machinery was 111.38, 3777.97, and 358.99 ha for rice transplanter, weeder, and combine harvester, respectively. This indicates that there is insufficient mechanization for rice weeding and that there is an urgent need to increase the machinery available for this agronomic operation in the paddy fields of Guilan Province.

Keywords:
Rice, paddy, power, Mechanization index, mechanization requirement

Department of Agronomy, College of Agriculture, Rasht Branch, Islamic Azad University, Rasht, Iran.
** Corresponding author's email: firoozi@iaurasht.ac.ir*

INTRODUCTION

Decreasing manual labor requirements is an important aspect of farm mechanization. Less number of labors is needed to complete the cultivation process by mechanized farm compared to traditional farm (Rahman *et al*., 2011). Manual cultivation of agricultural production is time-consuming and expensive. Mechanization of agricultural operations has been undertaken in many parts of the world to increase the income of farmers and promote the economic interests of agriculture (Singh, 2006). Agricultural mechanization is the application of agricultural machinery to land preparation, crop residue management, plant protection, harvesting, and threshing to produce crops. It adds value to the agricultural products and provides power for farm operations such as irrigation (Karale *et al.*, 2008). There is growing interest in mechanization of agricultural operations among farmers to decrease production costs and increase farm income (Singh, 2006). Decreasing the exertion of agricultural work and increasing its appeal, the area under cultivation and production per unit area are the other important benefits of agricultural mechanization (Almassi *et al*., 2006).

Of agricultural products, mechanization of rice as the world's second staple food is of special importance. Submerged cultivation and difficult working conditions in paddy fields make its cultivation labor intensive, expensive, and energy intensive. The total time required for traditional production of rice is estimated to be over 1780 man-hours ha^{-1} (Pateriya and Datta, 2012). Labor shortages during growing season and the low speed of manual cultivation compared with mechanized farming also are very important problems.

The mechanization of agriculture has been considered to be a major factor in the development and sustainability of rice production by the Iranian Ministry of Agriculture. Thus far, efforts in this area have not been sufficient and development of new approaches to solve this problem is still under consideration.

Codification of policies, plans, and models for agricultural development in different areas of the world requires detailed knowledge and analysis of the existing conditions. Without these studies, any planning, whether it is short, medium or long term, will fail (Loveimi and Almassi, 2003). An appropriate model can only be provided after all the possibilities and limitations for each district are studied. Cultural, geographic, economic and social differences mean that one model may not be effective for all parts of the world (Shahbazi, 1989).

Researchers have examined the quantity and quality of agricultural mechanization in different regions and countries and have developed and implemented several indicators. Ampratwum *et al.* (2004) found that the mechanization index (ratio of mechanical power to total mechanical, animal and human power) in Oman was 75% and the power per unit area was 1.1 kW ha^{-1}. They proposed the use of at least 916 double-axle tractors after 2002, to increase the mechanization index to 81% in Oman. The problems of mechanization of small hold farmers in the central strip in Nigeria were investigated by Yohanna *et al.* (2010). Their study showed that the level of mechanization (ratio of machine tools to total number of machines and manual tools for each cultivation operation) for cleaning products was 21.54%, land preparation was 24.62%, planting was 3.85%, spraying was 86.15%, fertilizing was 2.13%, weeding was 3.08%, harvesting was 40%, processing and storage of the product was 7.69%. Paman et al. (2012) studied the power available to small paddy fields of Riau province in Indonesia. They showed that, although access to farm power increased from 1997 to 2006, access to agricultural power in the region was low. The total power required to produce white rice, including plowing, preparing the nursery, planting, weeding, fertilizing, pest control, harvesting, threshing, transportation, cleaning, drying and converting the paddy to white rice was 418 kw (561 hp) per hectare on average. They recommended that farmers should be encouraged to use mechanical power for timely fulfillment of farm operation and increased productivity. Firouzi (2014) examined the mechanization indicators of paddy cultivation for five major districts in the city of Langarud in northern Iran. The results showed that the average mechanization level in the study areas was 1.37 $hpha^{-1}$. The mechanization requirements for land preparation was 0%, transplanting was 85.50%, weeding was 94.97% and harvesting was 43.20%. The mechanization of transplanting and weeding was deemed most important. Rasooli Sharabiani and Ranjbar (2008) investigated the status of farm mechanization in Sarab district in East Azerbaijan province in Iran. They showed that the overall power per unit of area in spring 2008 was 0.83 $hpha^{-1}$. The contribution to mechanization by animals was 1.24%, by humans was 2.23%, and by machinery 96.35% in the study area. Their results emphasized the importance of the use of machinery for agricultural activities in the region. Rasouli *et al.* (2010) studied the factors affecting farm mechanization of sunflower cultivation in Iran. Power per unit area was determined to be 0.5 $kWha^{-1}$ (1.67 $hpha^{-1}$). Shahraki *et al.* (2012) studied the quantity and quality of agricultural mechanization in Sistan and Baluchistan Province in Iran using statistics from the Agricultural Jihad Organization. Their study showed that the degree of mechanization of agricultural activities, especially for harvesting in Sistan and Baluchistan was much lower than for most developing countries. The power per unit area (0.68 $hpha^{-1}$) was also low. They recommended consolidation of agricultural land, provision of appropriate technology, availability of appropriate educational and promotional programs, and support for agricultural mechanization service companies for provision of farm machinery to promote agricultural mechanization in Sistan and Baluchistan Province. Loveimi and Almasi (2003) investigated the status of mechanization in the northern region of Ahwaz in Iran and found that the average power per unit area was 1.1 $hpha^{-1}$.

To increase agricultural mechanization, the quality and quantity of mechanization should be evaluated in detail for each region. Macroplanning can best be performed with knowledge of existing conditions. The present study investigated the status of mechanized power and self-propelled rice cultivation machinery in Guilan province in northern Iran.

MATERIALS AND METHODS

This study was conducted in Guilan Province in northern Iran. Rice, peanuts, tea, olives, kiwi fruit, and vegetables are the major agricultural and horticultural products in this province, but most farmers cultivate rice in paddy fields. Total rice cultivation is about 237000 ha, which ranks second among the rice-growing provinces of the country. Its annual production is estimated to be about 700,000 tons of white rice, which ranks first in Iran.

The present study investigated the mechanization level of agricultural operations (land preparation, planting, plant protection, and harvesting), area per self-propelled rice machinery, mechanization requirements for each agronomical operation, and power per unit of area in 16 cities in Guilan Province. Raw mechanization data from 2013 statistics of the Agricultural Jihad Organization of Guilan Province was employed to prepare this study.

Mechanization level

This is a quantitative index of the mechanized agronomic activity and equals the area under mechanized cultivation divided by the total area under cultivation (Lak and Almassi, 2011). This index is used to determine the ratio of mechanized operations at different agricultural stages. The mechanization level is often estimated for individual crops especial operations separately (Lak and Almassi, 2011):

$$ML = A_M / A_C \quad (1)$$

Where: ML =mechanization level (%); AM = mechanized cultivated area *(ha)*; AC = total cultivated area *(ha)*.

Power per unit of area

This is a qualitative index used in macroplanning and development of agricultural mechanization. It is the ratio of the total drawbar power available in a region to the total area under cultivation (Lak and Almassi, 2011):

$$PPA = P_d / A_c \ (hpha^{-1}) \quad (2)$$

Where: PPA = Power Per unit of Area *($hpha^{-1}$)*; P_d = total machinery power *(hp)*;
A_c = cultivated area *(ha)*.

The power per unit of area is similar to the

power per capita for agricultural land and shows the average power available per unit of cultivated agricultural land. The unit used to describe this index is horsepower per hectare *(hpha*$^{-1}$*)* or kilowatt per hectare *(kWha*$^{-1}$*)*. Because of the nature of rice cultivation and the fact that most equipment used in rice cultivation is self-propelled, all motorized power (two-wheeled and four-wheeled tractors, rice transplanters and weeders, self-propelled harvesters, rice combines) were included in the calculations. The actual power was calculated by multiplying the total rated power by 0.75.

Area per self-propelled rice machinery

This index determines the average area of land under cultivation worked by each self-propelled machine (transplanter, weeder, rice combine). This index was obtained by dividing the total area under cultivation by the total number of respective machines.

$$APSM=A_C/N_{SM} \quad (3)$$

Where: APSM= Area per self-propelled machinery *(ha machine*$^{-1}$*)*; A_C= total cultivated area *(ha)*; N_{SM}= number of self-propelled machinery.

Mechanization requirement

This index was calculated using the simple mathematical relation of 100 minus the mechanization level for each agronomical operation (Khambalkar *et al.,* 2010; Zangeneh *et al.*, 2010).

$$MR=100-ML \quad (4)$$

Where: MR= mechanization requirement (%); ML= mechanization level (%).

Analysis of data

After entering the raw data into Excel software, the means of the mechanization indices were calculated. The measure of standard error (SE) was calculated for each part of Guilan (east, west, central) and for the whole province. This statistical criterion indicates variation in the indices. The SE was calculated by dividing the standard deviation of the data by the square root of the number of data.

RESULTS AND DISCUSSION

The important indicators for power per unit of area and mechanization level were computed to determine the level of mechanization of

Table 1: Mechanization level (%) of farming various operations and power per unit of area *(hpha*$^{-1}$*)* in paddy fields of Guilan province.

Part of province	City	Mechanization level (%)				Power (unit area)$^{-1}$
		Tillage	Transplanting	weeding	harvesting	
WEST	Astara	100.00	32.28	1.09	29.06	2.00
WEST	Rezwanshahr	100.00	82.57	4.95	49.90	3.64
WEST	Talesh	100.00	51.00	0.18	29.81	3.26
WEST	Shaft	100.00	6.32	0.00	34.44	1.13
WEST	Somesara	100.00	9.60	0.23	69.59	1.98
WEST	Fuman	100.00	17.16	0.22	17.84	1.09
WEST	Masal	100.00	82.86	2.24	72.86	1.40
Average		100.00	40.25±12.35	1.01±0.68	43.36±8.05	2.07±0.38
CENTRAL	Anzali	100.00	21.06	0.13	76.45	2.83
CENTRAL	Rasht	100.00	8.99	0.00	53.08	1.91
CENTRAL	Roudbar	100.00	35.53	0.23	99.20	4.53
Average		100.00	21.86±7.67	0.79±0.07	76.24±13.31	3.09±0.77
EAST	Astaneh	100.00	21.48	1.57	59.19	2.64
EAST	Amlash	100.00	17.71	0.09	27.57	1.70
EAST	Roudsar	100.00	7.62	0.23	30.23	2.06
EAST	Siahkal	100.00	12.29	0.00	23.05	2.38
EAST	Lahijan	100.00	6.13	0.05	38.82	1.45
EAST	Langaroud	100.00	13.19	0.23	49.45	3.06
Average		100.00	13.07±2.39	0.36±0.24	38.05±5.70	2.22±0.24
Total (Mean)		100.00	26.61±6.26	0.72±0.34	47.53±6.18	2.32±0.29

Table 2: Area per unit of self-propelled rice machinery *(ha machine^{-1})* in the paddy fields of Guilan Province.

Part of Province	City	Transplanting	weeding	Reaper	Combine harvester
WEST	Astara	37.65	640.00	94.12	3200.00
WEST	Rezwanshahr	21.37	555.56	103.09	333.33
WEST	Talesh	70.43	5329.00	134.34	1141.93
WEST	Shaft	220.46	14330.00	114.64	3582.50
WEST	Somesara	147.22	4662.00	63.57	570.86
WEST	Fuman	167.11	6935.00	62.48	4623.33
WEST	Masal	32.11	7000.00	34.15	1166.67
Average		99.48	5635.94	86.63	2088.37
CENTRAL	Anzali	54.77	803.33	15.35	344.29
CENTRAL	Rasht	289.93	10389.33	82.02	1326.30
CENTRAL	Roudbar	88.87	3377.00	99.32	375.22
Average		144.53	4856.56	65.56	681.94
EAST	Astaneh	130.94	3928.33	179.92	130.94
EAST	Amlash	145.83	875.00	233.33	437.50
EAST	Roudsar	267.50	5350.00	356.67	205.77
EAST	Siahkal	113.69	2217.00	158.36	4434.00
EAST	Lahijan	243.02	23816.00	517.74	154.65
EAST	Langaroud	115.19	4550.00	700.00	100.00
Average		169.36	6789.39	357.67	910.48
Total (Mean)		111.38	3777.97	91.09	358.99

paddy fields in Guilan Province. All engine power including small and medium power tractors (average respective power of 35 and 60 *hp*), 2-wheeled tractors (power tillers with an average power of 7 *hp*), self-propelled rice machinery (rice transplanters, weeders, reapers and combine harvesters) were considered when calculating the mechanization level.

Table 1 show the mechanization level (ML) of land preparation in all sections of Guilan province Province is 100%. Farming operations generally include energy-intensive and control-dependent operations. Primary and secondary tillage are energy-intensive operations (Almassi *et al.*, 2006). This attribute, the limited time for tillage, and the negative effects of delay in land preparation on on time the other agronomical operations and the degree of mechanization of tillage for all cities in Guilan is 100%.

The lowest level of mechanization for rice transplanting was 6.13% in Lahijan in eastern Guilan followed by 6.32% in Shaft in western Guilan. The highest level of mechanization also was determined as 82.86% in Rezvanshahr followed by 82.57% in Masal in western Guilan, which represents a relatively desirable state of mechanization for rice transplanting in this region. The low level of mechanization for transplanting in Lahijan may be attributed to the small size, irregular geometry and dispersal of the paddy fields. Success in full implementation of land consolidation projects can improve the mechanization indicators for transplanting. The high cost of rice transplanting machinery and technical issues associated with their operating, problems in the preparation and supply of seedlings are the other important barriers for development of mechanized rice transplanting in Guilan province.

Table 1 show that the average mechanization level for rice weeding in Guilan is very low (about 0.72%). The mean values for mechanization level of rice weeding operation in western, eastern and central Guilan were determined to be 1.01, 0.36, and 0.79, respectively. The lesser mechanization level for rice weeding in comparison with other agricultural operations is unfavorable and requires more attention than the other agronomical operations.

It should be noted that mechanized rice weeding technically depends on the mechanization of transplanting. Transplanting should be mechanized before mechanizing the rice weeding operation. Table 1 show that the possibility of developing mechanization of rice weeding in Masal and Rezvanshahr are 82.86% and 82.57%, respectively. Indeed, these values are the mechanization levels of rice transplanting operation in Masal and Rezvanshahr, respectively. The shortage of rice weeding machinery in Guilan and the high price of rice weeders also effect the development of mechanization of rice weeding.

The average mechanization level of rice harvesting operation is desirable at 47.53%. The mechanization level of rice harvesting in central Guilan (76.24%) was higher than those of eastern and western Guilan (at 38.05% and 43.36%, respectively). This emphasizes the need to provide rice harvesting machinery to farmers in western and eastern Guilan. The physical strain associated with manual harvesting, high cost of manual harvesting, and the sensitivity to delays in rice harvesting caused by unstable weather at harvest time in northern Iran rank mechanization of harvesting as second to land preparation.

The average Power Per unit of Area (PPA) in paddy fields of Guilan province was 2.32 hpha-1 (Table 1). The power per unit area for central Guilan (3.09 *hpha*$^{-1}$) was higher than those of the eastern and western Guilan (2.22 and 2.07 hpha-1). These values are less than those reported by Ampratwum *et al.* (2004) in Oman, Rasooli Sharabiani and Ranjbar (2008) in Sarab, Shahraki *et al.* (2012) in Sistan and Baluchistan, and Loveimi and Almassi (2003) in northern Ahwaz. However, it should be noted that rice cultivation is more labor intensive than cultivation of other types of agriculture and requires more power.

The Average area Per Self-propelled rice Machinery (APSM) is shown in Table 2. According to the results, there is only one transplanter for every 111 ha of paddy fields in Guilan province. This index in Rasht is about 290 ha per machine. While manual transplanting provides uniform crop stand but it is quite expensive and needs a lot of labor besides involving a lot of drudgery (Manjunatha *et al.*, 2009). Singh *et al.* (1985) stated that rice transplanting needs about 250-300 man hours *ha*$^{-1}$ which is about 25% of the total labor requirement for rice cultivation. More consideration should be paid for this aspect of rice mechanization in Guilan.

There is only one rice transplanter for every 3778 ha of paddy fields (Table 2). Depending on the type and density of weeds in the paddy field, 10 to 15 workers per hectare are needed for rice weeding (Alizadeh, 2011). This illustrates the high labor force needed for manual compared to mechanical weeding and suggests that mechanization requires special consideration.

The mean area per rice combine was 359 ha, which is relatively good compared with weeding machinery. The higher area per combine harvester can be attributed to the independence of mechanization of rice harvest from the other operations (transplanting and weeding) and the difficult

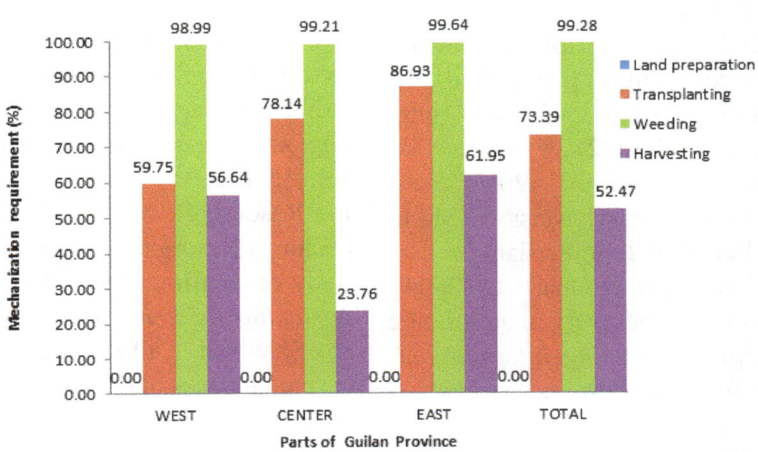

Figure 1: Mechanization requirement for various agronomic operations in paddy fields for different parts of Guilan province.

conditions for rice harvesting. The importance of timely harvesting of rice, limited time for harvest and cost-intensiveness of manual harvesting of paddy rice are additional factors.

Figure 1 shows the Mechanization Requirements (MR) for agronomic operations for paddy cultivation in Guilan Province. The mechanization requirement of land preparation equals zero.

The mechanization requirement of rice weeding operation in Guilan province is 99.28%. This figure underlines the urgency of removing barriers for development of mechanization of rice weeding in Guilan Province.

CONCLUSIONS

Results of this research indicated that except for tillage operation, there were considerable differences in mechanization level of various rice agronomical operations in three main regions of Guilan Province, Iran. The power per unit of area for central Guilan was higher than those of the eastern and western Guilan. The highest mechanization requirement was determined as 99.28% for rice weeding followed by 73.39% for rice transplanting operation. Therefore, increasing the machinery available for weeding operation needs to be given the topmost priority for mechanization of paddy cultivation in Guilan Province, Iran.

ACKNOWLEDGEMENTS

The author is grateful for the funding supplied by the Rasht branch, Islamic Azad University.

REFERENCES

1- Alizadeh, M.R. (2011). Field performance evaluation of mechanical weeders in the paddy field. *Scientific Research and Essays* 6(25), 5427-5434. http://dx.doi.org/10.5897/ SRE11.1412

2- Almassi, M., Loveimi, N. & Kiani, S. (2006). *Farm machinery management.* Ferdowsi University of Mashhad Press, Mashhad, Iran, P. 282. (In Persian).

3- Ampratwum, D., Dorvlo, A. & Opara, L. (2004). Usage of tractors and field machinery in agriculture in Oman. *Agricultural Engineering International: the CIGR Journal of Scientific Research and Development.* VI, 1-9.

4- Firouzi, S. (2014). An assessment of rice cultivation mechanization in Northern Iran (Langarud County in Guilan Province). *International Journal of Biosciences*, 5(3), 110-115.

5- Karale, D.S., Khambalkar, V.P., Bhende, S.M., Amale, S. & Wankhede, P. (2008). Energy economic of small farming crop production operations. *World Journal of Agricultural Science*, 4(4), 476-482.

6- Khambalkar,V.P. Jyoti, P., Sachin, K., Dipak, B. & Shilpa, D. (2010). Energy and economic evaluation of farm operations in crop production. *Journal of Agricultural Science,* 2(4), 191-200.

7- Lak, M.B. & Almassi, M. (2011). An analytical review of parameters and indices affecting decision making in agricultural mechanization. Australian *Journal of Agricultural Engineering*, 2(5), 140-146.

8- Loveimi, N. & Almassi, M. (2003). Investigation of the status of mechanization in north of Ahwaz county. *Iran Journal of Science and Technology of Agriculture and Natural Resources, Water and Soil Sciences,* 7(2), 227-239. (In Persian).

9- Manjunatha, M.V., Masthana Reddy, B.G., Shashidhar, S.D., & Joshi, V.R. (2009). Studies on the performance of self-propelled rice transplanter and its effect on crop yield. *Karnataka Journal of Agricultural Science*, 22(2), 385-387.

10- Paman, U., Inaba, S. & Uchida S. (2012). Power availability and requirements for small-scale rice farm operations: A case in Riau Province, Indonesia. *American Journal of Agricultural Science Engineering and Technology*, 1(2), 27-36.

11- Pateriya, R.N. & Datta, R.K. (2012). Design modifications of Mat Type Rice transplanter. *International Journal of Advanced Technology & Engineering Research* (IJATER), 2(6), 87-91.

12- Rahman, M.S., Monayem Miah, M.A., Moniruzzaman & Hossain, S. (2011). Impact of farm mechanization on labor use for wheat cultivation in Northern Bangladesh. *The Journal of Animal and Plant Sciences,* 21(3), 589-594.

13- Rasooli Sharabiani, V. & Ranjbar, I. (2008). Determination of the degree, level and capacity indices for agricultural mechanization in Sarab Region. *Journal of Agricultural Science Technology*, 10, 215-223.

14- Rasouli, F., Sadighi, H. & Minaei, S. (2010). Factors affecting agricultural mechanization: a case study on sunflower seed farms in Iran. *Journal of Agricultural Science Technology*, 11, 39-48.

15- Shahbazi, I. (1989). Agriculture advancement: the requirements of change and renewal of agriculture. [Translation.] Publication of Agriculture Dissemination Organization, Tehran, Iran. PP. 22-35. (In Persian)

16- Shahraki, J., Shahrakizad, A., Yaghoubi, M. &

Esfandiari, M. (2012). A survey on the level of mechanization development in Sistan and Baluchestan, Iran. *Journal of Applied Sciences Research,* 8(4), 2267-2271.

17- Singh, G. (2006). Estimation of a mechanization index and its impact on production and economic factors a case studies in India. *Biosystems Engineering,* 93(1), 99–106.

18- Singh, G., Sharma, T.R. & Bockhop, C.W. (1985), Field performance evaluation of a manual rice transplanter. *Journal of Agricultural Engineering Research,* 32, 259-268.

19- Yohanna, J.K., Fulani A.U. & Aka'ama, W. (2010). A survey of mechanization problems of the small scale (peasant) farmers in the Middle Belt of Nigeria. *Journal of Agricultural Science*, 3(2), 262-266.

20- Zangeneh, M., Omid, M. & Akram, A. (2010). Assessment of agricultural mechanization status of potato production by means of artificial neural network model. *Australian Journal of Agricultural Science,* 4(5), 372-377.

Preventive Factors to Found Walnut Production Cooperatives in Tuyserkan Township, Iran

Mohammad Abdolmaleky

Abstract

This study was conducted to seek the opinion of walnut producers on barriers affecting the success of producers to found production cooperatives regarding horticultural products in Tuyserkan Township, Iran. A sample of 234 walnut producers were randomly selected through multistage cluster sampling technique. Descriptive-correlation research method was used in this study which has been implemented through the questionnaire. For determining the validity of questionnaire, the face and content validity was used. Reliability for the instrument was estimated at 0.94. According to factor analysis, barriers to found and develop walnut production cooperatives were categorized into seven groups that explained 65.799% of the total variance of the research variables. The results also indicated that lack of cultural infrastructures, inadequate knowledge and incorrect opinions of producers and leaders, inadequate laws and vulnerability and infrastructural barriers had the most effect to found and develop walnut production cooperatives, respectively.

Keywords:
Barriers, Walnut Producers, Production Cooperatives, Iran

Assistant professor, College of Agriculture, Khoramabad Branch, Islamic Azad University, Iran
** Corresponding author's email: mabdolmaleky@yahoo.com*

INTRODUCTION

Changes of the recent age in the fields of economy, culture and social relations have led cooperatives to new conditions through which they couldn't go on without consistency. In industrialized countries, the government has paved effective steps through designing frameworks for the development of cooperatives in economic, social, and political parts. But developing countries face obstacles along the way such as lack of standardized environmental conditions, the mismatch of expectations and abnormal increases in cooperatives quantitatively. Consequently problems are facing cooperatives in developing countries, Iran is also faced with these problems (Abbasi *et al.*, 2009). Experts and sociologists believe that the current conditions of the hand (the slogan of the smaller government for a better future) are seriously considered, so more attention is needed to strengthen the cooperative sector and cooperatives for two reasons. First, strongly controlling is exerted on cooperative system, so responsibilities and controlling costs seem low to the government and finally supervision comes to sense widely more accurate. Second, there will be a new counterpart having important role on increasing the production quality and efficiency. However, during the past years, cooperatives have been able to practice their role and to find their real position in the economy of Iran (Kuhi, 2006). Based on further perspective 1404 and regarding this fact that, Iran should be at first scientific, economic and technology place among the oriental countries, while cooperatives are of successful socio-economic patterns being capable to establish instant development and justice. Considering the role of cooperatives, achieving this target will be feasible only by promoting human and social capitals; so, it is necessary to provide the efficient members in rural communities with a background to improve social and economic capitals in agricultural production cooperatives through one of important substrates of sustainable rural development and employment (Karami and Agahi, 2010a). Iran has a long history of rural solidarity and self-help efforts. However, formal cooperatives with the emphasis on the provision of agricultural services for their member were introduced by industrialized countries such as England to Iran only about 1940. The agricultural production cooperatives are considered to be the most important organizations that pay attention and try to support the rural development in general and the agricultural development in particular, through the activities and services offered for farmer welfare. It is generally believed that successfully managed agricultural cooperatives have a great potential in agricultural development in particular and rural development in general (Zarafshani *et al.*, 2010). Farmer organizations have a strong potential for building linkages to interventions/programmers in other sectors. The inter-sector linkages can enable the collective improvement of farmers' livelihoods from a number of government- and donor-supported interventions. Farmer organizations can facilitate a vertical exchange of information; enable farmers to access higher levels of management and contribution in decision-making process (Abdolmaleky, 2012).

The International Cooperative Alliance (ICA) defines a cooperative as "an autonomous association of persons united voluntarily to meet their common economic, social and cultural needs and aspirations through a jointly owned and democratically-controlled enterprise". In developed countries agricultural production cooperatives have significantly contributed to the mobilization and distribution of capital resources, entrepreneurship, constitute a forum for education and training, social welfare, poverty alleviation and other socio-economic problems (Tanzanian Federation of Cooperatives, 2006).

The most important agriculture cooperatives types in Iran are agricultural production cooperatives (Zarafshani *et al.*, 2010). In Iran, although, there are many cooperatives in rural area, but, it is alleged that cooperative does not play a significant role in rural development (Aref, 2011). Agricultural production cooperatives are certainly a major contributor to agricultural development in many countries. But, in Iran there are a number of barriers to use effectively agricultural production cooperatives as a tool for agricultural development. There are some barriers at national, local and organizational levels. Dependency of cooperatives to government and lack of cultural capacity for group collaboration, lack of resources, lack of leaders' knowledge, were the most important elements contributing to limit production cooperatives in agricultural and rural development (Allahdadi and Aref, 2011).

Cooperatives as a critical dimension of market structure in agriculture must periodically be assessed to determine the future viability of the cooperative form of business (Dunn *et al.*, 2002).

There are many studies regarding failure or success of agricultural cooperatives. Amini and Ramezani (2006) in the study of "evaluating effective factors on the success of poultry-farming cooperatives in Tehran province," concluded that the cooperatives have been successful to achieve desired goals in the statute and they indicated that some elements have the most effect on the success of agricultural production cooperatives including the performance of cooperatives ,managers' technical competences, membership in cooperative associations, the rate of participation in cooperative issues, profitability of cooperative members, quality of implemented training courses, number of training courses and managers' human talents, respectively. Indeed, there is a positive and significant relationship among these variables and the success of cooperative companies.

Zar'anzhad and Sharifi (2008) studied "an investigation on the attitude of managers and members of Ilam cooperatives towards effective factors for improving the productivity in cooperatives" they also pointed out that boosting financial power and using efficient forces influence promoting the productivity in these cooperatives.

Safari *et al.* (2009) found out, in a study named "effective factors on the success of cooperative companies", that certain elements affect the success of cooperatives including standards, trends, principles, methods of performance and stressing on performing them by employees and members, establishing a proper information system, buying, selling, updated information, continues training, holding educational courses in the fields of investment and financial affairs, marketing at level of directors and the board , active participation of members, determining criteria for evaluating the performance and reward, selecting the high experienced managers, sufficient initial capital, holding educational courses, applying the advanced methods of production ,creating a rich culture and supporting it, pondering needs appropriately at the beginning of establishing cooperatives, considering the demands of agents and providing them rapidly, modifying laws and regulations and executing them properly.

Karami and Agahi (2010b) in a study entitled "evaluating the role of creativity on the success of cooperatives", based on the viewpoints of respondents, and indicated that there are some factors contributing to success of cooperatives such as: the impact of innovation, access to information, effective and fast system of offers. Didi (2004) concluded that the homogeneity of members and stability of small groups had a considerable impact on the success of cooperatives, as well as he emphasized role of partnership on the utilization of resources and corporate success.

International Labor Organization (2004) considered social properties of human force in the company such as recognition, participation, knowledge, education level, quality of human resources and familiarity with the goals and governmental policies as effective factors on the success of cooperatives.

Bruynis *et al.* (2001) showed that the management and the establishment of formal condition a positive effect on the success of cooperatives. In this study, some factors help promoting cooperatives such as management quality, keeping touch with customers, members' interests, rate of members' trust, retaining members, conducting the board of directors.

Tuyserkan Township has suitable climate and environmental conditions for horticultural products especially for walnut production. Walnut production has the most important role in the rural and agricultural economy of Tuyserkan Township with total area of about 5100 ha which even lies in the first place in Iran (Jihad-e-Agriculture of Tuyserkan, 2013). Unfortunately, there are not effective walnut production cooperatives in Tuyserkan. Hence, this paper concerns with studying and investigating barriers contributing to the success for founding and developing walnut production cooperatives as perceived by walnut producers. Therefore, the primary purpose of the present study was to identify barriers to found walnut production cooperatives. The specific objectives of this study were to: (1) describe the demographic profile of walnut producers, (2) identify perceptions and attitudes of walnut producers to participate in cooperatives, and 3) identify barriers to success of walnut producers to found and develop production cooperatives.

MATERIALS AND METHODS

This study was conducted in the township of Tuyserkan, located in the west region of Iran, with total area of about 1464 square kilometers. Statistical sample in this study was drown randomly from 600 walnut producers. A number of 234 producers were selected through multistage cluster sampling based on their level of performance characteristics. The research design for this study was a survey design (Delavar, 2004). From a review of the literature, the researcher developed an instrument to collect data. The survey was divided into two sections. The first section was designed to gather data on personal characteristics of walnut producers, included gender, age, years of working experience, level of education, income and etc. The second section was designed to gather data about the producers' perceptions and attitudes with respect to the participation and membership in the walnut production cooperatives. Respondents were asked to rate their viewpoints concerning these variable on a five point Likert-type scale: (5=very much, 4=much, 3=moderate, 2=low and 1=very low). Face and content validity of the questionnaire were established using a panel of experts consisting faculty members in department of agricultural extension and education in Khoramabad Islamic Azad University and extension officers. Questionnaire reliability was estimated by calculating Cronbach's alpha coefficient. Reliability for the instrument was estimated at 0.94. The data was collected between March, 2013 and May, 2013. After gathering and encoding information of the questionnaires, data was obtained for analysis. Data was analyzed using the statistical package for the social sciences (SPSS, 19). Beside descriptive statistics, analytical statistics (factor analysis and ANOVA test) were employed for detailed analysis. Variables used in "factor analysis" included nineteen barriers on founding and developing walnut production cooperatives. Respondents were asked to rate their viewpoints concerning these variables on a five point Likert-type scale: (5=very much, 4=much, 3=moderate, 2=low and 1=very low).

RESULTS

Descriptive statistics

The first objective was to describe the demographic profile and socio-economic demographic characteristics of walnut producers of Tuyserkan Township, Iran. The findings showed that majority of the respondents were male (93.8%). Average age of respondents was 46.76 years. The minimum age of respondents was 27 and the maximum age was 65.Eighty-seven percent (87%) of them were married. Data showed that average household size in study area was 6 members per family. Regarding respondents' education levels, majority of the respondents (82.1%) were high School and illiterate. Fifty-three percent of the producers had more than 9 years of working experience and the rest under 9. Their average working experience was 12.48 years. Study also revealed that principal occupation of the27% of the respondents was "walnut garden holding". 41% of them were as "garden holder and farmer" and the rest were as "garden holder and rancher". Findings of the study also showed that their average garden land holding was 0.6 ha. Majority of them (93.6%) owned less than1 ha walnut garden land. Two-thirds (63%) of the producers sell their products in local markets. According to monthly income i.e. [(yearly income)/12], the average of their total monthly income was 9 million Rial per month. 70.5% of the respondents had less than 100 million Rial incomes yearly (Table 1).

The second objective of this study was to

Table 1: Some demographic characteristics of respondents

Feature	Percentage
Gender	
Male	93.8
Female	6.2
Age (years)	
Up to 35	12.8
36-45	37.6
46-55	39.3
Over 56	10.3
Average	46.76
Level of Education	
Illiterate	14.5
< High School	44.8
High School/GED	22.6
College	17.9
Work Experience	
Up to 9	47
10–19	29.5
20–29	11.5
30–39	12
Income (Million Rials)	
Up to 100	70.5
100 – 200	23.1
200 – 300	3.8
300 – 400	2.6

identify perceptions and attitudes of walnut producers to take part in cooperatives. To evaluate respondents' attitude about participation in agricultural production cooperatives, they were asked to rate their viewpoints on a five point Likert-type scale: (5=quite agree, 4=agree, 3=neutral, 2=disagree and 1=quite disagree) concerning twenty given items. According to the sum of scores of each respondent, they were placed in three classes as follows: negative, neutral and positive. Pondering revealed that 72.2% had positive attitude and interest for participation and membership in walnut production cooperatives. Results of the descriptive analysis have been shown in details (Table 2).

Table 2: Descriptive statistics on attitude of the respondents about participation in walnut production cooperatives

Respondents' attitude	Scores	F (%)
Negative	20 –47	6 (2.6)
Neutral	47–73	59 (25.2)
Positive	73– 100	169 (72.2)

The third objective was to identify barriers to success for founding and developing walnut production cooperatives. Exploratory Principal Component Analysis (PCA) was conducted to summarize the variables of the research to a smaller quantity and to determine the factors affecting the success of producers to found agricultural production cooperatives and the obtained factors were subjected to VARIMAX rotation. PCA is a form of factor analysis, which first looks for a linear combination of variables that extracts maximum variance from variables and then identifies a second linear combination to explain the remaining variance, leading to orthogonal, or uncorrelated, factors (Rehman *et al.,* 2007). The value of the Kaiser-Meyer-Olkin measure of sampling adequacy (KMO) was 0.72. Nelson and Thompson (2005) reported that KMO values of 0.6 and above are required for good factor analysis. Using the eigenvalue greater than one rule, the PCA suggested seven factors, which accounted for 65.799% variance in scores. The seven factors extracted and named in this study as follows: (1) lack of cultural infrastructures, (2) inadequate knowledge and incorrect opinions of producers and leaders, (3) inadequate laws and vulnerability, (4) infrastructural barriers, (5) insufficient capital resources and information delivery systems, (6) legal supports and insufficient profitability, (7) dependency to government and distrust to the structure of cooperatives (Table 3).

The first factor was called the "lack of cultural infrastructures". This factor had the most eigenvalue (3.802) among the other factors. Also this factor explained 19.011% of the total variances of the variables. "Lack of producers' cultural capacity for group collaboration "was the most important (M=4.68) cultural problems for the success to found walnut production cooperatives. Allahdadi and Aref (2011) revealed that there are some barriers at the national, local and organizational levels towards agricultural production cooperatives. Lack of cultural capacity for group collaboration was an important element contributing to limit production cooperatives in agricultural and rural development as well. Farmers' organizations may be im-

Table 3: Results of factor analysis for barriers to found walnut production cooperatives and the variable of each factor

Barriers to found production cooperatives	Mean± SD	Factor Loading	Eigen-values	Variance (%)	Cum (%)
Lack of cultural infrastructures			**3.802**	**19.011**	**19.011**
Unsuitable socio-cultural performance of existent cooperatives	4.61±0.92	0.782			
Reluctance of the producers to found cooperatives	4.07±1.28	0.706			
Lack of producers' cultural capacity for group collaboration	4.68±0.84	0.599			
Prevention of governmental organs to found cooperatives	4.58±0.84	0.505			
Inadequate knowledge and incorrect opinions of producers and leaders			**2.597**	**12.986**	**31.996**
Lack of confidence to the cooperatives	4.67±0.79	0.775			
Inadequate knowledge of the producers about cooperatives	4.55±1.04	0.675			
Lack of definite aims for cooperatives	4.44±0.88	0.653			
Inattention to the indigenous knowledge by mangers	4.64±0.77	0.522			
Inadequate laws and vulnerability			**1.755**	**8.777**	**40.773**
Time-consuming rules	4.54±0.87	0.801			
Low producers' rate of risk taking	4.42±1.02	0.592			
Infrastructural barriers			**1.465**	**7.326**	**48.099**
Inadequate education of cooperative members	4.62±1.05	0.823			
Trend of cooperatives to jobbing and broking	4.28±1.42	0.693			
Insufficient capital resources and information delivery systems			**1.265**	**6.327**	**54.426**
Lack of enough capital resources	4.58±1.82	0.855			
Inadequate producers' awareness on cooperatives	4.30±1.35	0.757			
Legal supports and insufficient profitability			**1.206**	**6.032**	**60.458**
Insufficient efficiency of cooperatives	4.28±1.36	0.855			
Lack of legal supports on cooperatives	4.12±1.48	0.753			
Dependency to government and distrust to the structure of cooperatives			**1.068**	**5.341**	**65.799**
Suspicion of members on board of directors and inspectors	3.88±1.36	0.749			
Suspicion of members on financial affairs	3.97±1.30	0.662			
Dependency to the government	4.04±1.30	0.620			

portant instruments for empowerment of farmers and encouragement of their participation (Hansen and Henriksen, 2004).

Also, "unsuitable socio-cultural performance of existent cooperatives" and "prevention of governmental organizations to found cooperatives "and" reluctance of the producers to found cooperatives "were the other barriers that limit founding cooperatives to help walnut producers to success in their operations.

In Iran, although, there are many cooperatives in rural area, but, it is alleged that cooperative does not play a significant role in agricultural development in particular and rural development in general. Allahdadi and Aref (2011) revealed that there are some of the problems faced by agricultural production cooperatives in Iran which have been common among the majority of the cooperatives such as, unclear and inadequate government policies on the development of agricultural cooperatives.

The second factor was called the "inadequate knowledge and incorrect opinions of producers and leaders". This factor that its eigenvalue was 2.597 explained 12.986% of the total variances of the variables. "Lack of confidence to the cooperatives",” inattention to the indigenous knowledge by mangers", and "inadequate knowledge of the producers about cooperatives" were the most important barriers in the above-mentioned factor.

Directors of the cooperatives should share producers' indigenous knowledge and consider their ideas and opinions towards improving the cooperative activities (Mokhtari *et al.,* 2012). Through using mass media such as radio and

TV educational programs, agricultural extension and education services in terms of importance, usefulness, aims and structure of the cooperatives, adequate knowledge and awareness should be delivered to the producers in order to enhance rate of their trust in cooperatives (Allahdadi and Aref, 2011; Prakash , 2002; Mokhtari *et al.*, 2012).

The third factor was called "inadequate laws and vulnerability". This factor that its eigenvalue was 1.755 explained 8.777% of the total variances of the variables.

The existence of clear and adequate government policies on the development of agricultural cooperatives ,for example about cooperative management, inputs supply, marketing etc., can facilitate and enhance a better trades for producers and enabling them to access higher levels of income (Aref, 2011; Prakash , 2002).

The forth factor was called "infrastructural barriers". This factor that its eigenvalue was 1.465 explained 7.326% of the total variances of the variables. In this factor," inadequate education of cooperative members "and" trend of cooperatives to jobbing and broking" are the mentioned barriers that prevent to develop and found walnut production cooperatives. Delivering needed information through proper trainings and extension-educational programs is an important factor to found and develop successful agricultural production cooperatives (Mokhtari *et al.*, 2012). In the rest factors," lack of enough capital resources" is the most important barrier for the success to develop production cooperatives (M=4.58). Lack of capital resources is one of the challenges of agricultural production cooperatives and one of the effective variables in the failure of these cooperatives in Iran (Allahdadi and Aref, 2011; Mokhtari *et al.*, 2012).

CONCLUSION AND RECOMMENDATIONS

The cooperatives in Iran are considered as the most effective organizations in rural agriculture. Similarly, because of their vast network and reach, the rural cooperatives are considered as the best promoters for rural development (Aref and Sarjit, 2009). This study was intended to seek the perception of walnut producers on barriers affecting the success to found production cooperatives in Tuyserkan Township. An important finding of the study was that many factors dealing with developing and founding cooperatives. Based on the findings of the present study cultural infrastructures such as "lack of producers' cultural capacity for group collaboration" and "inadequate knowledge and improper opinions of producers and leaders" were the most important factors contributing to develop production cooperatives.

The success of rural cooperatives is based on the ability of them to tackle with participatory and people-based approaches. So, it is concluded that government should strengthen legal framework of cooperatives and establish more effective people-based approaches to promote the cooperation among producers and their participation in the process of development of the country via cooperative proliferation.

Based on the opinion of the respondents as mentioned above, it is also recommended that responsible government ministries (cooperative and Jihad-e-agriculture) should make producers aware of the philosophy and capabilities of cooperation in order for drawing their participation through formal and informal educations, mass media, hold training courses for managers in the field of methods of participative management, deliver needed information through proper extension and educational programs , persuade them taking part in educational programs related to practical activities. Managers and directors of cooperatives should share and consider producers' indigenous knowledge and ideas for improvement of the cooperative activities.

It is concluded and recommended that cooperative ministry should support to establish new strong cooperatives through hold training courses for staff and members of cooperative companies existent in each city by educated agents who are representatives of these companies in province unions, organizing cooperatives in a frame of unions in order to improve institutional capacities of cooperative department, providing facilities and needed financial resources, consider as much as the presence of managers with abilities in management and planning in order to promote efficiency and effectiveness of cooperative activities with altogether, make frequently agricultural production cooperatives aware of new regulations through cooperative department, and be sure that the regulations are related and coordinated with given terms on cooperatives.

Government in Iran should establish strong cooperative unions for barter, providing internal and external marketing materials, hold training courses in the field of marketing inputs and outputs for managing directors and board of directors to improve the cooperative performance (Mokhtari *et al.,* 2012).

Factors identified in this study that contribute to found and develop walnut production cooperatives will help leaders and producers to be successful to found effective production cooperatives in Tuyserkan Township. The other states in Iran can also consider and adopt similar strategies in a frame of a handout to those are to establish agricultural production cooperatives so that they are familiarized with barriers to found agricultural production cooperatives and they can make necessary predictions on explanation plan to be able to found their own specified cooperative companies.

ACKNOWLEDGEMENT

This is to acknowledge of all respondents who answered to the questionnaire of the study. Also, thanks to the faculty members in college of agriculture of Khoramabad Islamic Azad University and extension experts in Jihad-e-Agriculture of Tuyserkan Township and all those who helped for implementing the research.

REFERENCES

1- Abbasi, R., Rasulzadeh, B., & Abbasi, P. (2009). Factors affecting success and failure of production cooperative of Ardebil province. *Journal of Economic, Social and Cultural Cooperation*, 20, 210 - 211.

2- Abdolmaleky, M. (2012). Preconditions of Small-Farmers' Empowerment to Success in Farm Operations in Lorestan Province, Iran. *World Applied Sciences Journal*, 19 (4), 523-529.

3- Aref, F. (2011). Agricultural cooperatives for agricultural development in Iran. *Life Science Journal*, 8(1), 82-85.

4- Aref, F., & Sarjit, S. G. (2009). Rural tourism development through rural cooperatives. *Nature and Sciences,* 7(10), 68-73.

5- Allahdadi, F., & Aref, K. (2011). Agricultural production cooperatives in Iran: Challenges and opportunities. *Journal of American Science,* 7(12), 471-474.

6- Amini, A. M., & Ramezani, M. (2006). Evaluating factors effective on the success of poultry-farm cooperatives in Tehran. *Journal of Agricultural and Development Economics,* 4, (55), 67- 89.

7- Bruynis, C., Goldsmith, P.D., Hahn, D. E., & Taylor, W. J. (2001). Critical success factors for emerging agricultural marketing cooperatives. *Journal of Cooperatives,* 16, 14-24.

8- Didi, B. (2004). Short communication fishing cooperatives participation in managing near shore resources: The case in Capiz, Central Philippians. *Fisheries Research,* 67, 81-91.

9- Delavar, A. (2004). *Theoretical and practical foundations of research in human sciences*, (3 ed., pp. 199-203), Tehran, Publication of Roshd Institution.

10- Dunn, J.R., Crooks, A.C., Frederick, D.A., Kennedy, T.L., &, Wadsworth, J.J. (2002). Agricultural cooperatives in the 21st Century. Retrieved December 2, 2011, from http://www.rurdev. usda.gov/rbs/pub/cir-60.pdf.

11- Hansen, J. R., & Henriksen, J. (2004). Farmer Empowerment. Experiences, lessons learned and ways forward. Technical Paper Vol. 1. Copenhagen: Danida, Technical Advisory Service. Retrieved December 12, 2011, from http:// www. diis. dk/ graphics/ Events/ 2004/ FarmerEmpowermentVol1Final.pdf.

12- International Labor Organization (2004). *Development and advancement of cooperatives* (Asghar Bayat, Trans.). Tehran: Paygan publication. Jihad-e-Agriculture of Tuyserkan (2013). Annual Statistics of the Agricultural Productions in Tuyserkan Township. Jihad-e-Agriculture, 50-55.

13- Karami, S.h., & Agahi, H. (2010a). Factors affecting the success of cooperatives: A case study off-season production cooperatives in Kermanshah province. *Journal of Rural Development*, 13(2), 31-60.

14- Karami, S.h., & Agahi, H. (2010b). Studying the role of creativity on the success of cooperatives. Paper Presented at the National Conference on Entrepreneurship and Development. University of Ilam, 15-16 May, 2010.

15- Kuhi, k. (2006). Pathological development of cooperatives in the province of East Azerbaijan. *Journal of Taavon*, 187-188.

16- Mokhtari, V. A., Ahmadpour, & Poursaeed, A. (2012). Regression analysis of factors in the agricultural success of production cooperatives of Ilam.

Trends in Advanced Science and Engineering, 4(2), 100-109. Retrieved from http://www.sciroad.com/ntase.html.

17- Nelson, S.J., & Thompson, G.W. (2005). Barriers perceived by administrators and faculty regarding the use of distance education technologies in pre-service programs for secondary agricultural education teachers. *Journal Agriculture Education,* 46, 36-48.

18- Prakash, D. (2002). Development of Agricultural Cooperatives-Relevance of Japanese Experiences to Developing Countries. Paper Presented at the14th ICA-Japan International Training Cource on "Strengthening Management of Agricultural Cooperatives in Asia". IDACA-Japan. Retreived: March 18, 2012, from http//: http:// www. uwcc. wisc. edu/ info/ intl/ daman_japan.pdf.

19- Rehman, T., McKemey, K., Yates, C.M., Cooke, R.J., Garforth, C.J., Tranter, R.B, Park, J.R., & Dorward, P.T. (2007). Identifying and understanding factors influencing the uptake of new technologies on dairy farms in SW England using the theory of reasoned action. *Agricultural Systems,* 94, 281-293.

20- Safari, H, Aryanfar K.h., & Ebrahimi, A. (2009). Factors influencing the success of cooperative companies. *Journal of Economic, Social and Cultural Cooperation*, 20 (212), 33-51.

21- Tanzanian Federation of Cooperatives (2006). Cooperatives and development in Tanzania. Retrieved January, 10, 2011, from http://www.hakikazi.org/papers/Cooperatives.pdf.

22- Zarafshani, K., Rostamitabar, F., Hosseininia, G.H., Akbari, M., & Azadi, H. (2010). Are agricultural production cooperatives successful? A case study in Western Iran. *American Eurasian Journal Agriculture & Environment Science*, 8(4), 482-486.

23- Zara'nezhad, M., & B. Sharifi (2008). Studying attitudes of members and managers of cooperatives of Ilam province to factors effective on improving productivity: (Case Study: Consumer cooperatives of Ilam province). *Journal of Humanity and Social Sciences, Especially For Management*, 8(29), 118 - 97.

Related to Agricultural Research, Distribution of Income and Agricultural Value Added

Zeinab Moinoddini [1,2]

Abstract

Agricultural Research is one of the most important factors in agricultural development, thus, the role and importance of research is increasing. Within the past years, all developed and consequently developing countries have engaged their most attention to promote their researches indexes. This study is investigating the effect of agricultural research on the distribution of income and agricultural value added in Iran, during the period of 1976 - 2012. Three SLS methods were used to determine the income distribution functions, value added and per capita income. The results showed that agricultural researches were effective on improvement of agricultural value added. The results also express that continuance in agricultural researches can increase per capita income and less inequality of income distribution. Effect of agricultural value added on inequality of income distribution was low but though increase in agricultural value added and decrease it.

Keywords:
Agricultural research, Agricultural value added, Per capita income, 3SLS

[1] *Invited Lecturer Payame Noor University of Kerman, Iran.*
[2] *PhD Student of Agricultural Economics Zabol University of Iran.*
* *Corresponding author's email: zmoinoddini@gmail.com*

INTRODUCTION

Economic growth is a necessary but not the sufficient condition for alleviation of menace of poverty. Economic growth coupled with equitable distribution of income makes possible the participation of the poor in economic activities that experience progress and expansion. A strong agriculture sector is source of accelerated growth but also has its robust poverty alleviating impact in the economy. However, this impact of agriculture sector on growth and poverty can be more effective if there is a strong production and consumption link between agriculture and other economic sectors. Economic growth is defined as development and expansion of resources and economic capacities in a specific period and income distribution considers the distribution of the same resources among the population of the specific society, too (Mehregan and Nassabian, 2010). Economic growth with equity and equality has been the prime objective of economists and policy makers. Economic growth is considered to be necessary for poverty alleviation both by economic theory and policy makers (Ali and Ahmad, 2014). Equal distribution of income provides the stability in an economy by providing the opportunities to the individuals of the society to equally benefit from the growth and development of the economy. In the process of growth, the productive capacity of the economy is increased. This growth of the economy would benefit everyone if the distribution of income is equal in the economy. On the other hand, with the higher income inequality the lesser portion of population would benefit from the process of development (Ali *et al.*, 2013). Poverty reduction is the result of growth within agriculture and not to the shift of inputs from low to high productivity sectors, associated with Kuznet's inverted "U" trajectory of inequality rising then falling in the course of development (Datt and Ravallion, 1996 and 1998). Agricultural research can bring about broad-based technological change in agriculture that benefits the poor in many different ways. First, it can help reduce poverty directly by raising the incomes. Second, technological change can help reduce poverty indirectly through the effects of adoption, by both poor and non-poor farmers, can have on the real income of others largely through lower food prices for consumers and increased employment and wage effects in agriculture and other sectors of economic activity through production, consumption, and savings linkages with agriculture (De Janvry and Sadoulet, 2002; Haggblade *et al.*, 2002).The agricultural sector in Iran plays an important role in the Iranian economy. The rapid economic growth achieved by certain Asian countries such as IRAN in the 1990s can be partially attributed to a rapid generation of knowledge and intellectual property, including new technologies. Developed countries tend to spend around 3 percent of their GDP on R&D (Khaksar Astane and Karbasi, 2007). In most developing countries, this average is much lower. Iran spends about 0.5 percent of its GDP on R&D. Nonetheless, its spending levels are well above the average of around 0.2 percent for Islamic world as a whole (Marashi, and Sheykhan, 2004). Fulginiti *et al.*, (2004) found that while the estimated rate of productivity change was 0.83% per year over the four decades, the average rate from 1985 to 1999 was a strong 1.90% per year. Similarly, Coelli and Rao (2005) reported that, over the period 1980–2000, productivity growth in Africa averaged 1.3% per year, much higher than the growth rate achieved by South America, and almost at par with that of Europe, which recorded a growth of 1.4%. Thirtle *et al.*, (2003) modeled such a pathway in assessing the impacts of research-led agricultural productivity growth on poverty in a sample of developing countries in Africa, Asia, and Latin America. In this paper, a simultaneous system of equations model that accounts for endogeneity of key macro-economic variables as well as for long term CGIAR operations was specified and estimated using three-stage least squares. The independent variables were selected based on the theoretical as well as empirical literature. Alston *et al.*, (1995) argue that the lag between the inception and completion of a line of research (i.e., research gestation lag) can be around 2–3 years for some crop-management types of research such as developing improved fertilizer recommendations.

A cross-country examination of the relationship between growth and poverty by Gallup et al. (1997) establishes that a one percent increase in agricultural GDP leads to a 1.61% increase in income of the poorest quintile, while the corresponding values for the manufacturing and services sectors are only 1.16% and 0.79%. Other cross-country studies (Bourguignon and Morrison, 1998; Timmer, 1997) provide further evidence of the pro-poor bias of agricultural growth, with only the results of White and Anderson contradicting this view. Timmer (1997) finds that manufacturing reduces poverty directly due to an increase in the income of employed workers, but it also worsens the distribution of income, reducing the effect on the poor, in contrast agricultural growth, which is not associated with worsening income distribution. Bourguignon and Morrison (1998), using a sample 38 small and medium size developing countries find that growth in agriculture and in basic services reduced poverty more than expanding industrial output. Ravallion (1995) used data from 36 developing countries, representing 78 percent of the population of the developing world, to assess the growth-poverty link during the 1980s. Growth reduces poverty, but has no systematic effect on inequality, suggesting that income gains were evenly spread. However, the econometric analysis left "a sizable unexplained variation in country performance at reducing poverty for a given rate of growth." In other words, clearly other factors matter too.

Bruno *et al.* (1996) reviewed the recent evidence and found that while income inequality differs significantly across countries, there is no discernable systematic impact over time of growth on inequality. However, there are exceptions, as a general rule sustainable economic growth benefits all layers of society roughly in proportion to their initial levels of living. However, countries that give priority to schooling, health and nutrition are more likely to see improving income distributions and higher average incomes over the longer term. A more equitable distribution of physical assets, notably land, can also help greatly.

Ali and Ahmad (2014) found that poverty alleviating impact of livestock sector, human capital, remittances from abroad and employment per factory in Pakistan. Results also show Household size increases poverty levels, poverty gap and poverty severity in the economy. The study also suggests some policy recommendation for the development of livestock sector in Pakistan. Considering noted Studies and the necessity of research, in this study is investigating the effect of agricultural research on the distribution of income and agricultural value added in Iran, during the period of 1976 - 2012 is discussed. Three SLS methods were used to determine the income distribution functions, value added and per capita income.

MATERIALS AND METHODS

A significant pathway of impact of agricultural research on poverty is in Iran. First, agricultural productivity is modeled as a function of lagged agricultural research expenditures and production factors, including fertilizer, labor, machinery, and irrigation. Second, GDP per capita is modeled as a function of agricultural productivity, land per agricultural worker, government expenditures, gross fixed capital investment, and the rural population as a percentage of total population. Finally, poverty is modeled as a function of income inequality, GDP per capita, government expenditures, and gross fixed capital investment and population growth rate (Alene and Coulibaly, 2009).

$$LVa=\alpha1+\alpha2LRe+\alpha3Re1+\alpha4LFer+\alpha5LLab+\alpha6LMac+\alpha7LIrr \quad (1)$$

FER: amount of fertilizer used (Kg / ha), LAB: labor, MAC: Tractor ha, IRR: Irrigation

$$LGDP=\alpha8+\alpha9LVa+\alpha10LLand+\alpha11LGoex+\alpha12LFci+\alpha13LRup \quad (2)$$

LAND: infield, GOEX: government expenditure, FCI: Agricultural Gross fixed capital, RUP: rural population,

$$LPove=\alpha14+\alpha15LGDP+\alpha16LGoex+\alpha17LFci+\alpha18LPogr \quad (3)$$

Poverty here means that the Gini coefficient, an indicator of inequality in income distribution is shown.

POVE : poverty, POGR: population growth

In another part of the study is to calculate return on investment in agricultural research (Alene and Coulibaly, 2009) $\frac{(\alpha_{re})(\overline{VA}/\overline{RE})}{1+ROR}=1$

α_{Re}: Sum of coefficients obtained from estimating the simultaneous equations $\overline{VA}$: average of value added, $\overline{RE}$: average of research expenditures, ROR: rate of return on investment.

The data is used in this study by a nation's official statistics, data from the Central Bank of the Islamic Republic of Iran and FAO website.

RESULTS

According to Table 1, in first equation Iranian Agricultural Research cost in the first year (Re) and value added of agriculture had a significant positive effect. Agricultural Research costs with one-year interrupt (Re1) had significant positive effect also and this effect was not significant in Iran. Due to the growth of Iran in recent years in research and development of research and Technology, expanding in recent years is justifiable. Variable effects of nitrogen (Fer) and irrigation (Irr) the agricultural value is meaningless, that due to the subsidized fertilizer in Iran. Variable labor force (Lab) has the significant negative effect on the value added and the reasons for this could be the simultaneous use of labor and machinery because labor productivity is reduced and ultimately cause is reduce the value-added agriculture and Variable machinery (Mac) significant positive effect has on the value added. Results of second equation indicate that the value added is positive and significant effect on per capita national income. Rural population has significant negative effect

Table 1: Simultaneous equation system estimates of the impact of agricultural research in IRAN.

Equation/variable	Parameter	Estimate	(t-value)
Value-added per hectare			
	α_1	27.51	1.62*
Constant	α_2	0.76	4.65***
Research expenditures (t)	α_3	0.35	2.07***
Research expenditures (t _ 1)	α_4	-0.02	-0.10
Fertilizer (kg per hectare)	α_5	-1.82	-1.68*
Labor (workers per hectare)	α_6	0.58	3.39***
Machinery (tractors per hectare)	α_7	-0.18	-0.30
Irrigation (% of crop land)	R^2	0.72	
GDP per capita	α_8	92.42	2.77***
Constant	α_9	0.57	1.99**
Value-added per hectare	α_{10}	-3.95	-4.57***
Land (hectares per worker)	α_{11}	0.96	1.57*
Government expenditures	α_{12}	0.44	0.98
(% of GDP)	α_{13}	-3.62	-2.10***
Fixed capital investment (% of GDP)	R^2		0.45
Rural population (% of total)			
Poverty	α_{14}	-0.11	-0.78
Constant	α_{15}	-0.06	-3.47***
GDP per capita	α_{16}	0.11	1.86**
Government expenditures (% of GDP)	α_{17}	0.09	1.76**
Fixed capital investment(% of GDP)	α_{18}	0.05	1.691*
Population growth(% per year)	R^2	0.63	

* p<0.1 ** p<0.05 *** p<0.01

Table 2: Elasticities linking agricultural research, productivity, income, and poverty.

	Elasticities	
1	Agricultural productivity—agricultural research	0.34
2	GDP per capita—agricultural productivity	0.57
3	Poverty—GDP per capita	-0.063
4	GDP per capita—agricultural research	=1*2=0.1938
5	Poverty—agricultural productivity	=2*3= -0.0359
6	Poverty—agricultural research	=1*2*3= -0.0122

on income per capita. In addition, the effect of gross fixed capital per capita income is meaningless. Land has a negative effect on per capita income that reduced land efficiency. Government expenditures are positive and significant effect on per capita income. Government expenditure, including costs such as education, health and ... increase agricultural production and ultimately increases per capita income. The third equation results show that income distribution is an inverse relationship with per capita income, that means is that increase in inequality in income distribution has led to increased poverty. The fixed gross capital investment and government costs have a positive effect on poverty reduction. R^2 in three equations are good that represents a good fit.

The percentage change for value-added agriculture, per capita income and income distribution as the elasticity coefficients are shown in Table 2 because as mentioned variables in logarithmic terms in the model. For example, elasticity than the agricultural value of agricultural products 1.1%, which is expressed that one percent increase in research costs, the agricultural added value increased by more than 1% . Results of second equation indicate that the value added is positive and significant effect on per capita national income. Rural population has significant negative effect on income per capita. In addition, the effect of gross fixed capital per capita income is meaningless. Land has a negative effect on per capita national income, that reduced land efficiency caused to it. Government expenditures are positive and significant effect on per capita income. Government spending, including costs, such as: education, health etc., increase agricultural production and ultimately increases per capita income. The third equation results show that income distribution is an inverse relationship with per capita income, that means is that increase in inequality in income distribution has led to increased poverty. The fixed gross capital investment and government Costs have a positive effect on poverty reduction. R^2 in three equations are good that represents a good fit.

Results of second equation indicate that the value added is positive and significant effect on per capita national income. Rural population has significant negative effect on income per capita. In addition, the effect of gross fixed capital per capita income is meaningless. Land has a negative effect on per capita national income, that reduced land efficiency caused to it. Government expenditures are positive and significant effect on per capita income. Government spending, including costs, such as: education, health and ... increase agricultural production and ultimately increases per capita income. The third equation results show that income distribution is an inverse relationship with per capita income, that means is that increase in inequality in income distribution has led to increased poverty. The fixed gross capital investment and government Costs have a positive effect on poverty reduction. R^2 in three equations are good that represents a good fit.

The percentage change for value-added agriculture, per capita income and income distribution as the elasticity coefficients are shown in Table 2 because as mentioned variables in logarithmic terms in the model. For example, elasticity than the agricultural value of agricultural products 1.1%, which is expressed that one percent increase in research costs, the agricultural added value increased by more than 1% .

CONCLUSION

Economic growth with equity and equality has been the prime objective of economists and policy makers. Economic growth is considered necessary for poverty alleviation by both economic theory and policy makers. Poverty reduction especially for the poorest can be greatly enhanced through distributional policies. All the evidence confirms that distribution is central to fighting poverty.

In this study, impact of agricultural research on the distribution of income and agricultural value added in Iran, during the period of 1976 - 2012 has been analyzed. The results of the study demand a robust agriculture policy focused on the development of agricultural researches. The results show that agricultural researches are effective on improvement of agricultural value added. The results also express that continuance in agricultural researches can increase per capita income and less inequality of income distribution. Effect of agricultural value added on inequality of income distribution is low but though increase in agricultural value added and decrease it. The results of the study reveal that rural population has significant negative effect on income per capita. The negative impact of this factor on income inequality reveals that the focus should be to improve the human capital situation in Iran. The Government expenditures are positive and significant effect on per capita income the focus of the government must be to increase the Government expenditures for better than income distribution.

We can be proposed increasing research expenditures, because due to the positive and significant effect it on the value added agricultural, we can by this factor increased Agricultural value added. Then with policy that increases of value added of agricultural, Increase the per capita income and low inequality in income distribution.

According these results Small changes in distribution can have a very large effect on poverty reduction. There are cases where inequality levels have changed relatively quickly. Moreover, the lack of knowledge of the determinants of inequalities and the relative neglect of distribution issues in recent decades may mean that there is untapped potential for reducing poverty through distribution changes.

REFERENCES

1- Alene, A. D., & Coulibaly. O. (2009). The Impact of Agricultural Research on Productivity and Poverty in Sub-Saharan Africa. *Food Policy*. 34 (2): 198–209.

2- Ali, S., & Ahmad, N. (2014). Livestock Development and Poverty in Pakistan: Evidence from the Punjab Province. *Journal of Basic and Applied Scientific Research*, 4(6), 269-276.

3- Ali, S., Mustafa, MQ. & Shahbazi, IA. (2013). Agriculture value added and income inequality in Pakistan: a time series analysis. *Research Journal of Economics, Business and ICT*, 8(2), 25-33.

4- Alston, J.M., Norton, G.W., & Pardey, P.G. (1995). *Science under scarcity: principles and Practice for agricultural evaluation and priority setting*. Cornell University Press, New York.

5- Bourguignon, F., & Morrison, C. (1998). Inequality and development: the role of dualism. *Journal of Development Economics*. 57: 233-257.

6- Bruno, M., Ravallion, M., & Squire, L. (1996). *Equity and Growth in Developing Countries: Old and New Perspectives on the Policy Issues*. Washington, D.C.: World Bank.

7- Coelli, T.J., & Rao, D.S.P. (2005). Total factor productivity growth in agriculture: a Malmquis index analysis of 93 countries, 1980–2000. *Agricultural Economics*. 32(1), 115–134.

8- Datt, G., & Ravallion, M. (1996). How important to India's poor is the sectoral composition of economic growth? *The World Bank Economic Review*, 10(1), 1-25.

9- Datt, G., & Ravallion, M. (1998). Farm Productivity and rural poverty in India. *Journal of Development Sutdies,* 34(4), 62-85

10- DeJanvry, A., & Sadoulet, E. (2002). World poverty and the role of agricultural technology: direct and indirect effects. *Development Studeis*. 38 (4), 1–26.

11- Fulginiti, L.E., Perrin, R.K., & Yu, B. (2004). Institution sand agricultural productivity in Sub-Saharan Africa. *Agricultural Economics*. 31, 169–180.

12- Gallup, J., Radelet, S., & Warner, A. (1997) "Economic Growth and the Income of the Poor". *Harvard Institute for International Development CAER II Discussion Paper*, 36, 30-61.

13- Khaksar Astane, H., & Karbasi, A.R. (2007).

Calculate the marginal rate of return on investment of Agricultural Research. *Journal of Agricultural Economics and Development*, 50, 35-46.

14- Marashi, N., & Sheykhan, N. (2004). Agricultural Research Program in Iran, *Journal of Agricultural Economics and Development.* 37, 279-301.

15- Mehrgan, N., & Nassabian, Sh. (2010). Agriculture Growth and Income Inequality in Iran. *America Eurasian Journal of Agricultural and Environmental Sciences,* 7(6), 637-643.

16- Ravallion, M. (1995). Growth and poverty: Evidence for developing countries in the 1980s. *Economics letters.* 48: 411-417

17- Ravallion, M., & Datt, G. (1999). When is growth pro-poor? Evidence from the diverse experience of India's states. *World Bank Policy Reseach Working Paper Series*, 2263.

18- Thirtle, C., Lin, L., & Piesse, J. (2003). The impact of research-led agricultural value added growth on inequality in income distribution reduction in Africa. *Asia and Latin American World Development.* 1959–1976.

19- Timmer, C. (1997). How well do the poor connect to the growth process? *Harvard Institute for International Development, CAER II Discussion Paper.* 17, 335-350.

20- Haggblade, S., Hazell, P., & Reardon, T. (2002) "Strategies for Stimulating Poverty-Alleviating Growth in the Rural Non-Farm Economy in Developing Countries". *Environment and Production Technology Paper.* 92, 127-141.

Examining Subsidy Polices on Maize Production in Iran (Panel Data Approach)

Negin Hosseingholizadeh [1], Jafar Haghighat [2] and Rassul Mohammadrezaei [3]*

Abstract

Among the agricultural important factors, inputs are the most significant in agricultural production. This article aimed to examine the impact of government subsidy policies on production of one of the most strategic products, namely maize, in Iran. To achieve this goal, panel data for the nine provinces of Iran's major producers of maize during the period of 1999-2007, is used. In this study, first the country's maize production function has been estimated by using data information for inputs: chemical fertilizer, labor, water, seeds and pesticides. And then, calculating the partial elasticity of production factors, sensitivity of production to changes in the value of inputs is evaluated. Also, using a methodology based on the maximum profit, inputs′ demand function is calculated.Results of analyzing government Subsidy Policy showed that, paying subsidy to chemical fertilizer decreases maize production 0.412 percent, because of low demand elasticity of this input. Also according to subsidy of seed, with regard to low demand elasticity of this input to its price, paying subsidy that decreases seed price, wouldn′t have so high effect on its consumption and consequently, on production growth, so that maize production only increased 0.478 percent due to paying subsidy to seed.

Keywords:
Subsidy, Production function, Input demand function, Panel Data, Maize, Iran

[1] *Graduate Student of Agriculture Economy, Iran-Tabriz University, Iran.*
[2] *Professor, Department of Agriculture Economy, Iran-Tabriz University, Iran.*
[3] *Professor, Department of Agriculture Economy, Iran-Tabriz University, Iran.*
* *Corresponding author's email: n.hosseingholizadeh@yahoo.com*

INTRODUCTION

Among the oil seeds in agriculture, maize play an important role as the main energy source for many people in the world and also for poultry. Therefore, the development of the area under cultivation and production is of special priority. During the recent years, the plan of increasing the production of maize in Iran has been associated with considerable successes. As, the amount of maize to produce, has reached from 250 thousand tons in 1992 to average 1,110 tons per year of 1999-2001. In the same period the area under cultivation of maize has increased from 60 thousand hectares to 181 thousand hectares and the yield, of 4100 kg to 6133 kg per unit area. In Iran, three provinces namely: Fars, Khuzestan and Kermanshah are in the first to third place respectively, from the point of area under cultivation and production. The importance of maize in oil and variety of food and starch industries, and many other products that are being used as human food is always emphasized and also, it must be said that approximately 65 to 70 percent of the poultry diets is consisted of maize. And maize is really considered as a strategic and determining good in our country's poultry industry. Various sources of statistics are predicting country's total maize consumption over 2 million and 700 thousand tons per year, and surely, with the increasing population and consumption of population, this figure will increase too. Considering the content expressed, the approaches to the ways of take advantage of available opportunities to achieve higher output at a time when Iran's accession to the WTO is discussed on one hand, and evaluating the technology on maize production and deployment of resources and production facilities of country due to severe lack of studies related to it on the other hand, highlighted the need to conduct this study, and further more researches.

Today, governments in all countries, especially Third World countries, play a crucial role in the growth process and economic development by economic policies. One of the government´s key economic policies is adopting appropriate supportive policies; why, many factors can lead to imbalances between different sectors of production, distribution, consumption and foreign trade.

In Iran the government has always involved in production and pricing of agricultural strategic commodities such as wheat and sugar beets and maize. One part of government policies that have persisted since the last years by now, has been the payment of production subsidies to producers of agricultural commodities that, this has done in terms of various supportive tools, including paying indirect subsidies to agricultural inputs such as fertilizer, pesticides, seed and machinery, to pay damages to agricultural producers, pay part of premiums of agricultural products, guarantee purchase of some basic products, provide services and agricultural researches freely, helping to invest in agro-industry units or payment of damages or exporting awards; that all of these factors can be accounted as production subsidies (Raymond and Renfro, 1992).

In total, according to government budget constraints and the effects of inexpensive and irregular use of the above mentioned inputs, especially chemical fertilizer, it´s necessary to address both the financial aspects of changes in the distribution pattern of the fertilizer input and its welfare effects, too.

In Iran, the payment of subsidies to agricultural inputs are done with the following two objectives (Parmeh, 2004):

1- Strengthen the agricultural sector and domestic capabilities of production to increase the quantity of production, increase competitiveness and improve the quality of the products in this section.

2- Support manufacturers by reducing production costs and finally keep prices of products low in this market.

Experiences and studies done about the use of inputs in agriculture and impact of subsidies shows that, one of the concepts and indicators of liberalization is elimination of subsidies, which in recent decades, this concept in our country, that means the elimination of subsidies as an economic goal of government, especially in the production of strategic crops such as maize has been considered which can be noted on the following selection of internal and external studies.

Debrah (2002) states the main motivation of assistance policies in agricultural sector of world in growth and economic development especially in rural areas, support jobs and investment, protect domestic production and reduce dependence on foreign counties and elimination and poverty reduction to achieve the proper conditions for life.

The general objective of the study was to examine the impact of government subsidy policies on production of one of the most strategic products, namely maize, in Iran.

MATERIALS AND METHODS

Data source

Data used in this study has been collected from the bank's cost of agricultural production from Ministry of Agricultural for the crop year 1999-2007. Statistics are estimated by panel data method. In order to estimate the model and doing the related tests of econometric, Evews6 software is used.

Model specification

Before discussing estimation, first stationary of variables is studied.

In this section, the stationary of whole provinces is tested by Unit Root in panel. In general, there are five tests to determine unit root in panel, most important of all; which are embedded in Eviews6 too, are as follows (Khazayi, 2008):

1- Levin, Lin and Chu test(LIC), (2003)

2- Im, Pesaran and Shin test(IPS), (2003)

3- Breitung test, (2000)

4- Fisher-type tests using ADF and FF tests, (Maddala and Wu (1999) and Choi(2001))

5- Hedri test

Estimating the production function

The importance of choosing the form of function for expressing the relationship between dependent and independent variables is doubled when the model´s estimated parameters to be the basis for new policies. For example, as in present study is considered too, when the purpose of estimating a production function is utilizing the parameters for calculating demand function of inputs and the elasticity of production and decisiding about how to use the inputs, so, carefully selection of correct form of function is of special importance. In the past, many tried to emphasize the importance of true selecting, particularly, correct selection of function form and show the sensitivity of parameters that reveal the structure and economic relations to the selection of functional form. Indeed, in economic literature, discussion and importance of selecting the functional form in production and consumption studies exists when Cobb-Douglas and Stone presented their functional form (Thompson, 1988).

Gujarati (1999) believes, fewer number of parameters, ease of interpretation, computational simplicity, well fit (R^2), power of generalization method and forecasting are other measures that can be useful in determining the econometrics model for the experimental works.

Consistence and compatibility of signs and parameters of function with economic theories are other criteria in superior pattern recognition from the viewpoint of Thompson (Thompson, 1988).

For this reason, with regard to discussed points, four types of flexible production function, included Transcendental logarithmic (Translog), Transcendental, Generalized power and Cobb-Douglas have been estimated by Pooled GLS method as the primary replacement for the relationship between factors of production and production of maize in Iran, and due to mentioned standards, Transcendental logarithmic(Translog) function was selected as the most appropriate maize production function of Iran.

In table 1 the general form of flexible functions estimated in the study, are described.

It should be noted that, although combining the data is of advantages, it creates some problems in model specification and the structure of residual, so that the residual includes errors associated with the time periods and cross sections or a combination of both, that in this connection, the following techniques have been introduced to reduce these problems (Azizan and Salami, 2005):

1- Combining data and using OLS estimation, that in these conditions, the classical assumptions about no correlation and Heteroscedasticity is violated.

2- Using fixed effects model, that is well known as Least Square Dummy Variable model (LSDV).

3- Using Error Component Model and apply GLS methods.

A model that is achieved at this step of our

study, is named as common effects model. But in Pool or common effects, intercept and coefficients of the variables are considered as the same for all provinces, but this limitation makes the true relationship between maize production and inputs become distorted and not well displayed. So, as several factors such as economic factors, climate and so on.. are different in various provinces, these differences have reviewed by Panel model in the next steps.

Therefore, Panel model against the Pool, estimates a separate intercept for each unit, so that the difference in intercept or individual effects can be expressed as dummy variables. In other words, a common approach in formulating panel data model, is based on the premise that the difference between units can be shown as the difference in intercept.

Thus, in equation (1) each xi is an unknown parameter that must be estimated. To assume that xi and yi consists of T observation for i th unit and ε_{it} is residual vector. γ_0 is intercept of source unit. The model described can be shown as follow:

$$Y_{it} = \gamma_0 + \sum_{i=1}^{n} \gamma_i D_i + \sum_{j=1}^{m} \beta_j X_{jit} + \varepsilon_{it} \quad (1)$$

D_i is dummy variable that is 1 for i th unit and otherwise have a value of zero.

Test of significance fixed effects

At this step, for choosing the method of estimating between common effects and fixed effects F-test is used.

In general, in panel data it´s necessary initially the homogeneous or heterogeneous of individuals (units) to be tested. If the units are homogeneous, Pooled Least Square method (common effects) can be used simply. otherwise, using fixed effects is necessary. In other words, it is necessary to demonstrate significance of fixed effects or simultaneous significance of dummy variables by using F-test that mentioned above (equation 2).

$$F_{(N-1,NT-N-K)} = \frac{(R^2_{UR} - R^2_{R})/(N-1)}{(1-R^2_{UR})/(NT-N-K)} \quad (2)$$

In equation (2), N is the number of cross sections. N-1, the number of limitations in restricted model or in other words, restricted model´s the degree of freedom. K is the number of explanatory variables and T, The number of observations over time.

The fixed effects is suggested as unrestricted model(UR) and pooled method as restricted model (R). thus, R^2_{UR} is the coefficient of determination of unrestricted model and R^2_R is the coefficient of determination of restricted model. So assumptions of the test can be defined as follows, that rejecting the null hypothesis indicates significance of fixed effects and using fixed effects method (Green, 2002).

H_0: provinces have the same intercept.

H_1: provinces have different intercept.

Then, the elasticity of production factors have been calculated and interpreted. Elasticity of inputs in flexible functions, is a function of the amount of the use of inputs and in some cases, production levels. According to the norm in such studies, mentioned elasticities are calculated in average of other factors.

As the production function of the study is translog form, the elasticity of the i-th input, also expressed in Table 1, is defined as the following form:

$$E_{Pi} = (\beta_i + \beta_{ii} (\ln X_i) + \sum_{j=2}^{n} (\ln X_i)) \quad (3)$$

Input demand function

The demand function of input is mathematical expression of the amount of applied input for business in various prices of inputs, product prices and income of firm. Input demand function, depending on including one or more variables and current and durable inputs, can be different shapes, that in this study, assuming a competitive market, input demand function will be obtained from the condition for profit maximization (Dejpasand, 1991). Although the monopoly of the corn market, and most agricultural products, using this relationship may not be correct, but due to lack of awareness maize buyers´ function using this approach is Justifiable.

Condition for profit maximization, in the case of allocating multiple inputs to produce a product in a competitive market is according to (4) equation:

$$VMP_X = P_X \rightarrow MP_X . P_m = P_X \quad (4)$$

For translog production function, equation (4) will be as follows:

$$P_y . \left(B_i + \beta_{ii} (\ln X_i) + \sum_{j=2}^{n} (\ln X_j) \right) = P_x$$

By displacing the elements of first order conditions of maximization(provided the second-order conditions), the input demand function is obtained that, with the placement of input and product prices in it, optimal amount of each input can be achieved. It should be noted, to calculate the demand functions, instead of other inputs, the average amount of inputs are used in the given years.

Finally, using the demand functions for inputs and assuming that the values of other inputs are consistent and the market is competitive, Price elasticity of inputs′ demand are calculated in accordance with equation (5):

$$E_{ii} = \frac{\partial \log X_i}{\partial \log P_i} \times 100 \quad (5)$$

RESULTS

In order to examine the impact of government subsidy policies in maize production, the elasticity of factors of production and then, price elasticity of demand were calculated. So, primly, maize production function is estimated by panel approach and then input demand functions are given.

Results of estimating production function

First, the stationary of the natural logarithm of all variables used in the model were tested. Results showed that all variables are stationary in levels(have unit root). According to the discussions, four types of flexible production function as expressed functions in table (1) are estimated by pooled GLS method, for five inputs including chemical fertilizer **(X_f)**, seed **(X_s)**, water **(X_w)**, labor force **(X_l)**, poison **(X_p)**, as the primary replacement for the relationship between factors of production and production of maize in Iran. The results of estimation along with t statistics obtained for each coefficients as well as R2 & D and F statistics associated with each of the equations are given in Table 2.

Translog production function for a single crop mode with five inputs (labor, seed, fertilizer, water and pesticides) are defined as follows:

$$\ln(Y) = \alpha + \sum_{i=1}^{5} \beta_i \ln(X_i) + 1/2 \sum_{i=1}^{5} \gamma_{ii} (\ln X_i)^2 + \sum_{i=1}^{1} \sum_{j=2}^{5} \gamma_{ij} (\ln X_i)(\ln X_j) \quad (6)$$

Considering F statistics, indicates that the regression is significant at a high level that the value of 56.20 for it, confirms this fact. This function′s determinant coefficient was obtained 0.82, this means that, 82 percent of changes in maize production are explained by exogenous variables. About Dourbin-Watson statistic, notifying this point is necessary that, despite getting the value of 1.16 for this statistic, although this figure doesn't represent the absence of serial correlation between the components of residuals of model, but basically serial correlation in pooled

Table 1: General form of flexible functions estimated in the study

Function name	Functional form	Elasticity of input i, (E_{xi})	The number of parameters
Transcendental logarithmic (Translog)	$\ln(Y) = \alpha + \sum_{i=1} \beta_i \ln(X_i) + 1/2 \sum_{i=1}^{n} \gamma_{ii} (\ln X_i)^2 + \sum_{i=1}^{n} \sum_{j=2}^{n} \gamma_{ij} (\ln X_i)(\ln X_j)$	$(\beta_i + \beta_{ii} (\ln X_i) + \sum_{j=2}^{n} (\ln X_j))$	1/2(n+1)(n+2)
Transcendental	$Y = \alpha \prod_{i=1}^{n} X_i^{\beta i} e^{\gamma i * xi}$	$((\beta_i / X_i + \gamma_i) * X_i$	2n+1
Generalized power	$\ln(Y) = \alpha + \sum_{i=1}^{n} \beta_i X_i + 1/2 \sum_{i=1}^{n} \gamma_{ii} (X_i)^2 + \sum_{i=1}^{n} \sum_{j=2}^{n} \gamma_{ij} (X_i)(X_j)$	$(\beta_{i+\gamma_{ii}} (x_i) + \sum_{j=2}^{n} \gamma_{ij(x_j)})(x_i / y)$	1/2(n+1)(n+2)
Cobb-Douglas	$Y = \prod_{i=1}^{n} X_i^{\beta^i}$	β_i	n+1

Table 2: The results of estimating production function as Pool, during 1997-2007

Parameters	Translog Function		Transcendental		Generalized power		Cobb-Douglas	
	Coefficient	t-statistic	Coefficient	t-statistic	Coefficient	t-statistic	Coefficient	t-statistic
INTERCEPT	-273.28* (0.0008)	-3.52	0.42 (0.98)c	-0.24	-68499.49 (0.77)	-0.28	*11.59 (0.000)	5.22
X_f	-	-	-0.001 (0.74)	-0.33	-2152.84 (0.58)	-0.58	-	-
X_s	-	-	-0.18 (cc0.68)	-0.40	8073.63 (0.40)	-0.83	-	-
X_w	-	-	-0.0007 (0.10)	-1.64	-242.48* (0.01)	-2.44	-	-
X_l	-	-	-0.0002 (0.10)	1.63	755.74 (0.55)	-0.59	-	-
X_p	-	-	-0.15* (0.0018)	3.24	140812.2* (0.000)	4.48	-	-
lnX_f	48.12* (0.0056)	2.86	2.55 (0.40)	-0.84	-	-	*-0.78 (0.01)	2.46
lnX_s	68.02* (0.0012)	3.38	-0.70 (0.70)	-0.37	-	-	-0.08 (0.69)	-0.39
lnX_w	9.50* (0.0007)	3.56	-0.49* (0.04)	-2.04	-	-	-0.93* (0.000)	-9.79
lnX_l	-11.58* (0.0003)	-3.82	-0.15 (0.32)	-0.98	-	-	-0.02 (0.83)	-0.20
lnX_P	1.25- (0.23)	-1.19	-0.14* (0.02)	2.33	-	-	*-0.31 (0.000)	5.50
$(1/2)(lnX_f)^2$	-2.63 (0.24)	-1.18	-	-	-	-	-	-
$(1/2)(lnX_S)^2$	3.18- (0.08)	-1.75	-	-	-	-	-	-
$(1/2)(lnX_W)^2$	*0.46- (0.005)	-2.87	-	-	-	-	-	-
$(1/2)(lnX_L)^2$	-0.03 (0.72)	-0.34	-	-	-	-	-	-
$(1/2)(lnX_P)^2$	*-0.10 (0.000)	5.63	-	-	-	-	-	-
$(1/2)(X_f)^2$	-	-	-	-	1.10* (0.02)	2.29	-	-
$(1/2)(X_S)^2$	-	-	-	-	-71.19 (0.48)	-0.70	-	-
$(1/2)(X_W)^2$	-	-	-	-	-0.24* (0.000)	5.52	-	-
$(1/2)(X_L)^2$	-	-	-	-	-0.09 (0.39)	-0.85	-	-
$(1/2)(X_P)^2$	-	-	-		-15922.2* (0.000)	-4.98	-	-
$(lnX_f).(lnX_S)$	-8.53* (0.0007)	-3.56	-		-	-	-	-
$(lnX_f).(lnX_W)$	-1.16* (0.0060)	2.83.	-		-	-	-	-
$(lnX_f).(lnX_L)$	1.74* (0.0005)	3.65	-		-	-	-	-
$(lnX_f).(lnX_P)$	-0.27 (0.08)	1.74	-		-	-	-	-
$(X_f).(X_S)$	-	-	-		-8.06 (0.49)	-0.68	-	-
$(X_f).(X_W)$	-	-	-		-0.19 (0.11)	-1.61	-	-
$(X_f).(X_L)$	-	-	-		-1.48 (0.50)	-0.67	-	-
$(X_f).(X_P)$	-	-	-		-68.23 (0.06)	-1.87	-	-
R^2	0.81		0.67		0.73		0.62	
D.W	1.16		1.09		1.06		1.23	
Prob (F-statistic)	0.000		0.000		0.000		0.000	

Table 3: Results of estimating production function as Panel

Explanatory variables	Coeficients	t-statistic
INTERCEPT	-147.99	-3.157
ln Xf	22.04	2.11
ln Xs	45.10	3.91
ln Xw	3.57	2.18
ln Xl	-0.33	-0.17
ln Xp	-0.22	-0.33
$(1/2)(\ln Xf)^2$	0.036	0.02
$(1/2)(\ln Xs)^2$	-2.05	-2.33
$(1/2)(\ln Xw)^2$	0.013	0.13
$(1/2)(\ln Xl)^2$	0.018	0.50
$(1/2)(\ln Xp)^2$	0.054	4.41
(ln Xf).(ln Xs)	-5.68	-0.88
(ln Xf).(ln Xw)	-0.57	-2.36
(ln Xf).(ln Xl)	0.040	0.13
(ln Xf).(ln Xp)	0.072	0.71

Fixed Effects (Cross)

_K-C	0.061367
_KSH-C	0.965216
_F-C	1.945700
_G-C	0.0194760
_H-C	0.715147-
_Y-C	1.329720-
_KH-C	1.0823061
_SI-C	0.958845-
_ES-C	1.070352-
R2	0.97
D.W	1.38
Prob(F-statistic)	0.000000000

data, has not so much problem, then, the results of model will be usable (Rashidghalam, 2010).

The model that is achieved at this stage, named common effects model, that its specification is according to equation (4):

$$\ln y = -273.28 + 48.12 \ln X_f + 68.02 \ln X_s + 9.50 \ln X_w - 11.58 \ln X_l - 1.25 \ln X_p - 2.63(\ln X_f)2 - 3.18 (\ln Xs)2 - 0.46 (\ln X_w)^2 + 0.03 (\ln X_l)^2 + 0.10 (\ln X_p)^2 - 8.53 (\ln X_f).(\ln X_s) - 1.16 (\ln X_f).(\ln X_w) + 1.74 (\ln X_l) + 0.27(\ln X_f).(\ln X_p) \quad (7)$$

But as mentioned before, in the pool, the coefficients of the variables and intercept are considered the same for all provinces, but this limitation makes the true relationship between maize production and production inputs become distorted and not be shown well. Therefore, as mentioned in the introduction, as several factors such as economic factors, climate and so on… vary in different provinces, at the next stage these differences will be checked by the Panel.

Thus, fixed effects model, which is our main model of panel, is as following form:

$$\ln Y = 0.06 + 0.96D_1 + 1.94D_2 + 0.19D_3 + ... - 1.07 D_8 + 22.04 \ln X_f + 45.10 \ln X_s + 3.57 \ln X_w - 0.33 \ln X_l - 0.22 \ln X_p + 0.36(\ln X_f)^2 - 2.05 (\ln X_s)^2 + 0.13 (\ln X_w)^2 + 0.18(\ln X_l)2 + 0.54(\ln X_p)^2 - 5.68(\ln X_f).(\ln X_s) - 0.57(\ln X_f).(\ln X_w) + 0.40(\ln X_f).(\ln X_l) + 0.72(\ln X_f).(\ln Xp) \quad (8)$$

As can be seen, in this estimate, a separate intercept is obtained for each province, the constant amount of intercept corresponds to Kerman province, with value 0.06, and if D1 = 1, the observations corresponds to Kermanshah, and is zero otherwise. For other dummy variables is interpreted like this. The question that comes to mind at this step, is that, which of the two estimated models are preferred? model (7) or model (8)…that is answered by F-test:

$$F = \frac{(0.97 - 0.81)/8}{(1 - 0.97)/58} = 38.66$$

Consequently, the null hypothesis that provinces have the same intercept, is rejected and fixed effects approach is accepted.

In this stage of research, one important point is necessary to be mentioned. In case of random effects estimation, the number of cross sections should be higher than the number of variables. Otherwise, we´ll face with an error message (Ashrafzadeh and Mehreghan, 2008). So, considering that in this study we have such a situation too, this case will occur. Thus, we stopped the work at this point and accepted the estimation by fixed effects method and we don´t need doing Hasmn test to choose between two methods of estimation, fixed and random effects.

Choosing fixed effects method has two important reasons c (Jhonston, 1997):

1- If the actual model is random effects and estimated by mistake by fixed effects, the estimation is consistent. But if the actual model is fixed effects and estimated wrong by random effects, the estimation is not consistent. Therefore, the precaution would be using fixed effects estimation method.

2- When the number of cross sections is less than the number of coefficients to be estimated, using random effects is not possible.

Calculate the partial elasticity of maize production

Results of calculating elasticity of production factors, according to equation (3), to determine the factors affecting the production of maize in country and show the amount of each input´s role in increasing or decreasing of production, are shown in table 4.

Results indicates that, production elasticity of chemical fertilizer is negative, that means 1 percent increase in consumption of chemical fertilizer, will decrease product 2.73 percent. This shows that, farmers are actually active in third area of production (the non-economic area), in other words, use this input more than optimal level in maize farms. The economical reason of this issue, is related to the prices of this input, that's why, chemical fertilizer is the subsidized input that, by receiving government subsidies, the price of it fell down and get away from its real price. So, farmers can buy and use it more than necessary. Therefore, it can concluded that as the input subsidy is reduced or eliminated, we´ll see optimal and economical use of it by farmers in second area of production. And finally, increased maize production in our country, in future.

Also, the numerical value of water elasticity is equal to (-0.51) which shows that, maize farmers are not doing economical in use of this input and are in third area or production. So, causes the indiscriminate loss of this valuable input into in farms.

Elasticity of seed was calculated 2.78, which indicates that one percentage change in consumption of seed will increase the production of seed 2.78 percent.

For labor, this value was obtained 0.38, that means a 0.38 percent increase of production due to one percent increase in applying the labor force.

Production elasticity of Poison is equal to 0.11 which shows, one percentage increase in consumption amount of this input, increases the production of maize by 0.11 percent. Poison is not included as subsidized inputs, so, optimal and appropriate use of it by farmers is not unexpected.

Results of calculating input demand functions

Demand for farm inputs, is a derived demand and its value largely depends on the demand for agricultural products. In general, demand for input in agricultural economics depends on following factors:

1- Price of product being produced

2- Input´s price

3- Price of substitute and complementary inputs in that exist in production function.

4- Technology coefficient or fixed parameters

Table 4: Results of calculating partial elasticity of production factors

Input	Fertilizer	Seed	Water	Labor	Poison
Elasticity	-2.73	2.78	-0.51	0.38	0.11

Source: Research findings

Table 5: Calculating input demand functions

Inputs	Input demand functions
Chemical fertilizer	$P_m.(22.04-0.36\mathit{ln}\ X_f-5.68\mathit{ln}\ X_s-0.57\mathit{ln}\ X_w+0.4\mathit{ln}\ X_1+0.72\mathit{ln}\ X_p).(M/X_f)=P_f$
Seed	$P_m.(45.10-2.05\mathit{ln}\ X_s-5.68\mathit{ln}\ X_f).(M/X_s)=P_s$
Water	$P_m.(3.57+0.13\mathit{ln}\ X_w-0.57\mathit{ln}X_f).(M/X_w)=P_w$
Labor	$P_m.(-0.33+0.18\mathit{ln}\ X_l+0.4\mathit{ln}\ X_f).(M/X_l)=P_l$
Poison	$P_m.(-0.22+0.54\mathit{ln}\ X_p+0.72\mathit{ln}\ X_f).(M/X_p)=P_p$

Table 6: results of calculating the price elasticity of demand for factors of production

Input	Fertilizer	Seed	Water	Labor	Poison
elasticity of demand	-0.151	-0.172	-0.189	-0.109	-0.047

of production function.

5- Under certain conditions, available budget for purchase of input that may be effective on input demand function .

Input demand functions are given in table (5).

Calculating price elasticity of demand

One indicator that can display the technology of a production unit as summary is Price elasticity of demand for production inputs. Own price elasticity of demand for a production input, shows percent of changes in quantity demand of that input as a result of changes in price of input. Which is defined as follows:

$$E_{ii} = \frac{\partial \log X_i}{\partial \log P_i} \times 100 \qquad (9)$$

In this study, using demand functions for inputs, obtained in the previous section, assuming that the values of other inputs are constant and the market is competitive, the price elasticity of demand for inputs are calculated as equation (9) and results are given in table 6.

According to table 6, all price elasticity of demand has proper and reasonable sign which are consistent with economic theories and show, there is an inverse relationship between prices and quantities of inputs. Also based on the results, the sensitivity of demand of corn farmers to all corn factors is smaller than one and therefore, the amounts of these inputs to their price has low elasticity. That means for every one percent change in input prices, the demand of inputs change less than one percent.

Also, the demand elasticity of labor to price changes is very small. Therefore, labor force does not adjust for changes in the wage and significant changes in the amount of labor, with an increase or decrease in wages, won't occur.

DISCUSSION

The purpose of subsidy policy and subsidizing, is to reduce production costs and increase farmer income, increase production, reduce imports and therefore reduce foreign ownership. On the other hand, considering the high cost of subsidizing chemical inputs, according to government subsidies should be completely removed and this cost of fertilizer should be used to produce better quality crops.

Our farmers just think to increase production per hectare through the overuse of chemical fertilizers than the quality of crops. If the price of inputs be too low, its use and application will increase without any restrictions, and this leads the farmer to use the input regardless of the ultimate efficiency. Thus, removal of subsidies, without negative effect on the production, can remove a heavy financial load of the government. Therefore, analysis and evaluation of factors affecting agricultural production and policy of price guarantees, easy access of producers to agricultural market and et al, is inevitable in politics.

For more detailed analysis of the effects of subsidy policies and examine the impact of subsidizing on maize production, Chart 1 is used.

Government's policy of subsidizing on input, will reduce its price. According to demand law, as the price and quantity have an inverse relationship to each other, the demand for it increases. If input demand be elasticity to price, reducing price, causes more consumption of

Table 7: Results of the impact of subsidy policy of inputs on maize production

Input	Chemical fertilizer	Seed
Elasticity of demand	-0.151	-0.172
Partial elasticity of production	-2.73	2.78
Percentage of changes in production, per one Percent change in input´s price	-0.412	0.478

input. Now if the partial elasticity of production to this input be negative, increasing consumption of it will reduce production. If the partial elasticity of production is between zero and one, increasing its consumption will increase production; but production rate increases less. If the partial elasticity of input production be greater than one, by increasing consumption of it, production will increase more. Based on Table 7, the partial elasticity of chemical fertilizer production is -2.73, increased consumption of this input which is the result of its reduced price through subsidy policy, reduces maize production by 0.412. So it can be concluded that subsidy policy of government to increase maize production has not been successful.

Also in relation to seed subsidy, with regard to low elasticity of this input to its price, it can be analyzed such as that, in return for subsidizing seed and reducing its price, according to own price elasticity of it that is equal to -0.172, the consumption of input increases much less than price reduction. As well as, production elasticity of input is positive and equals to 2.78. That means, farmers use this input in the first economic area. Increasing use of input, affects the production and increases the production to the rate of 0.478. It can be concluded that, demand and consumption of seed is inelastic to price changes and subsidies, which will reduce the price of seed, won't have so important effect on its consumption, in result of production increase. So here it is also clear that the policy of subsidy on corn production had little effect and only had great financial burden on the state (Table 7).

CONCLUSIONS AND RECOMMENDATION

1- According to numerical results, the production elasticity of fertilizer was obtained a negative value, indicating that: due to low price of this input because of subsidies, farmers do not use this input economically and use it in the third area of production(non-economic region).

2- According to the results, the best production function for corn production of Iran, based on test F, is Translog production function with fixed effects and coefficients of the inputs: fertilizer, seed, water and labor have a significant effect on the production of this strategic crop and estimated production function has diminishing returns to scale.

3- Results show that the elasticity marks are fully consistent with economic theories which express the right choice of production function and appropriate use of integrated data versus just cross-sectional or time series data.

4- In estimating the intended model of this study, based on provincial observations, the heterogeneity of provinces should be considered. As can be seen, if the production function is estimated regardless to this matter, estimates will be significantly different from the reality. (Comparison of the common effect method and fixed effects).

5- Results indicate that, demand elasticity of corn farmers to changes in price of all inputs is less than one, in other words, farmers are indifferent to changes in inputs.

6- As demand for chemical fertilizer is inelastic to its price(-0.151), on price reduction of this input due to subsidizing, increase level of its consumption, is less than the rate of price reduction. So changes in fertilizer price won't have any important effect in its consumption. Results showed that, per reducing the fertilizer price due to subsidy policy, corn production decrease to 0.412 percent. That means, government's policy of Subsidizing chemical fertilizer in corn production of country is not efficient.

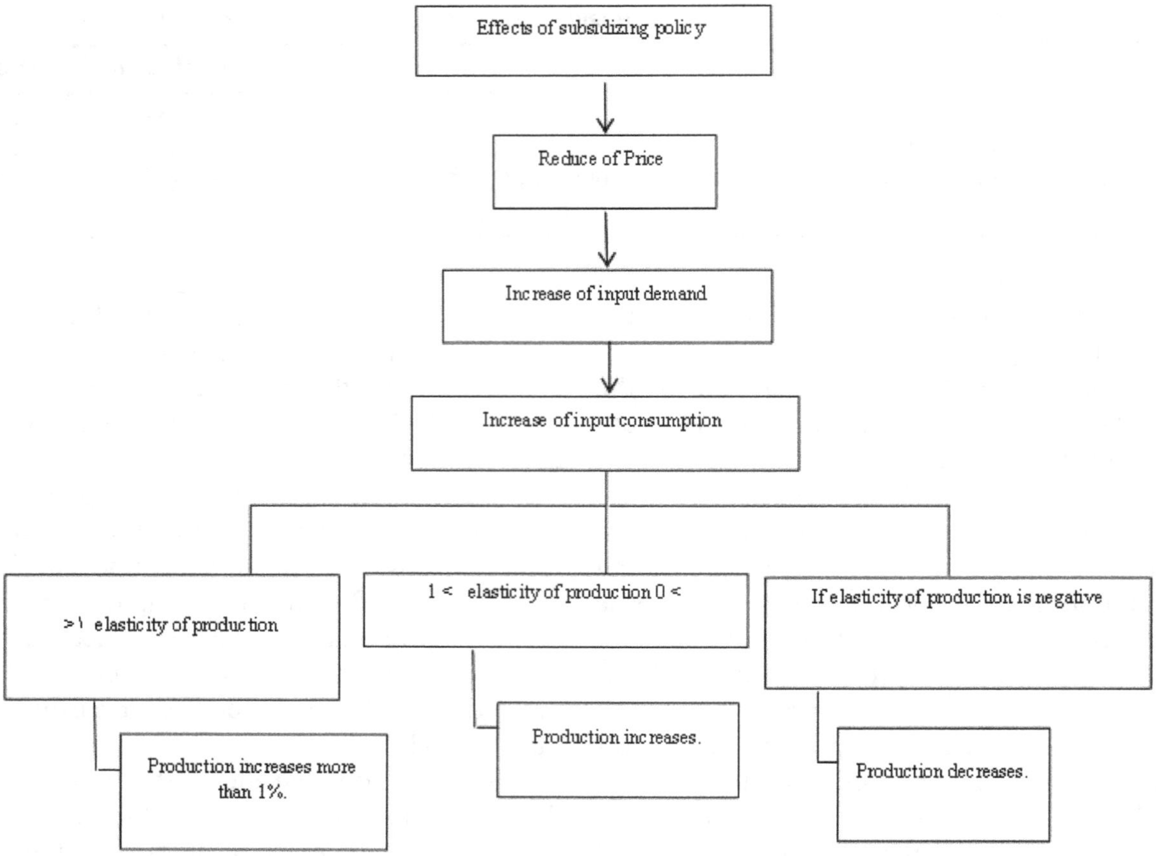

Chart 1: The effects of subsidizing production inputs′ policy
Resource: Rashid ghalam, 2009

Policy recommendations

7- Based on results, it′s revealed that corn farmers are using water too much which must be taken strategies to reduce the consumption of this valuable input. So, it is recommended, to better use of this input, using modern irrigation systems will provide necessary field of investment.

8- About two inputs: fertilizer and seed, price policy, including subsidies, or elimination of subsidies, will not cause so much changes in use of them. On the other hand, fertilizer subsidies reduce the production. So, it is recommended, eliminating subsidies and bringing prices to competitive prices, on one side a heavy financial burden of government is removed and on the other hand, environmental damages is prevented. But it should be kept in mind that, with sudden removing of subsidies, small farmers may be strongly affected and even, may lead them removed from production cycle.

9- Equivalent to the removed subsidies, either direct, cash and based on yield or Acreage, is better to pay to farmers. Moreover, removed subsidies can be allocated to other agricultural inputs or to indirect support policies such as reducing insurance premiums of farmers.

ACKNOWLEDGEMENTS

We would like to express our thanks to Dr. Jafar Haghighat and Rasul Mohammadrezaei, Professor of Department of Agriculture Economy of Tabriz University for their extensive help during the research doing.

REFERENCES

1- Ashrafzadeh, H., & Mehregan, M. (2008). Panel Data Econometrics, Institute of Cooperative Research of Tehran University, 38.

2- Azizan, A., & Salami, H. (2005). Choosing appropriate production function form of rice in Iran. Fifth Biennial International Conference Iranian Agricultural Economy, Zahedan, Sistan-Baluchestan University.

3- Debrah, k. (2002). Agricultural Subsidies In Sub-Saharan Africa: a Seflection, IFDC Africa Division

APIA Workshop, Lome.

4- Dejpasand, F. (1991). Review effect of fertilizer subsidy changes in production of (sugar beet), MS Thesis of Economics, Department of Economics and Political Science, Shahid Beheshti University.

5- Gujarati, D. (2003). Basic econometrics, Mc-Grow-hill.

6- Hosseinzad, J., & Salami, H. (2004). Select production function to estimate economic value of agricultural water (case study of wheat production). Agricultural economics and development, number 48.

7- Jhonston, J., & Dinardo, J. (1997). Econometric Methods, Mc Graw-Hill.

8- Khazayi, M. (2008). Foreign direct investment and value added of oil sector (case study, elected countries of OPEC), MS Thesis of Economics, Department of Economics, Tehran University.

9- Parmeh, Z. (2004). Review clear and estimate hidden subsidies in Iran Economy, Commercial reviews, 6: 32-42.

10- Rashidghalam, M. (2009). Effects of removing agricultural production factors subsidies on Sugar Beet production in Iran, Department of Agriculture, Tarbiyat Modaress University.

11- Raymond, Z., & Renfro, H. (1992), Fertilizer price and subsidy Polices in Bangladesh, World Development, 20(3): 437-455.

12- Thompson, C. D. (1988). Choice of flexible functional forms: Review and appraisal, Western Journal of Agricultural Economics , 13: 169-183.

25

Labour and Poverty: Empirical Relationship Using House Data from South Nigeria

Nsikak-Abasi A. Etim, Sunday Okon and Inibong A. Akpabio

Abstract

In Nigeria, most farming activities rely on family labor. However, rural-urban drift and the movement of young people away from agriculture are making labor increasingly pause. Thus, labor has become a major constraint to expanding the scope of production by small-scale resource poor farmers. This paper provides an empirical relationship between labor and poverty using data from households. Through a multi stage sampling procedure, 150 farming households were selected using questionnaire. Results of Foster, Greer and Thorbecke decomposition show that poverty incidence, depth and severity increase with increase in labor employed in farm operations implying that poverty is directly related to labor. Finding further reveals that the difference in poverty incidence of one of the sub-group (1–50 Vs 50–100) pair is statistically significant at ($P< 0.05$). Results suggest that the mandays of labor employed significantly affect the poverty incidence of farm households.

Keywords:
Labour, Poverty, Household, Nigeria.

Department of Agricultural Economics & Extension, University of Uyo, P.M.B. 1017, Uyo, Nigeria.
** Corresponding author's email: etimbobo@yahoo.com*

INTRODUCTION

Nigeria's agricultural production is highly labor intensive. Over 90% of all tasks in non–mechanized production systems depend on human labor, and for mechanized production systems depend on human labor, and for mechanized production systems, between 50 and 60% of the tasks depend on human labor (Olayide, 1981; Shaib *et al.,* 1997). According to Olayeni (1980) and Sharb *et al.,* (1997), family labor constitutes over 76% of farm labor. Generally, males are responsible for land clearing, ridging and weeding, while women undertake the processing and marketing of farm produce and they may also help in weeding, harvesting, bird scaring and the tending of cattle, sheep and goats. Hired labor has become important in farm operations especially during the peak periods of the various farming activities.

The rapid rural-urban migration of the youths and the resultant dwindling of an active farm labor force has become a major constraint to expanding agricultural production (Shaib *et al.,* 1997). Historically, the movement of labor away from agriculture as a result of rural-urban drift was relatively gradual between 1960 and 1970. During this period, young school leavers went into public and industrial sector employment, which was concentrated in the urban centers. In the 1970s, the deployment of earnings from petroleum export in urban construction industries created employment for unskilled labor. The resulting massive rural–urban migration depleted rural areas of farm labor. In the 1980s and 1990s when there were reverses in petroleum and construction activities, urban employment dropped. While some of the migrants returned to rural areas, many of them did not, and as a result labor shortages in the village have remained endemic. Consequently, farming sector wage rates have been on the increase, rising from a mere N0.45 per man-day in the 1970–1975 period to nearly N65.00 per man-day in the 1991–1994 period, making hired labor increasingly unaffordable to the small-scale farmer. More recently, these wage rates have increased astronomically between N1, 500 – N2, 500 per manday depending on the location and size of the plot.

Farm labor productivity has become increasingly low because farm households largely comprised fairly old people and very young children coupled with crude implements which impedes their ability to raise yield and income with subsequent reduction in poverty. According to Borlang and Dowswell (2010), no nation has been able to substantially reduce poverty and bring about economic growth and development without first markedly increasing the productivity of its agricultural and food systems. Like in many developing countries, poverty in Nigeria is essentially a rural phenomenon as most of the impoverished people live in the rural areas, where they derive their livelihood from farming (Etim and Ukoha, 2010). Investing in agriculture is a key to reducing poverty and hunger in developing countries and is essential element in addressing the current food price crises (Fan & Rosegrant, 2008; Etim & Ukoha, 2010). But agricultural production in the Akwa Ibom State is highly labor intensive as majority of the farming activities rely both on family and hired labor. Most of poverty studies are linked to agriculture (Canagarajah *et al.,* 1995; FOS, 1999; Khan

Table 1: Mean Household Expenditure (Adult Equivalent)

Item	Amount (₦) per month	Percentage Expenditure
Energy	1677.34	20.30
Clothing	1201.30	14.54
Health care/ Medication	1134.34	13.73
Education	2107.00	25.50
Food	2144.11	25.93
Total	8264.09	100.00

Mean 1652.82
2/3 1,101.88 (Poverty line)

Table 2: Labour Employed in all farm Enterprises

Family Labour (Mandays)	Frequency	Percentage
1 - 50	52	34.67
51 - 100	78	52
101 - 150	20	13.33
Total	150	

Table 3: Comparison of Poverty by Labour

Labour Employed (Mandays)	P0	P1	P2	Contribution to		
				P0	P1	P2
1–50	0.38	0.25	0.24	0.17	0.31	0.34
	(1.00)	(-0.25)	(-0.27)			
50–100	0.46	0.42	0.43	0.57	0.56	0.55
	(-0.20)	(-0.04)	(-0.08)			
100–150	0.66	0.71	0.84	0.26	0.13	0.11
	(2.11)**	(2.45)**	(2.57)**			
All	0.57	0.48	0.44	1.00	1.00	1.00

Figures in parentheses are t – values of Pα. ** Significant at 5%.

2001; Okunmadewa, 2001). This implies that a large proportion of the rural poor are engaged in one form of farming or the other. According to Mijindadi (1995), over 90% of the foods consumed in Nigeria are produced by these farm households. This study investigates the empirical relationship between family labor and poverty using data from farming households.

The concept of poverty dates back to 1899, when one of the earliest and most famous studies of poverty was conducted by Seebohm Rowntree in York. He used a concept of subsistence poverty and drew a poverty line in terms of a minimum weekly of sum of money, which was necessary to enable families secure the necessaries of a healthy life. According to Okunmadewa (2001), poverty is more easily recognized than defined. Hence, a universally acceptable definition of the term has remained elusive. Poverty is defined as total poverty as the expectation overtime of the poverty measured at each point in time. Poverty can be chronic (structural) or transitory, depending on how long poverty is expressed by an individual or a community. Chronic poverty is long term, persistent, the causes of which are largely structural and endemic, while transitory poverty is temporary, transient and short term in nature. Transitory poverty is defined as total poverty minus chronics poverty. Since the nineteenth century when rigorous studies in poverty began researchers have tried to establish fixed yardsticks against, which to measure poverty ideally, such a yardstick would be applicable to all societies and should establish a fixed level, usually known as the poverty line below; which poverty begins and above which it ends. A traditional measure of poverty stipulates that the number of people living on less than US$I per day. Although this traditional measure of poverty is commonly used, many in the development community have supported measures such as Millennium Development Goals (MDGs) that use a complex set of conditions as yardsticks in assessing the entire living situation of poor people (Rosegrant *et al.*, 2005). Absolute poverty is a situation of lack of access to resources required to obtain the minimum necessities required to maintain physical efficiency. Relative poverty, on the other hand, is the inability to attain a given minimum contemporary standard of living. Poverty can also be subjective. This refers to whether or not individuals or groups feel they are poor. Subjective poverty is closely related to relative poverty since those who are defined as poor in terms of standard of the day will probably see and feel themselves to be poor. The concept of subjective poverty is important since to degree, people act in terms of the way they perceive and define themselves. Poverty line is the threshold income below, which one is considered to be poor (Kakwani, 1993). It is the value of income or consumption expenditure necessary for a minimum standard of nutrition and other necessities. According to

Thorbecke (2004) there are currently two main methods of setting the poverty live i.e. Cost of Basic Needs (CBN) and the Food–Energy-Intake (FEI) methods.

The literature on aggregate measures of poverty and wellbeing is quite enormous. Many indices have been designed and developed to measure poverty and well-being. These comprise Sen index (1979); Foster-Greer-Thorbecke (FGT) poverty Index (1984); UNDP (1990), Integrated Poverty Index (IPI), Basic needs on balanced diet index, the Physical Quality of Life (PQLI) (Morris, 1994), Relative Welfare Index (IFAD, 1993), Index of Social Progress (Estes' 1984); Index of "Quality of Life" in nations (Slottje's 1991); Index of Quality of Life in metropolitan areas (Lui's 1977) This study however employs the Foster, Greer, Thorbecke weighted poverty measure for quantitative poverty assessment. This class of additively decomposable poverty measure is based on income/expenditure approach.

MATERIALS AND METHODS

Study area, sampling and data collection: The study was conducted in Akwa Ibom State, Nigeria. The state is located at latitude 4°33' and 5°53' North and longitude 7°25' and 8°25' East and occupies a total land area of 7,246 km^2. With an estimated population of about 3.9 million (NPC 2006), the state is bounded to the North by Abia State, to the East by Cross River State, to the West by Rivers State and to the South by the Atlantic Ocean. Administratively, the state is divided into 31 Local Government Areas and has 6 Agricultural Development Project (ADP) Zones viz: Oron, Abak, Ikot Ekpene, Etinan, Eket and Uyo.

The study area is in the rainforest zone and has two distinct seasons viz: the rainy and the short dry season. The annual precipitation ranges from 2000-3000 mm per annum. Most of the inhabitants of rural communities in the study area are farmers and the crops commonly cultivated include cassava, oil palm, yam, cocoyam, flitted pumpkin, okra, water-leaf, bitter-leaf, etc. In addition, some micro livestock are usually raised at backyards of most homesteads.

Primary data were used for this study. Farm-level intensive itinerary survey provided the basic cross-sectional data from 150 rural farming households in the study area. Data were collected from farm households using well structured questionnaire. Primary data included data on household income and expenditure, socio-economic characteristics of households and their heads, farm specific variables.

Multistage sampling technique was used for selecting the representative farm households that were used for this study. The first stage was the random selection of 3 out of the 6 Agricultural Development Project Zones in Akwa Ibom State. The second stage sampling was the random selection of 5 villages per ADP zone to make a total of 15 villages. Furthermore, a total of 10 households were randomly selected to make a total of 150 farming households.

Analytical techniques

The Foster, Greer and Thorbecke (FGT) weighted poverty index was used for the quantitative poverty assessment (Foster *et al.*, 1984). The reason for this choice is due to its decomposability of the overall population into sub-groups which allows for comparison. United Nations UN (2001) noted that the most important purpose of a poverty measure is to enable poverty comparisons.

The FGT measure for the subgroup $_i$th P α_i is given as:

$$Pi = n^{-1} \sum_{j=1}^{qi} \frac{z - Yji}{z,Omax}$$

Where $P_{\alpha i}$ is the weighted poverty index for the ith subgroup; n_i is the total number of households in the ith subgroup households in poverty; Y_{ji} is the per adult equivalent expenditure of household j in sub-group $_{ij}$, z is the poverty line and α is the degree of concern.

When α is equal to zero, it implies no concern and the equation gives the head count ratio for the incidence of poverty (the proportion of the farming households that are poor).

The poverty line used for this study is defined as the two-thirds of mean household expenditure adult equivalent. Adult equivalents were generated following Nathan and Lawrence (2005) as follows:

$$AE = 1 + 0.7\,(N1 - 1) + 0.5\,N2$$

Where AE = Adult Equivalent

N1 = Number of adults aged 15 and above

N2 = Number children aged less than 15

That is

$$Pi = ni^{-1} \sum_{j=1}^{qi} \frac{z - Yji}{z,Omax} = qi/ni$$

When α is equal to 1, it shows uniform concern and equation becomes

$$P_{1i} = ni^{-1} \sum_{j=1}^{qi} \frac{z - Yji}{z,Omax}^{1}$$

This measures the depth of poverty (the proportion of expenditure shortfall from the poverty line) according to Hall and Patrinos (2005), it is otherwise called the poverty gap-the average difference between the income of the poor and the poverty line.

When α is equal to 2, distinction is made between the poor and the poorest (Foster *et al.,* 1984, Assadzadeh and Paul, 2003). The equation become

$$P_{2i} = ni^{-1} \sum_{j=1}^{qi} \frac{z - Yji}{z,Omax}^{2}$$

The equation gives a distribution sensitive FGT index called the severity of poverty. It tells us the extent of the distribution of expenditure among the poor.

The FGT measure for the whole group or population was obtained using:

$$P = \sum_{i=1}^{m} \frac{P_1 n_1}{n}$$

Where P α is the weighted poverty index for the whole group, m is the number of subgroups while n and ni are the total number of households in the whole group and the $_i$th sub-group respectively.

The contribution (C_i) of each sub-group's weighted poverty measure to the whole group's weighted poverty measure was determined using;

$$Ci = \frac{n_i P \alpha_i}{n P\alpha}$$

The test of significance of $P\alpha_i$ (subgroup poverty measure) relative to the pα (whole group poverty measure) was given according to Kakwani (1993) by:

$$t = \frac{P\alpha_i - P_\alpha}{SE\,(P_{\alpha i})}$$

The above was used to test if significant difference exist between the Pα measures of a subgroup i with another j.

The weighted poverty measures (Pα) and their corresponding standard errors were calculated using the Microsoft Excel Package.

RESULTS AND DISCUSSION

The first step in the analysis of poverty is the determination of the poverty line. As stated in the methodology, the mean household expenditure adult equivalent was used to determine this threshold. Table I shows the average amount expended on basic consumption items of the households. The mean per adult equivalent household expenditure is ₦1, 652.82 and the poverty line is ₦1, 101.88.

The use of family labor is common in the study area. Findings however reveals that majority of the households (52%) have 50–100 mandays of family labor while 34.67% of the farm households employ less than 50 mandays of family labor.

Farm households with more than 100 mandays of family labor are 13.33%. Findings imply that majority (65.53%) of the households employ greater mandays of family labor suggesting that households had many members/dependants.

Labour Employed

Poverty was profiled among farm households using three labor sub-groups. Table reveals that 38, 46 and 66 percent of households with

Table 4: Poverty by Labour Employed (Mandays)

Labour Employed (Mandays)	P0	P1	P2
1–50 vs 50–100	-2.00**	-0.81	-0.79
1–50 vs 100–150	-0.88	-0.75	-0.81
50–100 vs 100-150	-0.33	-0.67	-0.82

** Significant at 5%.

labor less than 50, 50–100 and above 100 mandays respectively are poor. Their respective contributions to the whole group's poverty incidence are 17, 57 and 26 percent. Poverty incidence is significant (P<0.05) in households with more than 100 mandays of labor relative to that of the whole group. Poverty depth and severity follow similar pattern like the incidence of poverty.

Table 4 shows that the differences in poverty incidence in one of the three possible pairs (1-50 vs 50–100) mandays) is statistically significant (P<0.05). This means that the mandays of labor employed significantly affect the poverty incidence of farm households.

The level of poverty increases as the mandays of labor employed increases. This may be attributed to the fact that increase in household size which is caused by more younger household members, raises the dependency ratio and subsequently raises the poverty level. Thus, farm households with greater mandays of family labor have the propensity to be poorer than the households with smaller mandays of labor.

CONCLUSION

The incidence of poverty among farming households was 0.57 whereas the 0.48 constituted the proportion of expenditure shortfall from poverty line and 0.44 comprised the poorest of the poor. Results of FGT decomposition show that the level of poverty increases as mandays of labor employed increases. The analysis reveals that an increase in labor increases the probability of poverty. This is true because increased family labor results from larger household sizes and dependency ratios which tend to raise the level of poverty.

REFERENCES

1- Assadzadeh, A. and S. Paul (2003). Poverty Growth and Redistribution. A Case Study of Iran. In R. Van der Hoeven and A. Shorrocks (Eds). Perspective on Growth and Poverty. UNU/WIDER Helsinki, Finland.

2- Borlang, N. E. and C. Dowswell (2004). Food Security and Agricultural Development in Sub-Saharan Africa: The Challenges Ahead, in From Subsistence to Sustainable Agriculture in Africa (ed.) Steve A. Breth (Eds). Mexico City Sasakawa Africa.

3- Canagarajah, S. J. Ngwafon and S. Thomas (1995). Evolution of Poverty and Welfare in Nigeria 1985 – 92. PHRP West Central African Department, World Bank, Washington, DC.

4- Estes', R. (1984). The Social Progress of Nations. New York. Preager.

5- Etim, N. A. and O. O. Ukoha (2010) Analysis of Poverty Profile of Rural Households: Evidence from South South Nigeria. Agriculture and Social Science 6(3): 48 -52.

6- Fan, S, and M. W Rosegrant (2008). Investing in Agriculture to Overcome the World Crisis and Reduce Poverty and Hunger, IFPRI Policy Brief 3. International Food Policy Research Institute. Washington, DC.

7- FOS (Federal Office of Statistics) 1999. Poverty and Agricultural Sector in Nigeria. Federal Office of Statistics, Abuja, Nigeria.

8- Foster, J., J. Greer and E. Thorbecke (1984) A Class of Decomposable Poverty measures. Econometrica 52:761-5.

9- Hall, G. and H. A. Patrinos (2005) Latin America's Indigenous People. Latin America. A Time of Transition, Finance and Development. International Monetary Fund 42: 24.

10- IFAD (International Fund for Agricultural development) 1993. The State of World's Rural Poverty. A Profile of Latin America and the Carribean.

11- Kakwani, N. (1993) Measuring Poverty: Definition and Significance Test with Application to Cote d'lvoire. In Lipton, M. and J. Van der Gaag (eds), Including the Poor. The World Bank, Washington, DC.

12- Khan, M. H. (2001) Rural Poverty in Developing Countries. Implications for Public Policy. International Monetary Fund, Washington, DC.

13- Liu, B-C (1977) Economic and Non-Economic Quality of Life: Empirical Indicators and Policy Implications for Large Standard Metropolitan Areas, The American Journal of Economic and Sociology 36:225-40.

14- Mijindadi, N. S. (1995) Socioeconomic Constraints to Sustainable Agriculture in the moist Savannah of Sub-Saharan Africa, Procee-Cotonou, Benin Republic 19 – 23 September, 1994.

15- Morris, M. O. (1994). Measuring the Condition of the World's Poor. The Physical Life Index, New York. Perganon Press.

16- Nathan, O. F and B. Lawrence (2005). The Impact of Micro Finance on the Welfare of the Poor in Uganda. Final Report Submitted to African Economic Research Consortium (AERC) May, 2005.

17- NPC (National Population Commission) 2006. Population Census of the Federal Republic of Nigeria. Report at the National Level, National Population Commission, Abuja.

18- Okunmadewa, F. (2001). Poverty Reduction in Nigeria: A Four-Point Demand, Annual Guest Lecture, The House 20th Anniversary Lecture Delivered at the University of Ibadan.

19- Olayemi, J. K. (1980) Food Crop Production by Small Farmers in Nigeria. In Nigeria Small Farmers, S. O. Olayide, J. A. Eweka and V. E. Bello-Osagie eds. Centres for Agricultural Research and Development, University of Ibadan.

20- Olayide, S. O. (1981) Labour Demand in Agricultural Production in Nigeria.

21- Rosegrant, M. W., S. A Cline, W. Li, T. B. Sulser and R. A Valmonte-Santos (2005). Looking Ahead. Long-Term Prospects for Africa's Agricultural Development and Food Security. 2020 Discussion Paper 41. International Food Policy Research Institute, Washington, DC.

22- Sen, A. K (1976). Poverty An Ordinal Approach to Measurement. Econometrica 44(2): 219-231.

23- Shaib, B. A Aliyu and J. S. Bakshi (1997). Nigeria National Agricultural Research Strategy Plan: 1996 – 2010. Federal Ministry of Agriculture and Natural Resources.

24- Slottje, D. J (1991) Measuring the Quality of Life of Across Countries. The Review of Economics and Statistics 73: 684-93.

25- Thorbecke, E. (2004) Conceptual and Measurement Issues in Poverty Analysis. WIDER Discussion Paper. UNU/WIDER. Helsinki.

26- UN (United Nations) 2001. Indicators of Sustainable Development: Guidelines and Methodologies New York.

27- UNDP (United Nations Development Program) 1990. Human Development Report. Oxford University Press. New York.

PERMISSIONS

All chapters in this book were first published in IJAMAD, by Islamic Azad University; hereby published with permission under the Creative Commons Attribution License or equivalent. Every chapter published in this book has been scrutinized by our experts. Their significance has been extensively debated. The topics covered herein carry significant findings which will fuel the growth of the discipline. They may even be implemented as practical applications or may be referred to as a beginning point for another development.

The contributors of this book come from diverse backgrounds, making this book a truly international effort. This book will bring forth new frontiers with its revolutionizing research information and detailed analysis of the nascent developments around the world.

We would like to thank all the contributing authors for lending their expertise to make the book truly unique. They have played a crucial role in the development of this book. Without their invaluable contributions this book wouldn't have been possible. They have made vital efforts to compile up to date information on the varied aspects of this subject to make this book a valuable addition to the collection of many professionals and students.

This book was conceptualized with the vision of imparting up-to-date information and advanced data in this field. To ensure the same, a matchless editorial board was set up. Every individual on the board went through rigorous rounds of assessment to prove their worth. After which they invested a large part of their time researching and compiling the most relevant data for our readers.

The editorial board has been involved in producing this book since its inception. They have spent rigorous hours researching and exploring the diverse topics which have resulted in the successful publishing of this book. They have passed on their knowledge of decades through this book. To expedite this challenging task, the publisher supported the team at every step. A small team of assistant editors was also appointed to further simplify the editing procedure and attain best results for the readers.

Apart from the editorial board, the designing team has also invested a significant amount of their time in understanding the subject and creating the most relevant covers. They scrutinized every image to scout for the most suitable representation of the subject and create an appropriate cover for the book.

The publishing team has been an ardent support to the editorial, designing and production team. Their endless efforts to recruit the best for this project, has resulted in the accomplishment of this book. They are a veteran in the field of academics and their pool of knowledge is as vast as their experience in printing. Their expertise and guidance has proved useful at every step. Their uncompromising quality standards have made this book an exceptional effort. Their encouragement from time to time has been an inspiration for everyone.

The publisher and the editorial board hope that this book will prove to be a valuable piece of knowledge for researchers, students, practitioners and scholars across the globe.

LIST OF CONTRIBUTORS

Abdulazeez, Muhammad-Lawal and Adekunle, Dayo
Department of Agricultural Economics and Farm Management, University of Ilorin, P.M.B. 1515, Ilorin, Nigeria

Omotesho Kemi Funmi
Department of Agricultural Extension and Rural Development, University of Ilorin, P.M.B. 1515, Ilorin, Nigeria

Adekola, Olabisi Fatimat
Department of Agronomy, University of Ilorin, P.M.B. 1515, Ilorin, Nigeria

Yudi Ferrianta, Muhammad Fauzi Makki and Suprijanto and Rifiana
Lecturers in Faculty of Agriculture, Lambung Mangkurat University, Indonesia

Marzieh Panahzadeh Parikhani, Fatemeh Razzaghi Borkhani, Hossein Shabanali Fami and Naser Motiee
Department of Agricultural Management and Development, Faculty of Agricultural Economics and Development, University of Tehran, Karaj, Iran

Abotaleb Hosseinpoor
Assistant Professor in Agricultural Development Faculty of Agricultural, University of Mazandaran

Roya Eshraghi Samani
Department of- Agricultural Management, Ilam Branch, Islamic Azad University, Ilam, Iran

Alireza Poursaeed
Department of Agricultural Extension and Education, Ilam Branch, Islamic Azad University, Ilam, Iran

Asghar Bagheri
College of Agricultural Technologies and Natural Resources- University of Mohaghegh Ardabili

Fakher Kardoni
Former Graduate Student of Agro ecology, Department of Agronomy and Plant Breeding, Faculty of Agriculture, University of Birjand, Birjand, Iran

Majid Jami-Al Ahmadi
Associate Professor, Department of Agronomy and Plant Breeding, Faculty of Agriculture, University of Birjand, Birjand, Iran

Mohammad Reza Bakhshi
Assistant Professor, Department of Agronomy and Plant Breeding, Faculty of Agriculture, University of Birjand, Birjand, Iran

Kemi Funmilayo Omotesho
Department of Agricultural Extension and Rural Development, University of Ilorin, Ilorin

Azeez Muhammad-Lawal and Gafar Olawumi Iyiowu
Department of Agricultural Economics and Farm Management, University of Ilorin, Ilorin

Amir Ahmadpour
Department of Agricultural Extension and Education, Sari Branch, Islamic Azad University, Sari, Iran

Ali Sardar Shahraki
Ph.D. Candidate of Agricultural Economics, University of Sistan and Baluchestan, Iran

Javad Shahraki
Associate Professor, Agricultural Economics, University of Sistan and Baluchestan, Iran

Seyed Arman Hashemi Monfared
Assistant Professor, Civil Engineering, University of Sistan and Baluchestan, Iran

Parisa Alamdari and Setareh Amanifar
Assistant Professors of Soil Science, University of Zanjan, Iran

Azadeh Falsafian
Assistant Professor, Department of Agricultural Management, Extension and Education, Tabriz Branch, Islamic Azad University, Tabriz, Iran

Mohammad Reza Safari Motlagh
Department of Plant Protection, Faculty of Agriculture, Rasht Branch, Islamic Azad University, Rasht, Iran

Farzaneh Fallahpoor Salkooyeh
Graduate student of Agricultural Management, Rasht Branch, Islamic Azad University, Rasht, Iran

Nnamdi Ehirim, Stephen Onyeagocha, Glory Ben-Chendo, Ubon Essien and Melissa Okwara
Department of Agricultural Economics, Federal University of Technology, Owerri, Imo State, Nigeria

Emeka Osuji
Department of Agricultural Economics, Michael Okpara University of Agriculture, Umudike, Abia State, Nigeria

Lateef Lawal Adefalu, Oluwasogo David Olorunfemi and Yusuf Olatunji
Department of Agricultural Extension and Rural Development, University of Ilorin, Ilorin, Nigeria

LatifatKehinde Olatinwo
Department of Agricultural Economics and Extension, Federal University, Dutsin-Ma, Kastina State, Nigeria

Narges Mirzaie Gorji
M.Sc. Graduate of Rural Development, College of Agriculture and Natural Resources, University of Tehran, Iran

Hossein Shabanali Fami
Professors of Management and Agricultural Development, University of Tehran, Iran

Hooshang Iravani
Professors of Management and Agricultural Development, University of Tehran, Iran

Seyyed Milad Samizadeh
Department of Agricultural Management, Rasht Branch, Islamic Azad University, Rasht, Iran

Saeed Firouzi
Department of Agronomy, Rasht Branch, Islamic Azad University, Rasht, Iran

Mohsen Shafieyan
PhD Student, Industrial Management, Department of Management and Accounting, Rasht Branch, Islamic Azad University, Rasht, Iran

Mahdi Homayounfar and Mehdi Fadaei
Assistant Professor, Industrial Management, Department of Management and Accounting, Rasht Branch, Islamic Azad University, Rasht, Iran

Reza Esfanjari Kenari
Assistant Professor, Agricultural Economics, Faculty of Agricultural Sciences, University of Guilan

Zohre Karami
MSc, Agricultural Economics, Shiraz University, Shiraz, Iran

Seyadeh Sedighe Ahmadzade
Ph.D. Student, Agricultural Economics, Sari Agricultural Sciences and Natural Resources University

Arega Shumetie
Ph.D. Candidate in Haramaya University of Ethiopia and Makerere University of Uganda

Belay Kassa
Professor Pan African University, Africa Union, Addis Ababa, Ethiopia

Degye Goshu
Ph.D. Haramaya University, College of Agriculture and Environmental Science: School of Agricultural Economics and Agri-business: Department of Agricultural Economics, Ethiopia

Majaliwa Mwanjalolo
Ph.D. Makerere University, College of Agriculture and Environmental Science: School of Forestry, Environmental and Geographical Science: Department of Geography, Geo Informatics and Climatic Sciences, Uganda

Asghar Bagheri and Mojtaba Sookhtanlou
Faculty of Agriculture and Natural Resources, University of Mohaghegh Ardabili, Iran

Saeed Firouzi
Department of Agronomy, College of Agriculture, Rasht Branch, Islamic Azad University, Rasht, Iran

Mohammad Abdolmaleky
Assistant professor, College of Agriculture, Khoramabad Branch, Islamic Azad University, Iran

Zeinab Moinoddini
Invited Lecturer Payame Noor University of Kerman, Iran
Ph.D. Student of Agricultural Economics Zabol University of Iran

Negin Hosseingholizadeh
Graduate Student of Agriculture Economy, Iran-Tabriz University, Iran

Jafar Haghighat
Professor, Department of Agriculture Economy, Iran-Tabriz University, Iran

Rassul Mohammadrezaei
Professor, Department of Agriculture Economy, Iran-Tabriz University, Iran

Nsikak-Abasi A. Etim, Sunday Okon and Inibong A. Akpabio
Department of Agricultural Economics & Extension, University of Uyo, P.M.B. 1017, Uyo, Nigeria

Nalini Arumugam
Lecturer at Faculty of Agriculture and Biotechnology, UniSZA

Fatimah Mohamed Arshad
Director, Institute of Agricultural and Food Policies Studies at University Putra Malaysia

Eddie Chiew F.C and Zainalabidin Mohamed
Lecturers at Department of Agribusiness and Information Systems, Faculty of Agriculture, Universiti Putra Malaysia

Albert I. Ugochukwu
Department of Bioresource Policy Business and Economics, University of Saskatchewan, 51 Campus Drive Saskatoon, SK, S7N 5A8, Canada

Chuma I. Ezedinma
Root and Tuber Improvement Division, International Institute of Tropical Agriculture, P.M.B. 5320 Ibadan, Nigeria

Samaneh Abedi and Mohammad Kavoosi Kalashami
Ph.D. Student of Agriculture Economics, Department of Agriculture Economic, University of Tehran, Iran

Gholam Reza Peykani
Associate Professor, Department of Agriculture Economic, University of Tehran, Iran

Index

R

S

T

V

W

Z

www.ingramcontent.com/pod-product-compliance
Lightning Source LLC
LaVergne TN
LVHW060520210926
848485LV00005B/1112